Basic Methods for Microcomputer-Aided Analysis of Electronic Circuits

Basic Methods for Microcomputer-Aided Analysis of Electronic Circuits

M. Bialko

Technical University of Gdansk
Gdansk, Poland

R. Crampagne

E.N.S.E.E.I.H.T-INP
Toulouse, France

and

D. Andreu

E.N.S.E.E.I.H.T-INP
Toulouse, France

Prentice Hall

New York   London   Toronto   Sydney   Tokyo   Singapore

First published 1995 by
Prentice Hall International (UK) Limited
Campus 400, Maylands Avenue
Hemel Hempstead
Hertfordshire, HP2 7EZ
A division of
Simon & Schuster International Group

Typeset in Times 10 on 12pt
by Mathematical Composition Setters Ltd, Salisbury, Wiltshire

Printed and bound in Great Britain by
the University Press, Cambridge

Library of Congress Cataloging-in-Publication Data

Bialko, M. (Michal)
    Basic methods for microcomputer-aided analysis of electronic
circuits/M. Bialko, D. Andreu, R. Crampagne.
       p.   cm.
    Includes bibliographical references and index.
    ISBN 0-13-061284-7
    1. Electronic circuits—Mathematical models. 2. Electric circuit
analysis—Data processing. I. Andreu, D. II. Crampagne, R.
III. Title.
TK7867.B436   1994
621.3815′0285′5362—dc20   94-13684
                                         CIP

British Library Cataloguing in Publication Data

A catalogue record for this book is available from the British Library

ISBN 0-13-061284-7

---

1 2 3 4 5 99 98 97 96 95

Contents

Preface

There is no doubt that computer analysis and simulation are very important steps in the design and production of electronic circuits and systems. At present there exist several powerful computer simulators for these purposes such as SPICE, ECAP, NAP, etc., among which the best known is the SPICE simulator. The simulators offer excellent possibilities for the simulation and learning of properties of electronic circuits but do not enhance knowledge of the properties, advantages and limitations of various numerical methods used for circuit analysis, simulation and design.

In this book the basic methods of computer aided analysis of electronic circuits are presented. Each of the problems discussed is illustrated by a short computer program written in Turbo-Pascal which can be run on a small home or personal microcomputer. Hence the word 'microcomputer' is used in title of the book. The Pascal programs are written to be as simple as possible in order to provide the opportunity for readers to gain insight into the structure of algorithms and to follow the computational techniques. Running the programs provides the possibility for checking the solutions to problems speedily. The programs, which are collected on the diskette attached to the book, can easily be adapted to other similar problems.

The text is intended as an introduction to, or a short recapitulation of, the basic methods and numerical techniques that are frequently used in circuit analysis and simulation. The level of the book is acceptable for a wide group of potential readers, such as undergraduate students, practising engineers, advanced technicians or instructors, etc., who have no solid mathematical and advanced circuit theory background.

The material in the book is explained gradually starting from simple problems, simple numerical computation techniques and simple programs. Then, more and more complicated problems are discussed, and in the subsequent chapters, the material from previous chapters is used. Therefore, a description of electron devices models is presented first, followed by recapitulation of more important network properties; then the matrix generation is discussed, followed by sets of linear equation solutions, and so on. Finally, a new approach based on so-called artificial intelligence methods is introduced, together with descriptions of expert systems and their applications in the field of electronic circuits.

The text is divided into thirteen chapters. In Chapter 1 models of electron devices, including diode, bipolar, and MOS transistors, are described. In Chapter 2, which can be treated as introductory, some basic network theorems and transformations are collected. Also in this chapter some elementary numerical computations are performed, including definite and indefinite integral computations using such popular methods as rectangular, trapezoidal, and Simpson's methods. Chapter 3 gives the principles and examples of automatic nodal admittance (or mesh impedance) matrix generation. In Chapter 4 several numerical methods for a set of linear equation solutions are presented including: Gauss–Jordan, Gauss elimination and Lower/Upper matrix decomposition methods. Also, the inversion of matrices using the LU decomposition method is described. Chapter 5 deals with numerical generation of derivatives of network functions. Based on the Tellegen theorem, an adjoint network is defined and sensitivities of the transfer function for circuit element variations are computed. An application of the adjoint network for Thevenin equivalent circuit generation is also shown. In Chapter 6 non-linear resistive circuits are considered. Solutions of non-linear equations using the Newton–Raphson algorithms for circuits with diodes and transistors are described. Circuit representations of Newton–Raphson methods are also introduced to convert non-linear circuit branches into equivalent linear circuits (in a single iteration step). Chapter 7 explains the basic methods for the solution of first-order differential equations, including extrapolation and interpolation, single-step and multi-step algorithms. Algorithm stability and accuracy problems are discussed. Companion models for capacitor and inductor associated with different interpolation algorithms are described. In Chapter 8 transient analysis methods for higher-order linear and non-linear dynamic circuits are described and the simultaneous solution of a set of equations is presented. For non-linear dynamic circuits, linearized models for non-linear resistive elements and associated models for reactive elements are used and the Gauss elimination method is applied to the solution. Chapter 9 is devoted to the automatic generation of state variable equations and their solutions in the time domain. The intermediate steps in the procedure, such as: automatic generation of the hybrid matrix, transformation of the initial state and output equations into the normal form, and their solutions for different excitation functions are described. In Chapter 10 the transient response of the circuit using the inverse Laplace transform for a given transmittance $T(s)$ is presented. Intermediate steps leading to residue and time response computation are also described. An application of the convolution integral to find the time response of the circuit described by its impulse response is also explained and used in the computer program. In Chapter 11 a sinusoidal steady state analysis is performed leading to the magnitude and phase frequency characteristics computations. This analysis is based on the semisymbolic representation of the transfer function. The intermediate steps of the method: generalized, or modified, admittance matrix generation and unit circle polynomial interpolation methods to find the coefficients of the numerator and denominator of the transfer function are described in detail. Chapter 12 presents an application of Fourier series to an analysis of electronic circuits excited by non-sinusoidal periodic signals. Examples of Fourier series coefficient computation are presented. Reconstruction of the time function $f(t)$ based on the Fourier series coefficient knowledge is also described. The last chapter 13 gives an introduction to artificial intelligence methods for analysis and design problem solutions in the area of electronic circuits. It gives a short description of the characteristics of artificial intelligence methods based on symbolic, non-algorithmic and heuristic approaches; the properties and structure of expert systems are also described and compared with classical algorithmic programs. Examples of heuristic programming are presented and, finally, applications of popular expert system shells, CLIPS and EXSYS, for expert systems development are given.

All programs written in Turbo-Pascal 4.0 are collected on diskette attached to the book: providing the reader with the set of example programs (along with the explanations given in the text) will facilitate speedy understanding of algorithm applications. Running the programs with different data entered by the user may not only explain the operation and behaviour of circuits in different situations but may also explain the delicate (sometimes) aspects of numerical computations such as accuracy and stability of some algorithms.

This book is a result of multiple-year cooperation (intensified over the last three years) between the Technical University of Gdansk, Poland, and the Ecole Nationale Supérieure d'Electrotechnique, d'Electronique, d'Informatique et d'Hydraulique de Toulouse, France. The majority of the topics and programs given here have been taught and demonstrated in the classroom (using a computer screen-wall projector) in the Department of Electrical Engineering and Electronics at the Technical University of Gdansk, Poland, Universitaet Karlsruhe, FR Germany, and ENSEEIHT, France.

The authors would like to express their thanks and appreciation to the Authorities of the Institut National Polytechnique at Toulouse and also to the Director of ENSEEIHT, Professor D. Amoros, for their support. This support has allowed the authors to meet together many times for discussions and final preparation of the text.

Professor Michal Bialko
Toulouse/Gdansk 1993

1 Models of Electronic Components

1.1 Model generation of electronic elements

A model allows us to describe the behaviour of a component used in a circuit simulation software package. It is called an equivalent circuit and is developed from basic circuit theory elements and numerical component parameter values.

The complete model would, in fact, be able to simulate the component in any application (DC, transient, small signal analysis etc., ...). However, the complete model, if it exists, is often very complex and so a part of the model is merely adapted to a given situation.

The six minimum elements for modelling are represented in Fig. 1.1.

1.1.1 Resistors

Resistors are defined by the current–voltage relation $i(t) = f[v(t)]$. If this function is a straight line passing through the origin, the resistance has a constant value: the resistor is linear. If not, the resistor is non-linear:

1. $v(t) = Ri(t)$: resistor characterized by its resistance R.

2. $i(t) = Gv(t)$: resistor characterized by its conductance $G\left(G = \dfrac{1}{R}\right)$.

3. $v(t) = f[i(t)]$: current controlled resistor.
4. $i(t) = g[v(t)]$: voltage controlled resistor.

(See Fig. 1.1(a).)

1.1.2 Capacitors

The function defining capacitors relates the charge $q[v(t)]$ to the potential difference at their terminals, $v(t)$. If this is a proportional function, that is, $q(t) = Cv(t)$, then the capacitor is linear. Its capacitance is constant and has value C. If this is not true then the capacitor is non-linear.

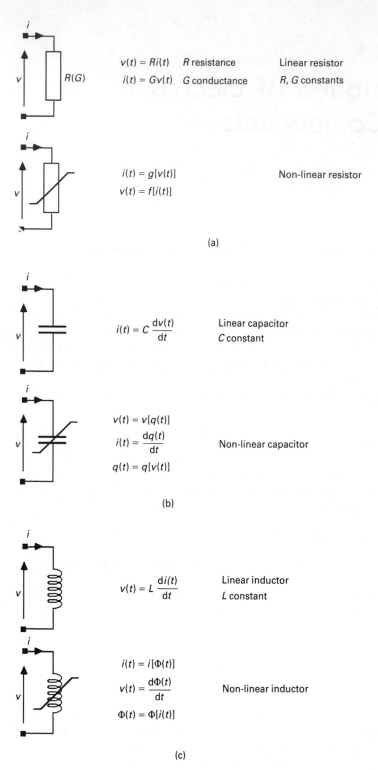

$v(t) = Ri(t)$ R resistance   Linear resistor
$i(t) = Gv(t)$ G conductance R, G constants

$i(t) = g[v(t)]$ Non-linear resistor
$v(t) = f[i(t)]$

(a)

$i(t) = C \dfrac{dv(t)}{dt}$ Linear capacitor
                                      C constant

$v(t) = v[q(t)]$
$i(t) = \dfrac{dq(t)}{dt}$ Non-linear capacitor
$q(t) = q[v(t)]$

(b)

$v(t) = L \dfrac{di(t)}{dt}$ Linear inductor
                                      L constant

$i(t) = i[\Phi(t)]$
$v(t) = \dfrac{d\Phi(t)}{dt}$ Non-linear inductor
$\Phi(t) = \Phi[i(t)]$

(c)

Figure 1.1   (continued)

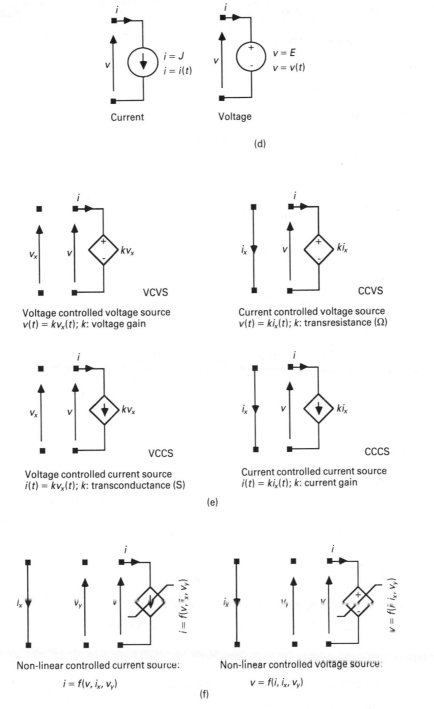

Figure 1.1 Basic elements used in a circuit simulator. (a) Linear and non-linear resistors. (b) Linear and non-linear capacitors. (c) Linear and non-linear inductors. (d) Independent sources. (e) Linear controlled current and voltage sources. (f) Non-linear controlled current and voltage sources

There are two possibilities:

1. When

$$v(t) = f[q(t)]$$

and therefore

$$\frac{dv(t)}{dt} = \frac{df[q(t)]}{dt} = \frac{dq(t)}{dt}\frac{df[q(t)]}{dq}$$

giving

$$i(t) = \frac{1}{\dfrac{df(q)}{dq}}\frac{dv(t)}{dt} = C[q(t)]\frac{dv(t)}{dt}$$

Thus, a capacitance varying as a function of charge is defined

2. When

$$q(t) = f[v(t)]$$

and therefore

$$i(t) = \frac{dq(t)}{dt} = \frac{df[v(t)]}{dt} = \frac{df[v(t)]}{dv}\frac{dv(t)}{dt} = C[v(t)]\frac{dv(t)}{dt}$$

Thus, a capacitance varying as a function of voltage is defined. (See Fig. 1.1(b).)

1.1.3 Inductors

The function defining inductors relates magnetic flux to the current flowing through them. In the same way as for capacitors, linear inductors ($\Phi(t) = Li(t)$) having constant inductance L, and non-linear inductors, denoted by the relations $i(t) = f[\Phi(t)]$ or $\Phi(t) = f[i(t)]$, are defined. (See Fig. 1.1(c).)

1.1.4 Independent current and voltage sources

The sources are characterized either by a current, J, or by a voltage, E, which do not depend on the state of the circuit. J and E can have constant or non-constant values. (See Fig. 1.1(d).)

1.1.5 Linear controlled current and voltage sources

The values of linear controlled current and voltage sources are proportional either to the current flowing through a controlling branch or to the voltage between two nodes of the network. (See Fig. 1.1(e).)

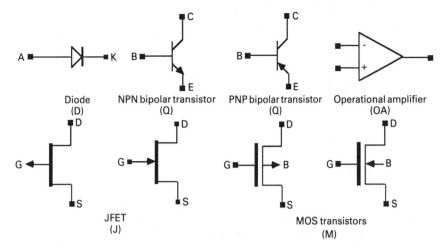

Figure 1.2 Semiconductor devices used in electronic circuits

1.1.6 Non-linear controlled current and voltage sources

The values (current or voltage) of non-linear controlled current and voltage sources are non-linear functions of currents flowing through branches and voltages between any two nodes. These sources can be defined as a circuit element such as previous basic elements in some simulators (PSPICE for example). They will be used in the following sections of this chapter in order to model the semiconductor devices (diode, transistors, ...). (See Fig. 1.1(f).)

1.1.7 Semiconductor devices

Models of semiconductor devices are more complex because they are described by the six preceding basic elements: the main devices are the diode (symbol D), the bipolar transistor (symbol Q), the JFET (symbol J) and the MOS transistor (symbol M) and sometimes op. amp. (Fig. 1.2). When the simulator finds one of these elements, it makes reference immediately to a model. The main models are described in the following sections of this chapter.

1.2 The models of the diode

1.2.1 The ideal DC model

The basic structure of the diode or PN junction is represented in Fig. 1.3(a). The P-region is called the anode (A) and the N-region is called the cathode (K). When

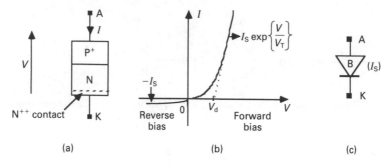

Figure 1.3   (a) The basic diode structure. (b) Its *I–V* characteristics. (c) Symbolic representation used for the ideal Boltzmann diode

the diode is biased, a current flows through the structure. Ideally, this current is represented by the well-known Boltzmann equation

$$I = I_S \left[\exp\left(\frac{V}{V_T}\right) - 1 \right] \tag{1.1}$$

where V is the anode–cathode potential difference $V_A - V_K$ (Fig. 1.3(a)), V_T is the thermal voltage $\{kT/q\}$, and I_S is the diode saturation current. The current I is equal to $-I_S$ for negative voltages ($V < -3V_T$) and the diode is said to be reverse biased. For positive voltages, the current I can be approximated by the equation $I \approx I_S[\exp(V/V_T)]$ and the diode is said to be forward biased.

The different approaches of diode modelling are presented in this section.

The ideal diode (the Boltzmann diode) is most easily modelled by equation (1.1). In this expression, I_S appears as an electrical parameter which can be extracted directly from the *I–V* characteristics (Fig. 1.3(b)). In fact, this parameter is very strongly dependent on the technology, the type of material, the component geometry and the temperature.

The *I–V* characteristic shows, in forward bias, a 'cut-off' voltage V_d (diode threshold voltage). This value, calculated for the standard magnitude currents (1 mA to 1 A), depends strongly on I_S. Although very often used by circuit designers, V_d does not appear as a parameter in the diode model. The order of magnitude of I_S for three types of material per cm^2 of active surface is

1. Germanium: $I_S = 1.2$ mA.
2. Silicon: $I_S = 0.2$ nA.
3. Gallium arsenide: $I_S = 0.75 \times 10^{-16}$ A.

The Boltzmann diode will be represented throughout this work by the symbol shown in Fig. 1.3(c).

1.2.2 Diode secondary effects: the real diode

The properties of semiconductor materials and the technology associated with component fabrication do not allow the construction of diodes strictly having the characteristics described by the Boltzmann equation due to the following:

1. The current and voltage in a semiconductor are not infinite. The maximum current is of the order of 100 A/cm^2. The maximum operating voltage (breakdown voltage) depends on the fabrication technology of the diode.
2. The semiconductor bulk materials have a resistance depending on their doping, which is taken into account by introducing resistors in diode models.
3. The basic theory assumes that the junction is flat and infinite (in area). A diode has a very small surface (several tenths μm^2). Secondary characteristics (carrier recombination, leakage current) are, therefore, taken into account by the introduction of an emission coefficient or a second Boltzmann diode in diode models.

According to the Boltzmann equation, the current I is nearly equal in magnitude to the saturation current I_S for reverse-bias voltage values that are relatively very large compared to $-V_T$. In reality, when a junction is in reverse-bias mode, the current changes can be described by the following empirical equation (avalanche effect):

$$I = \frac{-I_S}{1 - \left(\frac{|V|}{V_{bd}}\right)^n} \qquad -V_{bd} < V < -10\ V_T \quad \text{and} \quad 3 < n < 6 \tag{1.2}$$

V_{bd}, the breakdown voltage, depends on the doping level and the size of the P and N regions.

Equation (1.2) is difficult to implement numerically (very rapid transition from a zero current region to an infinite current region: the denominator goes to zero when $V = -V_{bd}$).

Many different models are used by simulators.

1. *Piecewise model* (MICROCAP II):

$$V > -V_{bd}: \quad I = I_s\left[\exp\left(\frac{V}{V_T}\right) - 1\right]$$

$$\tag{1.3}$$

$$V \leqslant -V_{bd}: \quad I = \frac{V + V_{bd}}{R_z}$$

The reverse-bias operating region is modelled by a voltage source V_{bd} (see Fig. 1.4) in series with a small resistance R_z allowing, thereby, a rapid variation of current for $V < V_{bd}$ (the forward-bias mode is always modelled by Boltzmann's equation).

2. *Non-linear model* (SPICE). The reverse-bias operating region is modelled by an exponential function. In reality, the model uses the overall equation

$$I = I_S\left[\exp\left(\frac{V}{nV_T}\right) - 1\right] - I_{bd}\exp\left(\frac{-V - V_{bd}}{n_{bd}V_T}\right) \qquad 1 < n_{bd} < 2 \tag{1.4}$$

Equation (1.4) demonstrates two currents which don't interact together. If we make

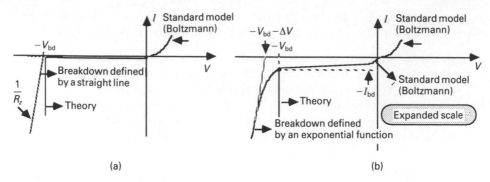

Figure I.4   Various models of the reverse-bias operating region of a diode. (a) MICROCAP model. (b) SPICE model

the realistic assumption that V_{bd} is often greater than a hundred V_T, we can write:

(a) For $V > -V_{bd} + 100\ V_T$ $I = I_S[\exp(V/nV_T) - 1]$ if I_{bd}/I_S is smaller than 10^{+6}. This is the well-known functioning of a Boltzmann diode.

(b) For $V < -V_{bd} + 10 \times V_T$, taking into account the former assumptions, $I = -I_{bd}\exp((-V - V_{bd})/n_{bd}V_T)$; the effect of the Boltzmann diode is equal to zero. In this case, the breakdown region is only modelled. I_{bd} is called the break-down current (value of I at $V = -V_{bd}$). This parameter value can be defined by the user. The current I, in the reverse-bias mode, only has a significant value for $V = -V_{bd} - \Delta V, \Delta V$ representing a threshold voltage (similar to V_d in forward bias − see Fig. 1.4(b)). The breakdown potential is, therefore, equal to $-V_{bd} - \Delta V$.

   Note that in equation (1.4), the current I is not equal to zero for $V = 0$. This is an approximate model. Fig. 1.4 shows the variations of the diode current in the reverse-bias mode, depending on the model described.

Semiconductor materials have a certain resistivity, inversely proportional to the doping levels, represented by resistors in series with the terminals of the ideal component. Therefore, the voltage on the real diode can be decomposed as

$$V_{\text{real diode}} = V_{\text{ideal diode}} + R_S I \tag{1.5}$$

where $R_S I$ represents the voltage drop.

   The real diode is therefore represented by a Boltzmann diode in series with a series resistance R_S (Fig. 1.5(b)), taking into account the resistance of the P and N semiconductor regions. This resistance, of the order of several ohms for a low voltage diode, will induce, if a current of 100 mA flows, a voltage fall of several hundred millivolts, which is comparable to the threshold voltage V_d (0.6 V for a silicon diode). A high voltage diode will have a much more significant series resistance value and therefore an even higher voltage drop.

   In the reverse-bias mode, even with small bias ($V < -5V_T$), the current I is never equal to $-I_S$. This is due to the parasitic phenomena which lead to an inverse conduction. This additional current is modelled by a high value ($\cong$ several MΩ)

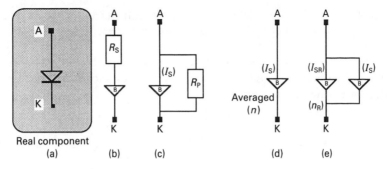

Figure 1.5   Various models for a real diode. (a) The physical component. (b, c) Introduction of series + parallel resistors. (d, e) Introduction of the emission coefficient (modelled with one or two Boltzmann diodes)

parallel resistor between the anode and the cathode. The equivalent circuit, taking this resistance into account, is shown in Fig. 1.5(c).

In practice, the forward-bias diode equation is as follows:

$$I = I_S \left[\exp\left(\frac{V}{nV_T}\right) - 1 \right] \qquad (1.6)$$

where n is called the emission coefficient; it is a function of the current level and is close by 1 for a medium current value. At very low current levels, however, this parameter value has a tendency to increase between 1 and 2. To take this variation into account, some models use two Boltzmann diodes with different n (Fig. 1.5 (e)) or use only one diode with an 'average n' coefficient (Fig. 1.5 (d)).

In the model containing two diodes, the 'n' parameters are different and the 'I_S' values are different, too. In general, the parameters of the diode modelling low injection are named I_{SR} and n_R; I_{SR} is larger than I_S in PSPICE.

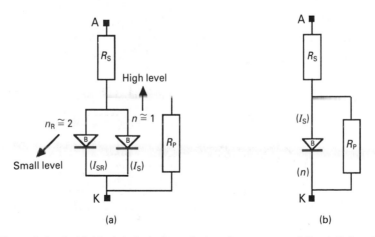

Figure 1.6   (a, b) Models including all the phenomena of Fig. 1.5 (b–e)

Model shown in Fig. 1.6 includes all the phenomena indicated in Fig. 1.5.

An example of a model that takes all the effects described in this section into account is shown in Fig. 1.7.

Note: because of the insertion of the 'n parameter', the Boltzmann diode symbol is modified as is shown in Fig. 1.5(d). If n is not mentioned on the symbolic representation of the diode, then $n = 1$.

1.2.3 The dynamic model

If the voltage across a diode varies rapidly, the junction presents a reactive behaviour. The following two different physical mechanisms give rise to such behaviour:

1. The first takes place in the quasi-neutral region (far from the junction).
2. The second is located at the depletion layer (at the proximity of the junction).

The analysis of these effects attempts to model this dynamical behaviour of the diode by two capacitors, connected in parallel.

When the diode's polarity varies rapidly, the distribution of the associated charge carriers tends towards a new distribution after a brief transient period, that is, there is a charge storage phenomenon. The quantity of associated charge is proportional to the current crossing the junction:

$$Q_D = \tau I(V) \tag{1.7}$$

The constant τ is a time constant which represents the minimum time required to either store or remove the charge; it is called the *transit time*, and is a function of the geometric and physical characteristics of the diode.

The quantity of charge Q_D is associated with a non-linear capacitance C_D

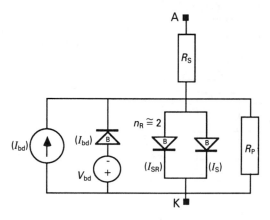

Figure 1.7 Model taking into account the main static phenomena

(diffusion capacitance) in parallel with the Boltzmann diode:

$$C_D = \frac{dQ_D}{dV} = \tau \left.\frac{dI}{dV}\right|_V = \tau \frac{I_S}{V_T} \exp\left[\frac{V}{V_T}\right] \tag{1.8}$$

and $C_D = \tau g$ (g is the dynamic conductance, $g = dI/dV$).

When a varying reverse voltage is applied across the junction, the electrostatic potential difference V_J that supports the depletion zone in the junction region changes as well; this fact creates a variation in the charge Q_J stored in that region and a modification of the electric field (as a result of a change in the voltage). This phenomenon is accounted for by a junction capacitance C_J in parallel with the Boltzmann diode, of the form

$$C_J = \frac{C_{J0}}{\left[1 - \dfrac{V}{\Phi_0}\right]^m} \tag{1.9}$$

where m (called the grading coefficient) is a constant depending on the doping distribution level ($m = 0.5$ for an abrupt junction and $m = 0.33$ for a diffused junction); C_{J0} is a constant representing the junction capacitance for zero external voltage; Φ_0 represents the junction potential.

All the dynamic phenomena are modelled by two parallel capacitances (Fig. 1.8); in this diagram, for simplicity, the breakdown phenomena are omitted.

Therefore, a total non-linear capacitance is obtained:

$$C = C_D + C_J \tag{1.10}$$

In equation (1.9), C_J tends toward infinity as V tends toward Φ_0. This equation no longer reflects a physical reality and is therefore no longer valid. A more accurate analysis of the capacitance C_J is represented by Fig. 1.9. The values of C_J can be extracted from experimental data. The following approximation has been found in

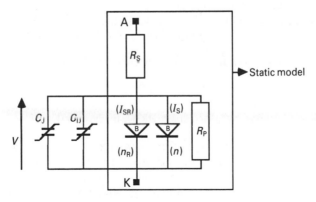

Figure 1.8 Dynamical model of the diode

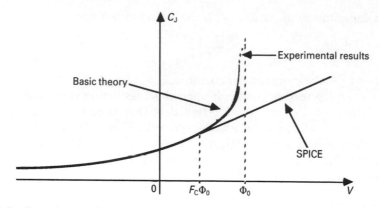

Figure 1.9   Capacitance C_J as a function of applied voltage: the different approaches

a number of simulators:

1. If $V < F_C\Phi_0$, C_J is represented by equation (1.9).
2. If $V > F_C\Phi_0$, the relation $C_J(V)$ is linearized apart from $V = F_C\Phi_0$. F_C is called foward-bias depletion capacitance coefficient.

In general, F_C is taken to equal 0.5. The following equations are used in SPICE:

$$C = \tau\,\frac{\mathrm{d}I}{\mathrm{d}V} + C_{J0}\left[1 - \frac{V}{\Phi_0}\right]^{-m} \qquad V < F_C\Phi_0 \qquad (1.11\mathrm{a})$$

$$C = \tau\,\frac{\mathrm{d}I}{\mathrm{d}V} + \frac{C_{J0}}{[1 - F_C]^{1+m}}\left[1 - F_C(1 + m) + \frac{mV}{\Phi_0}\right] \qquad V > F_C\Phi_0 \qquad (1.11\mathrm{b})$$

In practice, for both reverse bias and weak forward bias, the capacitance C_J is dominant. Moreover, for a strongly forward-biased diode ($V \gg V_T$), C_D becomes dominant with respect to C_J.

Some software packages use other approximations of C_J. For example, we can model the capacitance by using equation (1.9) if V is negative; if V is greater than zero, the capacitance has a constant value: C_{J0}.

1.2.4 Small signal model

For small variations around an operating point fixed by a constant source (small signal mode of operation), the non-linear characteristics of a diode can be linearized.

From the non-linear model shown in Fig. 1.8, after linearization, the corresponding incremental diode conductance is obtained using the following equation:

$$g = \left.\frac{dI}{dV}\right|_{V^*} = \frac{I_S}{nV_T} \exp\left[\frac{V^*}{nV_T}\right] + \frac{I_{SR}}{n_R V_T} \exp\left[\frac{V^*}{n_R V_T}\right] \tag{1.12}$$

V^* represents the operating point voltage.

The value of g varies to a great extent. It is zero for reverse bias and proportional to I for forward bias.

Also, the total charge stored by the diode ($Q = Q_D + Q_J$) undergoes a variation of ΔQ, so that

$$C_D = \left.\frac{dQ_D}{dV}\right|_{V^*} \quad \text{and} \quad C_J = \left.\frac{dQ_J}{dV}\right|_{V^*} \tag{1.13}$$

Then, we obtain:

$$C = \left.\frac{dQ}{dV}\right|_{V^*} = \frac{\tau}{nV_T} \exp\left[\frac{V^*}{nV_T}\right] + C_{J0}\left[1 - \frac{V^*}{\Phi_0}\right]^{-m} \qquad \text{for } V < F_C\Phi_0 \tag{1.14a}$$

$$C = \left.\frac{dQ}{dV}\right|_{V^*} = \frac{\tau}{nV_T} \exp\left[\frac{V^*}{nV_T}\right] + \frac{C_{J0}}{[1 - F_C]^{1+m}}$$

$$\times \left[1 - F_C(1+m) + \frac{mV^*}{\Phi_0}\right] \qquad \text{for } V \geqslant F_C\Phi_0 \tag{1.14b}$$

Thus, the small signal model of a diode is represented in Fig. 1.10.

1.2.5 Models used in typical circuits

The models used in most cases in the following chapters are represented in Fig. 1.11(a, b). For convenience, the expressions for the various elements used are recalled: $\Phi_0 = 0.7$ V; $m = 0.5$; $F_C = 0.5$.

Figure 1.10 Diode small-signal model

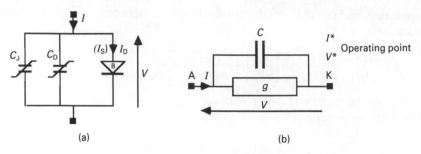

Figure 1.11   Models used in classical circuits. (a) 'Large signal' model. (b) 'Small signal' model

For the large signal model we use the following equations:

$$I_\mathrm{D} = I_\mathrm{S}\left[\exp\left(\frac{V}{V_\mathrm{T}}\right) - 1\right] \text{ (cf. equation (1.1))}$$

$$C = C_\mathrm{D} + C_\mathrm{J} \text{ (cf. equation (1.11))}$$

$$C_\mathrm{D} = \tau\,\frac{I_\mathrm{S}}{V_\mathrm{T}}\exp\left[\frac{V}{V_\mathrm{T}}\right] \text{ (cf. equation (1.8))}$$

$$C_\mathrm{J} = C_{J0}\,\frac{1}{\sqrt{1 - \dfrac{V}{0.7}}} \qquad V < 0.35 \text{ volt}$$

$$C_\mathrm{J} = 2.83\ C_{J0}\left[0.25 + \frac{V}{1.4}\right] \qquad V > 0.35 \text{ volt}$$

and for the small signal model:

$$g = \frac{I_\mathrm{S}}{V_\mathrm{T}}\exp\left[\frac{V^*}{V_\mathrm{T}}\right] = 0 \qquad \text{reverse bias}$$

$$g = \frac{I^*}{V_\mathrm{T}} \qquad \text{forward bias}$$

$$C_\mathrm{D} = \tau\,\frac{I^*}{V_\mathrm{T}} \qquad \text{forward bias}$$

$$C_\mathrm{J} = C_{J0}\,\frac{1}{\sqrt{1 - \dfrac{V^*}{0.7}}} \qquad \text{reverse bias}$$

1.3 The bipolar transistor

Fig. 1.12 summarizes the different regions according to the polarization of the base–emitter and base–collector junctions of an NPN transistor (represented in

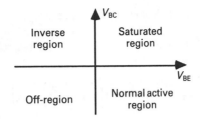

Figure 1.12   Regions of operation for a bipolar NPN transistor

Fig. 1.2):

1. Normal active region: these are the conditions necessary to produce the transistor effect, that is, $V_{BE} > 0$ and $V_{BC} < 0$. For these conditions, we can write $I_C = \beta_F I_B$.
2. Saturated region: $V_{BE} > 0$ and $V_{BC} > 0$. The two junctions are forward biased and inject charge carriers into the base. The current gain β is reduced.
3. Inverse region: $V_{BE} < 0$ and $V_{BC} > 0$. Collector and emitter are interchanged. Here, there is a transistor effect as in the normal active region but much weaker (this is due to the different properties and characteristics of the two junctions): $I_E = \beta_R I_B \ (\beta_R \ll \beta_F)$.
4. Off region: $V_{BE} < 0$ and $V_{BC} < 0$. The two junctions being reverse biased, the current is practically equal to zero: $I_C \approx I_B \approx I_E \approx 0$.

The transistor characteristics are frequently represented by the $I_C(V_{CE})$ functions for a given base current (Fig. 1.13). Only the active and saturated regions are represented.

   The bipolar transistor is considered as two connected diodes and the approach to model generation is fundamentally the same as that of the diode; classically, two basic models are used: the Ebers–Moll model and the Gummel–Poon model.

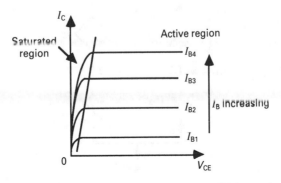

Figure 1.13 $I_C(V_{CE})$ characteristics of an NPN transistor

1.3.1 The basic static model: Ebers–Moll equations

The basic expressions of the transistor currents as a function of base–emitter and base–collector voltages are as follows:

$$I_C = \alpha_F I_{ES}\left[\exp\left(\frac{V_{BE}}{V_T}\right) - 1\right] - I_{CS}\left[\exp\left(\frac{V_{BC}}{V_T}\right) - 1\right] \tag{1.15}$$

$$I_E = -I_{ES}\left[\exp\left(\frac{V_{BE}}{V_T}\right) - 1\right] + \alpha_R I_{CS}\left[\exp\left(\frac{V_{BC}}{V_T}\right) - 1\right] \tag{1.16}$$

$$I_B = -\{I_C + I_E\} \tag{1.17}$$

In these expressions, α_F and α_R are, respectively the large signal forward and reverse current gain of a common base bipolar transistor. These equations can be represented by the Ebers–Moll equivalent circuit shown in Fig. 1.14, where the base–emitter diode is characterized by the saturation current I_{ES} and the base–collector diode by I_{CS}. The two controlled non-linear sources I_{CC} and I_{EC} are functions of diode currents (I_F and I_R):

$$I_F = \frac{I_{CC}}{\alpha_F} \tag{1.18}$$

$$I_R = \frac{I_{EC}}{\alpha_R} \tag{1.19}$$

$$I_{CC} = \alpha_F I_{ES}\left[\exp\left(\frac{V_{BE}}{V_T}\right) - 1\right] = \alpha_F I_F \tag{1.20}$$

$$I_{EC} = \alpha_R I_{CS}\left[\exp\left(\frac{V_{BC}}{V_T}\right) - 1\right] = \alpha_R I_R \tag{1.21}$$

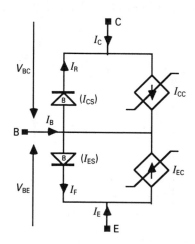

Figure 1.14   Basic Ebers–Moll model

This model has four parameters: I_{CS}, I_{ES}, α_R, α_F. However, these parameters are not independent and are related as follows: $\alpha_F I_{ES} = \alpha_R I_{CS} = I_S$.

I_S is called the transistor saturation current. Thus, the Ebers–Moll model uses three parameters: I_S, α_R and α_F. The preceding equations become

$$I_F = \frac{I_S}{\alpha_F}\left[\exp\left(\frac{V_{BE}}{V_T}\right) - 1\right] \tag{1.22}$$

$$I_R = \frac{I_S}{\alpha_R}\left[\exp\left(\frac{V_{BC}}{V_T}\right) - 1\right] \tag{1.23}$$

$$I_{EC} = I_S\left[\exp\left(\frac{V_{BC}}{V_T}\right) - 1\right] \tag{1.24}$$

$$I_{CC} = I_S\left[\exp\left(\frac{V_{BE}}{V_T}\right) - 1\right] \tag{1.25}$$

There are two approaches for the Ebers–Moll model: the injection model (Fig. 1.15(a)) and the transport model (Fig. 1.15(b)). Mathematically, the two models are identical. Their difference comes from the reference currents used. In the injection model, the reference currents I_F and I_R are the currents crossing the diodes. In the transport model, the reference currents are the current sources I_{CC} and I_{EC}.

Generally, the α_F and α_R parameters are not used for transistor modelling. They are replaced by the more familiar β_F and β_R parameters:

$$\beta_F = \frac{\alpha_F}{1 - \alpha_F} \text{ and } \beta_R = \frac{\alpha_R}{1 - \alpha_R} \tag{1.26}$$

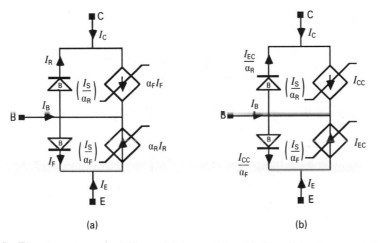

(a)   (b)

Figure 1.15   The two types of Ebers–Moll models. (a) The injection model. (b) The transport model

β_F and β_R are, respectively, the large signal forward and reverse current gains of a common emitter transistor. The Ebers–Moll model can be modified by replacing I_{CC} and I_{EC} by a single source using the following expression:

$$I_{CT} = I_{CC} - I_{EC} = I_S\left[\exp\left(\frac{V_{BE}}{V_T}\right) - \exp\left(\frac{V_{BC}}{V_T}\right)\right] \qquad (1.27)$$

This model is represented in Fig. 1.16. The model has a different topology; the base–emitter and base–collector diodes are modified, too. Now, they don't represent the physical characteristics of the junctions since their saturation currents are different: I_S/β_F for base–emitter diode and I_S/β_R for base–collector diode.

   This is the simplest model usually used in transistor circuit simulators. In fact, it only requires three parameters β_F, β_R, I_S.

1.3.2 The static model: secondary effects

The semiconductor material used and the technology associated with the fabrication of the bipolar transistor do not permit the components to be represented solely by the Ebers–Moll equations. As in the diode case, the following phenomena need to be taken into account: resistivity of the materials, breakdown phenomenon, Early effect and others.

   The material resistivity of emitter, base and collector regions are represented as resistors on each of the base, emitter and collector contacts shown in Fig. 1.17. Their values are theoretically a function of the semiconductor parameters (size, doping levels, etc., ...). The emitter and collector resistances are usually considered constant. Their values ($\cong 1\ \Omega$ for emitter, $\cong 10\ \Omega$ for collector) influence the simulation minimally. The base resistance is also assumed constant in the Ebers–Moll model. In fact, this resistance value depends on the level of the current. This

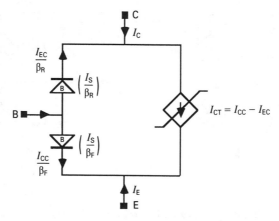

Figure 1.16   The modified Ebers–Moll model used in simulators

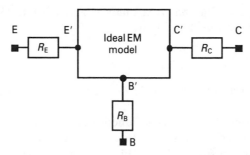

Figure 1.17 Ebers–Moll model taking the terminal resistances into account

variation is taken into account in the Gummel–Poon model (described in the next section). In SPICE, for example, two types of base resistance are used:

1. A constant external resistance R_{B0} representing the contact resistance.
2. An internal resistance of the active base region (base region physically under the emitter) varying as a function of the base current; this resistance is defined by two parameters R_{BM} and I_{RB} and causes a non-uniform polarity of the base–emitter junction.

The variation of this resistance as a function of the base current I_B is shown in Fig. 1.18 and is represented by the equation

$$R_B = R_{BM} + 3(R_{B0} - R_{BM})\left(\frac{\tan z - z}{z\,(\tan z)^2}\right) \qquad \text{where: } z = \frac{-1 + \sqrt{1 + \dfrac{144\,I_B}{\pi^2 I_{RB}}}}{\dfrac{24}{\pi^2}\sqrt{\dfrac{I_B}{I_{RB}}}}$$

$$(1.28)$$

In the normal active region, the collector current I_C has the following expression:

$$I_C = I_S \exp\frac{V_{BE}}{V_T} - I_S \exp\frac{V_{BC}}{V_T} - \frac{I_S}{\beta_R}\left[\exp\frac{V_{BC}}{V_T} - 1\right] \qquad (1.29)$$

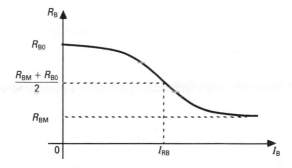

Figure 1.18 Variation of the base resistance as a function of the current I_B

However, in this region V_{BC} has a strongly negative value ($V_{BC} \ll -V_T$), so the preceding expression can be simplified:

$$I_C \cong I_S \exp \frac{V_{BE}}{V_T} \tag{1.30}$$

For a given base current ($V_{BE} = $ const), I_C is constant, too. In fact I_S depends on physical base dimensions.

Strong reverse polarization of B–C junctions causes an increase of I_S; therefore the current I_C increases equally and the $I_C(V_{CE})$ characteristics are no longer horizontal as is shown in Fig. 1.19(a). This phenomenon is called the Early effect. This feature can be accounted for by introducing the Early voltage V_A, which is shown in Fig 1.19(b).

The slope of the characteristics can be calculated simply:

$$\frac{\partial I_C}{\partial V_{CE}} = \frac{\partial I_C}{\partial V_{CB}} = \frac{I_{C0}}{V_A} \text{ (for } V_{BE} = \text{constant), with } I_{C0} = \beta_F I_B \tag{1.31}$$

Thus, we obtain the linearized characteristic:

$$I_C = I_{C0}\left[1 + \frac{V_{CB}}{V_A}\right] \tag{1.32}$$

Very strong reverse bias of the B–C junction causes an increase of the collector current I_C due to the avalanche phenomenon. The increase of I_C is represented, as in the diode, by the following expression:

$$I_C = \frac{I_{C0}}{1 - \left[\dfrac{V_{CB}}{BV_{CB}}\right]^n} \tag{1.33}$$

BV_{CB} represents the breakdown voltage of the collector–base junction; I_{C0} is the value of the current I_C without the avalanche phenomenon. This phenomenon is not modelled in PSPICE.

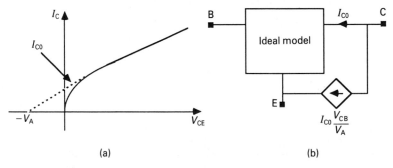

(a)   (b)

Figure 1.19 $I_C(V_{CE})$ characteristics. (a) Early effect. (b) Model taking into account the Early effect

1.3.3 The Gummel–Poon static model (the variations of β with current)

The Ebers–Moll model doesn't take into account the second-order phenomena existing in the majority of actual components: the effects of weak and strong injection. When the currents are weak, an additional current due to carrier recombination must be taken into account: this current decreases the gain. The effect of large currents equally decreases the gain and is modelled by modification of the saturation current I_S. The curve of the gain $\beta(I_C)$ with $V_{BC} = \text{const}$ shown in Fig. 1.20(b) represents the following three regions:

1. Region 1 (weak current level): β increases as a function of I_C.
2. Region 2 (medium current level): β is a constant (equal to β_{FM}).
3. Region 3 (high current level): β decreases as a function of I_C.

Let us describe in more detail the variation in each region.

1. *Medium current: region 2.* The gain in this region is called β_{FM}. The two currents I_B and I_C then have the following expressions (for $V_{BC} = 0$):

$$I_C = I_S\left[\exp\left\{\frac{V_{BE}}{V_T}\right\} - 1\right]$$

$$I_B = \frac{I_S}{\beta_{FM}}\left[\exp\left\{\frac{V_{BE}}{V_T}\right\} - 1\right] \tag{1.34}$$

The values of β_{FM} and I_S can be obtained directly from the graph of Fig. 1.20(a) for $V_{BC} = 0$. In this case, β_{FM} is equal to β_F defined for the Ebers–Moll model.

2. *Low current: region 1.* The decrease of β is caused by carrier recombination creating an additional current. This phenomenon is modelled by two Boltzmann diodes, one between the base and emitter, the other between the base and collector

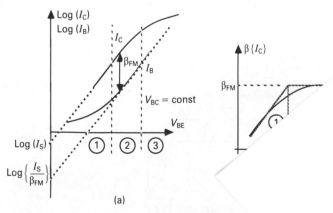

Figure 1.20   (a) Base and collector currents variation.
                    the current gain β

(as diode modelling in Section 1.2). This modelling requires four parameters:

(a) I_{SE} and I_{SC}: saturation currents of base–emitter and base–collector diodes modelling low injection effects;
(b) n_{EL} and n_{CL}: corresponding non-ideality factors.

These diode currents are added to the ideal diode currents, thus modelling the transistor. Fig. 1.21 represents the bipolar transistor model with the recombination components.

From the measurement of $\beta(I_C)$, one can determine the value of the current I_L (Fig. 1.20(b)). I_L represents the limit between regions 1 and 2. It is represented by the following equation:

$$I_L = I_S \left(\frac{\beta_{FM} I_{SE}}{I_S} \right)^{\frac{n_{EL}}{n_{EL} - 1}} \qquad (1.35)$$

3. *Strong injection: region 3.* The phenomenon of strong injection is modelled by recalculating the parameter I_S. The I_S general expression, which is a function of the voltage, can be represented by I_{S0} (I_{S0} is the I_S value in region 2 which is also the parameter in the Ebers–Moll model when $V_{BC} = 0$). I_{CT} (see equation (1.27)) can then be written in the following way:

$$I_{CT} = \frac{I_{S0}}{q_b} \left[\exp\left\{ \frac{V_{BE}}{V_T} \right\} - \exp\left\{ \frac{V_{BC}}{V_T} \right\} \right] \qquad (1.36)$$

where $q_b = \dfrac{Q_B}{Q_{B0}}$

The coefficient q_b representing the ratio between the total majority base charge,

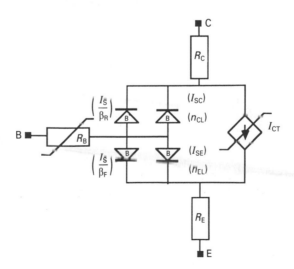

Figure 1.21   Equivalent circuit of the Gummel–Poon model

Q_B and zero-bias majority base charge Q_{B0} takes into account the following features in its analytical equation:

(a) the Early effect (through the intermediary voltage V_A);
(b) the phenomenon of strong injection (through the 'knee' currents I_{KF} and I_{KR} (for the reverse region) corresponding to the start of the drop in gain represented in Fig. 1.20(b)).

Note: For $V_{BC} < 0$, in the medium current region, the β gain is larger than β_{FM} because the Early effect increases the I_C current and thus the β gain.

1.3.4 The dynamic 'large signal' model

Having defined the static models, we are now going to consider the effects of stored charges in these models. The importance of these phenomena has already been presented in the section relating to the diode model. They are identical. This model is obtained by introducing the following three capacitors, connected between electrode junctions:

1. Two capacitors between base–emitter and base–collector terminals.
2. One substrate capacitor: integrated circuit transistors have junctions between the collector and the substrate of the chip, often reverse biased and introducing a junction capacitor.

The dynamic large signal model described here corresponds to the Ebers–Moll model.

The modelling of B–E and B–C capacitors is similar to that of the diode capacitor (see Section 1.2.3). They are divided in two types:

1. Junction capacitances modelled by the following parameters:

   (a) V_{JE}, Φ_E, C_{JE0}; M_{JE} for base–emitter,
   (b) V_{JC}, Φ_C, C_{JC0}; M_{JC} for base–collector,
   (c) F_C for the two junctions.

2. Diffusion capacitances which are function of I_{EC} or I_{CC} currents:

$$C_{DE} = \tau_F \frac{d(I_{CC})}{dV_{BE}} = \frac{I_S}{V_T} \tau_F \exp \frac{V_{BE}}{V_T} \tag{1.37}$$

$$C_{DC} = \tau_R \frac{d(I_{EC})}{dV_{BC}} = \frac{I_S}{V_T} \tau_R \exp \frac{V_{BC}}{V_T} \tag{1.38}$$

where τ_F and τ_R represent the forward and reverse transit time, respectively. These capacitors are represented in Fig. 1.22.

For integrated circuit transistors, it is also necessary to take into account the substrate capacitor C_{JS} from an identical expression as for the junction capacitor. The substrate capacitor is connected between collector and substrate for the NPN

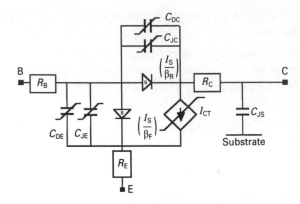

Figure 1.22   Transistor dynamic large signal Ebers–Moll model

transistor, as shown in Fig. 1.22. As a typical junction capacitance, it requires the following parameters: C_{JS0}, Φ_S, m_S, F_C.

Note that, in Fig. 1.22, the Boltzmann diode and the two capacitors C_J and C_D can be associated to a real diode with the following parameters:

1. $I_{Seq} = \dfrac{I_S}{\beta_F}$, $\tau_{eq} = \beta_F \tau_F$ and $C_{J0} = C_{JE0}$ for the B–E junction.

2. $I_{Seq} = \dfrac{I_S}{\beta_R}$, $\tau_{eq} = \beta_R \tau_R$ and $C_{J0} = C_{JC0}$ for the B–C junction.

In the case of the dynamic Gummel–Poon model, the non-ideal diodes don't modify the values of the capacitors C_J and C_D.

1.3.5 The small signal model

This model is built directly from the dynamic transistor model. It is shown in

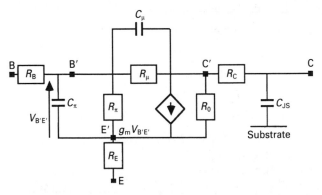

Figure 1.23   Small signal transistor model

Fig. 1.23. R_B, R_E and R_C are the series resistances described in Section 1.3.2. R_π, C_π and R_μ, C_μ are the capacitances and dynamic resistances of the diodes. A resistor of value R_0 is added to take into account the Early effect. C_{JS} represents the substrate capacitance.

The following expressions define the small signal parameters:

$$g_m = \left.\frac{\mathrm{d}I_C}{\mathrm{d}V_{BE}}\right|_{(V_{BE}^*)}; \qquad I_C \text{ is the DC collector current} \qquad (1.39a)$$

$$R_\pi = \left.\frac{\mathrm{d}V_{BE}}{\mathrm{d}I_{BE}}\right|_{(V_{BE}^*)}; \qquad I_{BE} \text{ is the effective current of the base–emitter}$$
$$\text{diode} \quad (1.39b)$$

$$R_\mu = \left.\frac{\mathrm{d}V_{BC}}{\mathrm{d}I_{BC}}\right|_{(V_{BC}^*)}; \qquad I_{BC} \text{ is the effective current of the base–collector}$$
$$\text{diode} \quad (1.39c)$$

$$C_\pi = C_{BE}|_{(V_{BE}^*)} = C_{DE}|_{(V_{BE}^*)} + C_{JE}|_{(V_{BE}^*)} \qquad (1.39d)$$

$$C_\mu = C_{BC}|_{(V_{BC}^*)} = C_{DC}|_{(V_{BC}^*)} + C_{JC}|_{(V_{BC}^*)} \qquad (1.39e)$$

$$R_0 = \left.\frac{\mathrm{d}V_{CE}}{\mathrm{d}I_{CT}}\right|_{(V_{CE}^*)} \qquad (1.39f)$$

where V_{BE}^*, V_{BC}^* and V_{CE}^* are quiescent (operating) point voltages.

1.3.6 Simplified models used in typical circuits

The models used in the following chapters are shown in Figs 1.24 and 1.25. For

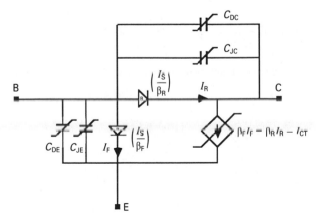

Figure 1.24 Simplified 'large signal' transistor model

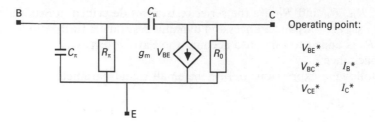

Figure 1.25   Simplified 'small' signal transistor model

convenience, as in the diode models, the expressions for the various elements used are recalled. We assume that $\Phi_E = \Phi_C = 0.7$ V, $m_E = m_C = 0.5$ and $F_C = 0.5$.

1.   *'Large signal' model* (Fig. 1.24)

$$I_F = \frac{I_S}{\beta_F}\left[\exp\left(\frac{V_{BE}}{V_T}\right) - 1\right] \qquad I_R = \frac{I_S}{\beta_R}\left[\exp\left(\frac{V_{BC}}{V_T}\right) - 1\right]$$

$$I_{CT} = \beta_F I_F - \beta_R I_R = I_S\left[\exp\left(\frac{V_{BE}}{V_T}\right) - \exp\left(\frac{V_{BC}}{V_T}\right)\right]$$

$$C_{DE} = \tau_F \frac{I_S}{V_T}\exp\left[\frac{V_{BE}}{V_T}\right] \qquad C_{DC} = \tau_R \frac{I_S}{V_T}\exp\left[\frac{V_{BC}}{V_T}\right]$$

$$C_{JE} = C_{JE0}\frac{1}{\sqrt{1 - \dfrac{V_{BE}}{0.7}}} \qquad V_{BE} < 0.35 \text{ volt}$$

$$C_{JE} = 2.83 C_{JE0}\left[0.25 + \frac{V_{BE}}{1.4}\right] \qquad V_{BE} > 0.35 \text{ volt}$$

$$C_{JC} = C_{JC0}\frac{1}{\sqrt{1 - \dfrac{V_{BC}}{0.7}}} \qquad V_{BC} < 0.35 \text{ volt}$$

$$C_{JC} = 2.83\, C_{JC0}\left[0.25 + \frac{V_{BC}}{1.4}\right] \qquad V_{BC} > 0.35 \text{ volt}$$

2.   *'Small signal' model* (Fig. 1.25)

$$R_\pi = \left[\frac{I_S}{\beta_F V_T}\exp\left[\frac{V_{BE}^*}{V_T}\right]\right]^{-1} = \frac{V_T}{I_B^*}$$

$$g_m = \frac{I_S}{V_T}\exp\left[\frac{V_{BE}^*}{V_T}\right] = \frac{\beta_F}{R_\pi}$$

$$C_\pi = \frac{\tau_F}{V_T} \exp\left[\frac{V_{BE}^*}{V_T}\right] = \beta_F \tau_F I_B^*$$

$$C_\mu = C_{JC0} \frac{1}{\sqrt{1 - \dfrac{V_{BC}^*}{0.7}}}$$

$$R_0 = \frac{V_A}{V_T g_m}$$

Typically, for an AC simulation, the bipolar transistor is in the normal active region ($I_C = \beta_F I_B$) and so $V_{BE} \gg V_T$ and $V_{BC} \ll -V_T$.

1.4 The MOS transistor

There exist two types of transistor: the N channel MOS and the P channel MOS (see Fig. 1.2). For N channel MOS, the normal polarization conditions are where the drain and the gate are positively biased with respect to the source. The voltage V_{GS} determines the concentration of carriers in the channel. When V_{GS} is greater than V_{Th}, the threshold voltage, a current flows between source and drain. This operating region can be divided into the following two parts:

1. For low voltages $V_{DS} < V_{Dsat}$ (saturation voltage), the current I_D increases with the voltage V_{DS} (ohmic region).
2. For $V_{DS} > V_{Dsat}$, the current I_D becomes independent of V_{DS} ('pinch-off' or saturation region).

These characteristics are represented in Fig. 1.26. For a gate–source voltage equal to zero, some transistors are in the on-state; these transistors are called depletion

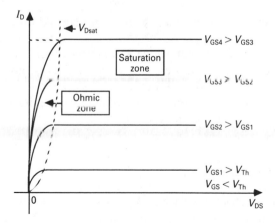

Figure 1.26 $I_D(V_{DS})$ characteristics of an N channel MOS transistor

MOS transistors. A voltage of opposite sign to that of normal conditions has to be applied to cancel the conduction.

The MOS transistor is a very widely used component in integrated circuits and so its modelling is very important. Models of different complexity are presented in this section. The study will always use an N-channel MOS transistor. The conversion to a P-channel MOS can be made by changing the signs of polarities and currents. We present here the following two main types of models:

1. The first-level model or basic model defined from the theory of the MOS transistor and only valid for 'long channel' devices.
2. The second-level model or complete model defined with physical secondary phenomena included.

1.4.1 The basic static model

The threshold voltage of N MOS transistors. The threshold voltage V_{Th} above which the transistor conducts depends on the doping density of the substrate N_B, the thickness of the gate oxide t_{ox} and additional physical parameters characterizing the metal–oxide and oxide–semiconductor interfaces. It can be shown that V_{Th} is given by the following equation:

$$V_{Th} = V_{FB} + \Phi_S + \gamma\sqrt{\Phi_S - V_{BS}} \qquad (1.40)$$

where: V_{FB}–the flat band voltage characteristic of the material (voltage which, when applied to the semiconductor–oxide junction, cancels out the effect of the stored charge at this interface)

Φ_S – diffusion potential

γ – bulk threshold parameter $= \dfrac{\sqrt{2\varepsilon_S q N_B}}{C_{ox}}$ \qquad (1.41)

ε_S – dielectric constant of silicon and C_{ox} represents the oxide capacitor (for an area of 1 cm^2).

Note that in expression (1.40) V_{Th} is a function of the bulk–source potential V_{BS}. Often, this potential is considered to be zero. This is, however, not the case in integrated circuits. Frequently, all the transistors are made on the same substrate and if the sources are not at the same potential there will be, inevitably, a bulk effect on some transistors. Most models take V_{BS} into account. The threshold voltage V_{Th0} corresponds to $V_{BS} = 0$.

The drain current characteristics.

1. Ohmic or non-pinched zone: when $V_{GS} > V_{Th}$, the electrons accumulate near to the oxide surface, allowing a current to flow. The calculation principle of drain current I_D of a MOS structure comes from an integration of Ohm's

law, along the conduction channel MOS structure. The following expression can be obtained:

$$I_D = \mu_0 C_{ox} \frac{W}{L}\left(V_{GS} - V_{Th} - \frac{V_{DS}}{2}\right) V_{DS} \tag{1.42}$$

where μ_0 represents the carrier mobility at the surface and W and L represent the width and length of the gate electrode, respectively.

2. Pinched or saturated zone: equation (1.42) is valid as long as the pinch-off conditions are not reached, that is, as long as there is a channel all along the gate. In effect, the conduction channel does not depend on V_{GS} alone but also on V_{DS}. When the voltage, V_{DS}, is greater than V_{Dsat}, the channel next to the drain disappears; the current remains constant:

$$I_D = \mu_0 C_{ox} \frac{W}{2L} (V_{GS} - V_{Th})^2, \; V_{DS} \geqslant V_{Dsat} = V_{GS} - V_{Th} \tag{1.43}$$

The condition $V_{DS} = V_{Dsat}$ is the limit of the pinch-off.

The basic static model of a MOS transistor is represented by a non-linear current source connected between the source and drain and is controlled by the gate–source voltage (Fig. 1.27). The current source I_D has the following expressions:

$$I_D = 0 \qquad V_{GS} < V_{Th} \qquad \text{off region} \tag{1.44a}$$

$$I_D = \beta\left(V_{GS} - V_{Th} - \frac{V_{DS}}{2}\right)V_{DS}, \; V_{GS} \geqslant V_{Th} \text{ and } V_{GS} - V_{DS} > V_{Th}$$
$$\text{ohmic region} \quad \text{(1.44b)}$$

$$I_D = \frac{\beta}{2} (V_{GS} - V_{Th})^2, \qquad V_{GS} \geqslant V_{Th} \text{ and } V_{GS} - V_{DS} \leqslant V_{Th}$$
$$\text{pinched region} \quad \text{(1.44c)}$$

where $\beta = \mu_0 C_{ox} \dfrac{W}{L} = K_p \dfrac{W}{L}$ is called the transconductance parameter.

Note that the gate–source input resistance is physically infinite because of the gate oxide insulation.

Starting from these equations, the $I_D(V_{DS})$ equation can be written for both transistor types, N-channel and P-channel. Table 1.1 summarizes the bias conditions of MOS transistors

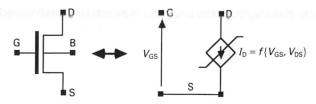

Figure 1.27 Basic static model of a MOS transistor

Table 1.1   Bias conditions of MOS transistors

N-type enhancement MOS	$V_{Th} > 0$	$V_{DS} > 0,\ V_{GS} > 0$
N-type depletion MOS	$V_{Th} < 0$	$V_{DS} > 0,\ V_{GS} > 0$
P-type enhancement MOS	$V_{Th} < 0$	$V_{DS} < 0,\ V_{GS} < 0$
P-type depletion MOS	$V_{Th} > 0$	$V_{DS} < 0,\ V_{GS} < 0$

1.4.2 Extensions of the basic static model

For integrated circuit transistors, it is necessary to develop a much more sophisticated static model to take into account phenomena due to reduced dimensions such as output resistance, terminal resistances and parasitic diodes.

Channel length modulation (output resistance). When the transistor enters in pinchoff mode, the channel disappears near to the drain. Little by little, as the voltage V_{DS} increases, the real length of the channel decreases and so the current rises. To a first approximation, this can be modelled by introducing a parameter λ, called the channel-length modulation factor, which modifies the current equation in the following way:

$$I_D = I_{D0}(1 + \lambda V_{DS}) \tag{1.45}$$

where I_{D0} is the value of the drain current calculated by equation (1.44). From this equation, we obtain a constant output conductance g_{ds}:

$$g_{ds} = g_{out} = \frac{dI_D}{dV_{DS}} = \lambda I_{D0} \tag{1.46}$$

Two terminal resistors R_S and R_D next to the source and drain, respectively, shown in Fig. 1.28, represent the material's resistivity. They are generally of the order of several ohms.

The two diodes, respectively connected between the drain and the bulk and the source and the bulk, model the transistor parasitic junctions between drain, source and substrate (bulk). They normally operate in the reverse-bias mode and therefore have a current of the same order as the saturation current I_S. The basic level 1 static model is shown in Fig. 1.28. All the electrical parameters of the model can be calculated simply from the physical, geometric and technological parameters:

$$\beta = \mu_0 C_{ox} \frac{W}{L} = K_p \frac{W}{L} \quad \text{and} \quad C_{ox} = \frac{\varepsilon_{ox}}{t_{ox}} \tag{1.47}$$

Only λ is defined from electrical measurements (a physical calculation will be proposed in the level 2 model).

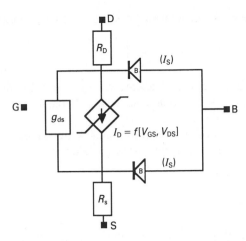

Figure 1.28   The basic level 1 static model

1.4.3 The complete static model (Level 2)

The complete model of the MOS transistor is derived from a more rigorous calculation for the drain current I_D and the different electrical parameters contained in this expression. Its topology is identical to that of the basic level 1 model shown in Fig. 1.28.

The calculation principle for the drain current is the same as for the basic model. Ohm's law is integrated along the conduction channel but without making an approximation. Of course, more complex relations are obtained, but these are still analytical ones.

Ohmic region.

$$I_D = K_P \frac{W}{L} \left\{ \left(V_{GS} - V_{FB} - \Phi_S - \frac{V_{DS}}{2} \right) V_{DS} \right.$$

$$\left. - \frac{2}{3}\,\gamma \left[(V_{DS} - V_{BS} + \Phi_S)^{3/2} - (-V_{BS} + \Phi_S)^{3/2} \right] \right\} \qquad (1.48)$$

Pinched region. The following value for $V_{DS} = V_{Dsat}$ is obtained:

$$V_{Dsat} = V_{GS} - V_{FB} - \Phi_S + \gamma^2 \left[1 - \sqrt{1 + \frac{2}{\gamma^2}\,(V_{GS} - V_{FB})} \right] \qquad (1.49)$$

and therefore $I_D = I_{Dohmic}\,(V_{Dsat})$ where $I_{Dohmic}\,(V_{Dsat})$ corresponds to equation (1.48).

The mobility represented by the parameter K_P ($K_P = \mu_0 C_{ox}$) is assumed constant in expression (1.48). However, the mobility decreases as the gate voltage increases.

SPICE proposes an empirical relation to simulate this phenomenon:

$$K_P' = K_P \left[\frac{\varepsilon_S}{\varepsilon_{ox}} \frac{U_c t_{ox}}{V_{GS} - V_{Th} - U_t V_{DS}} \right]^{U_e} \tag{1.50}$$

where U_c is the critical field coefficient, U_e the mobility degradation exponent smoothing and U_t the transverse field coefficient. Typical values of these parameters are $U_c = 10^4$; $U_t = 0$; $U_e = 0.1$.

The channel length modulation parameter λ described in the basic model is constant. To be more precise it is necessary to calculate it as a function of physical parameters:

$$\lambda = \frac{L - L'}{L V_{DS}} = \frac{\Delta L}{L V_{DS}} \tag{1.51}$$

where L represents the physical channel length and L' the electrical channel length. The quantity ΔL is a function of the drain–source voltage and the substrate doping N_B:

$$\Delta L = X_D \left[\frac{V_{DS} - V_{Dsat}}{4} + \sqrt{1 + \left\{ \frac{V_{DS} - V_{Dsat}}{4} \right\}^2} \right] \text{ and } X_D = \sqrt{\frac{2\varepsilon_S}{q N_B}} \tag{1.52}$$

1.4.4 The 'large signal' dynamic model

The 'large signal' dynamic model is shown in Fig. 1.29. This model contains, in

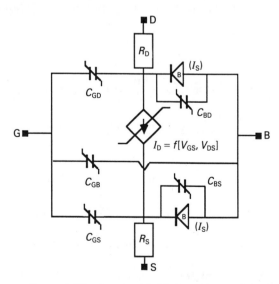

Figure 1.29 'Large signal' dynamic model

addition to the elements described in the static model, two types of capacitors:

1. the diode junction capacitors,
2. the gate capacitors, due to the oxide layer.

The junction capacitors of the drain–bulk and source–bulk diodes are denoted C_{JBD} and C_{JBS}, for which expressions are derived in Section 1.2 relating to the diode. Each capacitor is, however, divided into the following two parts:

1. a capacitance C_J due to the active surface of the diode (proportional to the drain or source area), modelled by the parameters C_{J0}, m_J, Φ_{J0} and F_C,
2. a 'side wall' capacitance C_{JSW} due to the junction created by lateral diffusion (proportional to the drain or source perimeter), modelled by the parameters C_{J0SW}, m_{JSW}, Φ_{J0SW} and F_C.

The total capacitance is given by the relation $C_T = AC_J + PC_{JSW}$, where A is the junction area and P the junction perimeter. These capacitances can also be modelled using C_{J0} = total capacitance at $V = 0$, denoted, respectively, C_{BD} and C_{BS}.

There are three gate capacitors: the gate–source, gate–bulk and gate–drain. They are composed of the following two types of capacitors:

1. the characteristic oxide capacitor (function of the parameter C_{ox}),
2. the overlap capacitors due to the part of the oxide covering the source, drain and bulk.

This model involves the following parameters:

1. $C_{ox} = \varepsilon_{ox}/t_{ox}$: capacitance of the gate oxide,
2. C_{GS0} and C_{GD0}: overlap capacitances of the gate–source and gate–drain, respectively (per unit length of W),
3. C_{GB0}: gate–bulk overlap capacitance (per unit length of L).

Concerning C_{GB}, this capacitance in the linear or saturated region is only a parasitic capacitor (overlap capacitor). It is important in the off region:

$$C_{GB} = C_{ox}WL + C_{GB0}L \qquad V_{GS} < V_{Th}$$

$$C_{GB} = C_{GB0}L \qquad V_{GS} > V_{Th}$$

Capacitances C_{GS} and C_{GD} are important in the linear and saturation regions. In the off region, they can be considered as parasitic capacitors (overlap capacitors). In the linear region these capacitances are equal, so

$$C_{GS} = C_{GD} = \frac{1}{2} C_{ox}WL \text{ (the total capacitance under the gate being}$$
$$C_{ox}WL)$$

In the saturated region, C_{GD} consists only of an overlap capacitance (C_{GD0}). C_{GS} can be calculated using the total charge stored in the channel and

$$C_{GS} = \frac{2}{3} C_{ox}WL$$

Of course, we have to add the overlap parasitic capacitances. In conclusion:

$$C_{GS} = C_{GS0}W; \ C_{GD} = C_{GD0}W; \qquad V_{GS} < V_{Th}$$

$$C_{GS} = \frac{1}{2} C_{ox}WL + C_{GS0}W; \qquad V_{GS} > V_{Th}; \ V_{GD} > V_{Th}$$

$$C_{GD} = \frac{1}{2} C_{ox}WL + C_{GD0}W; \qquad V_{GS} > V_{Th}; \ V_{GD} > V_{Th} \qquad (1.53)$$

$$C_{GS} = C_{GS0}W; \qquad V_{GS} > V_{Th}; \ V_{GD} < V_{Th}$$

$$C_{GD} = \frac{2}{3} C_{ox}WL + C_{GD0}W; \qquad V_{GS} > V_{Th}; \ V_{GD} < V_{Th}$$

1.4.5 The 'small signal' model

The MOS transistor 'small signal' model is shown in Fig. 1.30. It is obtained directly from the 'large signal' model (Fig. 1.29). R_D and R_S are the terminal resistances modelled in Section 1.4.2. g_{bd}, C_{BD}, g_{bs} and C_{BS} are the capacitances and dynamic conductances of the diodes, respectively. The I_D current source is linearized and is represented by two linear controlled current sources $g_m V_{GS}$ and $g_{mbs} V_{BS}$ and a conductance g_{ds}. Note the following expressions for these components:

$$g_{ds} = \frac{dI_D}{dV_{DS}} \qquad (1.54a)$$

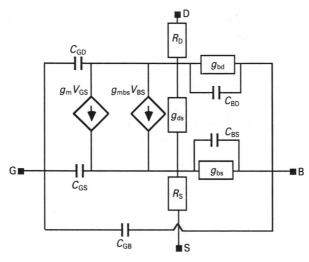

Figure 1.30   The 'small signal' model of the MOS transistor (operating point: V_{GS}^*, V_{BS}^*, V_{DS}^*)

$$g_m = \frac{dI_D}{dV_{GS}} \tag{1.54b}$$

$$g_{mb} = \frac{dI_D}{dV_{BS}} \tag{1.54c}$$

$$g_{bd} = \frac{dI_{BD}}{dV_{BD}} \qquad I_{BD} \text{ is the bulk–drain diode current} \tag{1.54d}$$

$$g_{bs} = \frac{dI_{BS}}{dV_{BS}} \qquad I_{BS} \text{ is the bulk–source diode current} \tag{1.54e}$$

$$C_{BD} = C_{JBD}(V_{BD}^*); \qquad C_{BS} = C_{JBS}(V_{BS}^*) \tag{1.54f}$$

$$C_{GD} = C_{JGD}(V_{GD}^*); \qquad C_{GB} = C_{JGB}(V_{GB}^*); \; C_{GS} = C_{GS}(V_{GS}^*) \tag{1.54g}$$

1.4.6 Models used in typical circuits

The standard models used in the following chapters are shown in Fig. 1.31(a,b). In the following chapters, only the basic models are used, with electrode bulk diodes being reverse biased: their currents are neglected. The simplified large signal model is presented in Fig. 1.31(a). The values of I_D are given by equations (1.44a)–(1.45c). Note that the different capacitors, in reality, have very complex expressions. In this book, we use only average values. So all the capacitors are linear. As in the 'large signal' model, the diodes are reverse biased. So the diode conductances g_{bd} and g_{bs} can be neglected. Then, we obtain the model represented in Fig. 1.31(b), where the

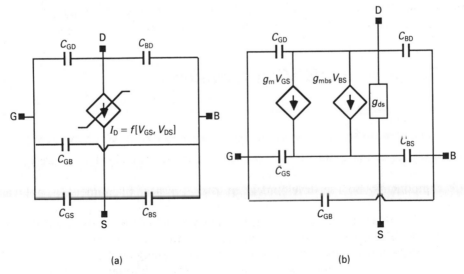

(a)   (b)

Figure 1.31 Models used in typical circuits. (a) Large signal. (b) Small signal

different conductance elements have the following expressions:

$$g_{ds} = \frac{dI_D}{dV_{DS}}\bigg|_{V^*_{DS},\; V^*_{GS}} = 0 \qquad V^*_{GS} < V_{Th}$$

$$g_{ds} = \frac{dI_D}{dV_{DS}}\bigg|_{V^*_{DS},\; V^*_{GS}} = \beta[V^*_{GS} - V_{Th} - V^*_{DS}] \qquad \begin{array}{l} V^*_{GS} \geqslant V_{Th}; \\ V^*_{GD} \geqslant V_{Th} \end{array}$$

$$g_{ds} = \frac{dI_D}{dV_{DS}}\bigg|_{V^*_{DS},\; V^*_{GS}} = 0.5\lambda\beta(V^*_{GS} - V_{Th})^2 \qquad V^*_{GS} \geqslant V_{Th};\; V^*_{GD} < V_{Th}$$

$$g_m = \frac{dI_D}{dV_{GS}}\bigg|_{V^*_{DS},\; V^*_{GS}} = 0 \qquad V^*_{GS} < V_{Th}$$

$$g_m = \frac{dI_D}{dV_{GS}}\bigg|_{V^*_{DS},\; V^*_{GS}} = \beta V^*_{DS} \qquad V^*_{GS} \geqslant V_{Th};\; V^*_{GD} \geqslant V_{Th}$$

$$g_m = \frac{dI_D}{dV_{GS}}\bigg|_{V^*_{DS},\; V^*_{GS}} = \beta(V^*_{GS} - V_{Th})(1 + \lambda V^*_{DS}) \qquad \begin{array}{l} V^*_{GS} \geqslant V_{Th}; \\ V^*_{GD} < V_{Th} \end{array}$$

$$g_{mbs} = \frac{dI_D}{dV_{BS}}\bigg|_{V^*_{DS},\; V^*_{GS}} = 0 \qquad V^*_{GS} < V_{Th}$$

$$g_{mbs} = \frac{dI_D}{dV_{BS}}\bigg|_{V^*_{DS},\; V^*_{GS}} = \frac{\beta\gamma V^*_{DS}}{2\sqrt{\Phi_S - V^*_{BS}}} \qquad V^*_{GS} \geqslant V_{Th};\; V^*_{GD} \geqslant V_{Th}$$

$$g_{mbs} = \frac{dI_D}{dV_{BS}}\bigg|_{V^*_{DS},\; V^*_{GS}} = \beta\gamma\, \frac{(V^*_{GS} - V_{Th})(1 + \lambda V^*_{DS})}{2\sqrt{\Phi_S - V^*_{BS}}} \qquad \begin{array}{l} V^*_{GS} \geqslant V_{Th}; \\ V^*_{GD} < V_{Th} \end{array}$$

where V^*_{DS}, V^*_{GS} and $V^*_{GD} = V^*_{GS} - V^*_{DS}$ correspond to the operating point. The different capacitors have the same values as in the 'large signal' model.

1.5 The junction field effect transistor

1.5.1 Introduction

The JFET (Junction Field Effect Transistor) presented in Fig. 1.2 is rarely used. They are sometimes encountered in silicon integrated circuits and also act as parasitic components. So the development of their physical phenomena is not very important for users. Thus, in this section we present only the equations, not the physics. The JFET models are included in simulators but only basic equations are considered. SPICE for example, contains only a one level model.

The JFET devices are components of the same family as the MOS devices. So their current–voltage equations have the same form, although the physics is not exactly the same. (The JFET devices have no gate oxide.) For JFET, V_P is the pinch-off voltage and is equivalent to V_{Th} for MOS transistors ($I_D = 0$ for $V_{\mathrm{GS}} < V_P$ in the case of a N.JFET). V_P is negative for N.JFET (and positive for P.JFET).

The large signal phenomena are typically modelled by two junction capacitors: C_{GD} and C_{GS}, formed by two reverse biased diodes.

1.5.2 Models used in typical circuits

The standard models used are presented in Fig. 1.32 (a, b). For convenience, as in the bipolar transistor models, the expressions for the various elements used are recalled. We assume that $\Phi_0 = 0.7$ V; $m = 0.5$; $F_C = 0.5$ for the two junctions.

1.5.2.1 'Large signal' model

The components represented in Fig. 1.32(a) have the following expressions:

$$I_D = 0 \qquad V_{\mathrm{GS}} < V_P$$

$$I_D = [\beta\,[2(V_{\mathrm{GS}} - V_P) - V_{\mathrm{DS}}]\,V_{\mathrm{DS}}](1 + \lambda V_{\mathrm{DS}}) \qquad \text{ohmic zone}$$
$$V_{\mathrm{GS}} > V_P;\ V_{\mathrm{GS}} - V_{\mathrm{DS}} > V_P$$

$$I_D = [\beta(V_{\mathrm{GS}} - V_P)^2]\,(1 + \lambda V_{\mathrm{DS}}) \qquad \text{pinched zone}$$
$$V_{\mathrm{GS}} > V_P;\ V_{\mathrm{GS}} - V_{\mathrm{DS}} < V_P$$

$$C_{\mathrm{GD}} = C_{\mathrm{JD0}}\ \frac{1}{\sqrt{1 - \dfrac{V_{\mathrm{GD}}}{0.7}}} \qquad V_{\mathrm{GD}} < 0.35\ \text{volt}$$

$$C_{\mathrm{GD}} = 2.83\ C_{\mathrm{JD0}}\left[0.25 + \frac{V_{\mathrm{GD}}}{1.4}\right] \qquad V_{\mathrm{GD}} > 0.35\ \text{volt}$$

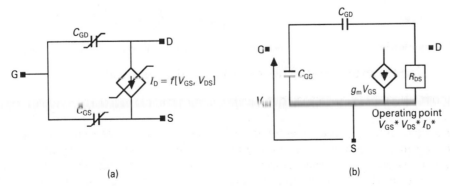

(a)   (b)

Figure 1.32   Models used in classical circuits. (a) 'Large signal' model. (b) 'Small signal' model

$$C_{GS} = C_{JS0} \frac{1}{\sqrt{1 - \dfrac{V_{GS}}{0.7}}} \qquad V_{GS} < 0.35 \text{ volt}$$

$$C_{GS} = 2.83 C_{JS0} \left[0.25 + \frac{V_{GS}}{1.4} \right] \qquad V_{GS} > 0.35 \text{ volt}$$

Note that, in these expressions, the β parameter is not the same as in the MOS model. In fact, $\beta_{JFET} = 1/2\beta_{MOS}$. This is the same approach as in the PSPICE simulator.

1.5.2.2 'Small signal' model

The model obtained is presented in Fig. 1.32(b):

$$g_m = 0 \qquad V_{GS} < V_P$$

$$g_m = 2\beta V_{DS}^*(1 + \lambda V_{DS}^*) \qquad \text{ohmic zone}$$

$$g_m = 2\beta(V_{GS}^* - V_P)(1 + \lambda V_{DS}^*) \qquad \text{pinched zone}$$

$$R_{DS} = \infty \text{ (open circuit)} \qquad V_{GS} < V_P$$

$$R_{DS} = [2\beta\{V_{GS}^* - V_P - V_{DS}^*\}(1 + \lambda V_{DS}^*)$$
$$\qquad + \beta\lambda\{(2(V_{GS}^* - V_P) - V_{DS}^*)V_{DS}^*\}]^{-1} \qquad \text{ohmic zone}$$

$$R_{DS} = [\beta\lambda(V_{GS}^* - V_P)^2]^{-1} \qquad \text{pinched zone}$$

$$C_{GD} = C_{JD0} \frac{1}{\sqrt{1 - \dfrac{V_{GD}^*}{0.7}}} \qquad V_{GD}^* < 0.35 \text{ volt}$$

$$C_{GS} = C_{JS0} \frac{1}{\sqrt{1 - \dfrac{V_{GS}^*}{0.7}}} \qquad V_{GS}^* < 0.35 \text{ volt}$$

1.6 Circuit modelling: the operational amplifier

Macromodelling or behavioural modelling permits, from circuit characteristics (static, dynamic, ...), the development of an equivalent circuit that is simpler but has the same functioning. However, this equivalent model does not represent the circuit physically. These macromodels, as in the case of discrete components, can have different degrees of complexity and accuracy. In this section, we present the op.amp. macromodel that is generally used.

1.6.1 Modelling of basic parameters of an op.amp.

The main characteristics of an op.amp. are as follows:

1. $A = V_{out}/V_{in}$ voltage gain defined by the maximum gain A_0 (DC gain or gain at zero frequency) and cut-off frequencies,
2. Finite output swing,
3. Slew rate limiting,
4. Input and output impedances.

Let us give more details of these characteristics.

A typical Bode plot of the amplitude response of op.amp. is shown in Fig. 1.33. Every cut-off frequency presents a real negative pole with the terms $1/(s + \omega_1)$ and $1/(s + \omega_2)$. The total gain is

$$A = \frac{A_0 \omega_1 \omega_2}{(s + \omega_1)(s + \omega_2)} \tag{1.55}$$

The purpose is to represent this function using an equivalent circuit. Typically, the kind of function $1/(s + \omega_1)$ can be associated with an RC two-pole in parallel with a controlled source, as shown in Fig. 1.34.

Thus, we obtain

$$\frac{V_{out}}{V_{in}} = \frac{g_m R}{1 + sRC} = \frac{A_0}{1 + s/\omega_1} \tag{1.56}$$

with

$$A_0 = g_m R; \quad \omega_1 = \frac{1}{RC}$$

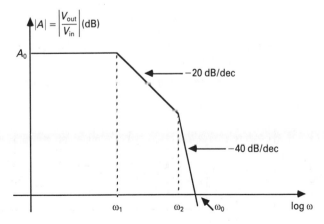

Figure 1.33 Amplitude Bode plot of an op.amp

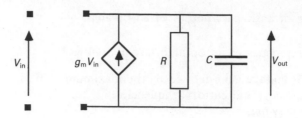

Figure 1.34   Modelling of a function $\dfrac{1}{s + \omega_1}$

The transfer function is realized on this basis. Then, the different two-ports are connected in cascade corresponding to each pole. The voltage controlled sources permit the isolation of different two-ports. The total gain should be distributed over stages. The values of R and C are not single.

Example. Let us consider an op.amp. characterized by a DC gain $A_0 = 10^5$ and two cut-off frequencies $f_1 = 100$ Hz and $f_2 = 10^6$ Hz. Determine an equivalent circuit modelling this gain function; assume that the op.amp. output resistance R_0 is equal to 75Ω.

The voltage gain of the op.amp. is represented by the following transfer function:

$$A = \frac{10^5}{\left(1 + \dfrac{s}{2\pi \times 100}\right)\left(1 + \dfrac{s}{2\pi \times 10^6}\right)}$$

Hence, this transfer function can be modelled by the equivalent circuit shown in Fig. 1.35.

Hence, we obtain

$$A = \frac{g_1 R_1 g_2 R_2 g_3 R_0}{(1 + R_1 C_1 s)(1 + R_2 C_2 s)}$$

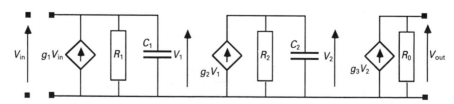

Figure 1.35   Equivalent electric circuit modelling the voltage gain

and we can deduce

$$
\begin{cases}
f_1 = \dfrac{\omega_1}{2\pi} = \dfrac{1}{2\pi R_1 C_1} \\[2mm]
f_2 = \dfrac{\omega_2}{2\pi} = \dfrac{1}{2\pi R_2 C_2} \\[2mm]
A_0 = g_1 g_2 g_3 R_1 R_2 R_0
\end{cases}
$$

and

$$
\begin{cases}
(R_1 C_1)^{-1} = 2\pi \times 10^2 \ \text{s}^{-1} \\
(R_2 C_2)^{-1} = 2\pi \times 10^6 \ \text{s}^{-1}
\end{cases}
$$

taking, for example,

$$
\begin{cases}
R_1 = 10 \ \text{k}\Omega \\
R_2 = 100 \ \Omega
\end{cases}
$$

Thus

$$
\begin{cases}
C_1 = \dfrac{1}{2\pi \times 10^2 \times 10^4} = 0.16 \ \mu\text{F} \\[4mm]
C_2 = \dfrac{1}{2\pi \times 10^6 \times 10^2} = 1.6 \ \text{nF}
\end{cases}
$$

and

$$
g_1 g_2 g_3 = \frac{A_0}{R_1 R_2 R_0} = \frac{10^5}{10^6 \times 75} = 1.33 \times 10^{-3}
$$

Thus, for example,

$$
g_1 = 1 \ \text{S}; \quad g_2 = 0.1 \ \text{S}; \quad g_3 = 0.013 \ \text{S}
$$

The output voltage swing limitation in op.amp. is usually modelled by the output stage of the macromodel. For a voltage greater than $|V_{0\text{Max}}|$, the output stage is saturated and so the amplifying function is modified. The I–V function representing this phenomenon is shown in Fig. 1.36(a). It is a non-linear function which can be modelled as follows:

1. If $V_{\text{out}} < |V_{0\text{Max}}|$, $I = 0$ (the op.amp. presents the standard output resistance R_0).
2. If $V_{\text{out}} \geqslant |V_{0\text{Max}}|$, $I = (V - V_{0\text{Max}})G$. The voltage output is practically constant and equal to $V_{0\text{Max}}$.

This non-linear description can easily be modelled by a diode circuit as shown in Fig. 1.36(b).

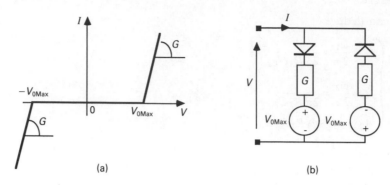

Figure 1.36   Output voltage swing limitation modelling. (a) I–V characteristic. (b) Model

The slew-rate phenomenon represents the maximum voltage variation speed (dV/dt) without signal distortion. This limit is due to the too small values of transistor currents of the input stage (differential stage) which charge the circuit capacitors. This parameter is very important, specially in high speed circuits.

In order to model this transistor current limitation, the non-linear function shown in Fig. 1.37(a) is used. I_M represents the maximum current of the input stage and V_{IMax} represents the corresponding voltage. Such a non-linear current source connected in parallel with a capacitor C allows us to model the slew rate. Indeed, if we consider the loading speed of the capacitor shown Fig. 1.37(b), we obtain

$$\left|\frac{dV_C}{dt}\right| = \frac{1}{C}|I_C| = \frac{I_M}{C} \text{ for } V_C \geq V_{IMax} \text{ (constant current load)}$$

$$\left|\frac{dV_C}{dt}\right| = \frac{1}{C}|I_C| = \frac{g_m V_{in}}{C} \text{ for } V_C < V_{IMax} \text{ (linear controlled current load: } I = g_m V_{in})$$
$$(1.57)$$

I_M/C represents the S_R parameter: maximum speed of capacitor loading.

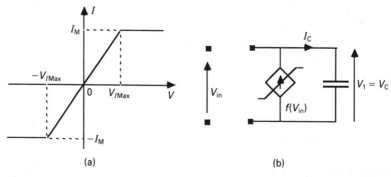

Figure 1.37   'Slew-rate' modelling. (a) I–V characteristic. (b) Model subcircuit

The 'slew rate' is modelled by a stage including a non-linear controlled source $f(V_{in})$ and a capacitor C.

The controlled current source is linear for $|V| \leqslant |V_{IMax}|$, $(I = g_m V_{in})$. This stage is also used for gain modelling (DC gain value A_0 and first cut-off angular frequency ω_1). A resistor R must be added in parallel with capacitor C in order to obtain $\omega_1 = (RC)^{-1}$. Thus, we obtain the circuit shown in Fig. 1.38.

In the linear operating range, the following relation can be written

$$\frac{V_1}{V_{in}} = \frac{g_m R}{1 + RCs} = \frac{A_0}{1 + RCs} \tag{1.58}$$

The following expressions also have to be considered:

$$\frac{I_M}{C} = S_R, \ \omega_1 = \frac{1}{RC} \ \text{and} \ R \gg \frac{1}{C\omega_1} \tag{1.59}$$

Another expression can be obtained taking ω_0 into account, the unit gain angular frequency (in the case where the slope of the Bode plot is -20 dB/Dec, up to the frequency ω_0, with $\omega_0 < \omega_2$). In such a situation $\omega_1 = \omega_0/A_0$. Then, the transfer gain function for $\omega = \omega_0$ can be approximated by

$$V_1 = g_m V_{in} \frac{1}{j\omega_0 C} \tag{1.60}$$

and

$$\left| \frac{V_1}{V_{in}} \right| = 1 = \frac{g_m}{\omega_0 C}$$

thus

$$g_m = \omega_0 C = \frac{A_0}{R} \tag{1.61}$$

Finally, the last parameters to be defined are I_M and V_{IMax}. We obtain

$$I_M = \frac{S_R}{\omega_1 R} \ \text{and} \ V_{IMax} = \frac{I_M}{g_m} \tag{1.62}$$

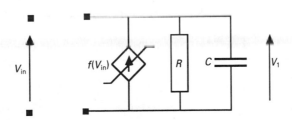

Figure 1.38 Model subcircuit for slew-rate, A_0 gain and angular frequency ω_1 modelling

1.6.2 The amplifier macromodel used in the PSPICE simulator

In the PSPICE simulator, the proposed amplifier macromodel is much more complicated. This macromodel, shown in Fig. 1.39, takes into account the static, dynamic and frequency characteristics.

The circuit presented in Fig. 1.39 consists of the following three stages:

1. The input stage is modelled by two basic bipolar transistors (simple Ebers–Moll model), sources and linear passive components. This stage models the common mode and differential mode characteristics. It is designed in order to take current and voltage offset into account. Its gain is equal to unity (an assumption permitting the simplification of calculation parameters). The capacitor C_E models the second-order effect of the slew rate. The capacitor C_1 models the second-order effect of the phase response. Note that the input stage is modelled by a differential circuit as in real amplifers.
2. The interstage permits differential mode and common mode voltage gain modelling. This stage consists of only linear elements. The dominant time constant is modelled by capacitor C_2.
3. The output stage provides the proper DC and AC output resistance and models voltage and current limitation.

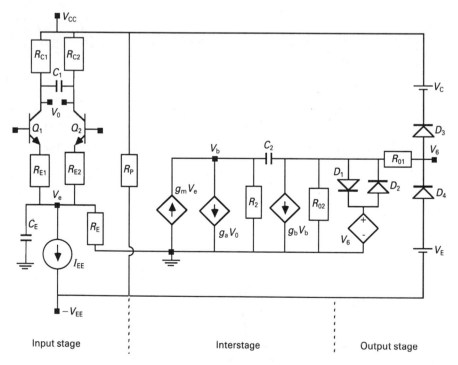

Figure 1.39 Example of macromodelling: the op.amp. model used in PSPICE

In Fig. 1.39, the bipolar transistors Q_1 and Q_2 are modelled by two parameters (Ebers–Moll model with $\beta_R = 1$): β_{F1} and I_{S1} for Q_1, β_{F2} and I_{S2} for Q_2. The diodes D_1, D_2, D_3 and D_4 are modelled only by their saturation currents denoted, respectively, $I_{SD1}, I_{SD2}, I_{SD3}$ and I_{SD4}. The other components have the following expressions:

$$I_{S1} = I_{SD3} = I_{SD4} = 8 \times 10^{-16} \text{ A} \tag{1.63a}$$

$$R_2 = 100 \text{ k}\Omega \tag{1.63b}$$

$$C_E = \frac{2I_{C1}}{S_R} - C_2 \text{ where } I_{C1} = I_{C2} = \frac{C_2}{2} S_R^+ \tag{1.63c}$$

$$I_{B1} = I_B + \frac{I_{BOS}}{2} \text{ and } I_{B2} = I_B - \frac{I_{BOS}}{2} \tag{1.63d}$$

$$\beta_{F1} = \frac{I_{C1}}{I_{B1}} \text{ and } \beta_{F2} = \frac{I_{C2}}{I_{B2}} \tag{1.63e}$$

$$I_{EE} = \left(\frac{\beta_{F1} + 1}{\beta_{F1}} + \frac{\beta_{F2} + 1}{\beta_{F2}} \right) I_{C1} \tag{1.63f}$$

$$R_E = \frac{200}{I_{EE}} \tag{1.63g}$$

$$I_{S2} = I_{S1} \left(1 + \frac{V_{0S}}{V_T} \right) \tag{1.63h}$$

$$R_{C1} = R_{C2} = \frac{1}{2\pi f_{0 \text{ dB}} C_2} \tag{1.63i}$$

$$R_{E1} = R_{E2} = \left(\frac{\beta_{F1} + \beta_{F2}}{\beta_{F1} + \beta_{F2} + 2} \right) \left(R_{C1} - \frac{1}{g_{m1}} \right) \text{ where } \frac{1}{g_{m1}} = \frac{V_T}{I_{C1}} \tag{1.63j}$$

$$C_1 = \frac{C_2}{2} \tan \Delta\phi \tag{1.63k}$$

$$R_P = \frac{(V_{CC} + V_{EE})^2}{P_d - V_{CC}(? I_{C1}) - V_{DD} I_{DD}} \tag{1.63l}$$

$$g_a = \frac{1}{R_{C1}} \tag{1.63m}$$

$$g_m = \frac{1}{R_{C1} \text{ (CMRR)}} \tag{1.63n}$$

$$R_{01} = R_{0-\text{ac}} \text{ and } R_{02} = R_{\text{out}} - R_{01} \tag{1.63o}$$

$$g_b = \frac{a_{VD} R_{C1}}{R_2 R_{02}} \tag{1.63p}$$

$$I_{SD1} = I_{SD2} = I_X \exp\left(-\frac{R_{01} I_{SC}}{V_T}\right) \text{where } I_X = (2I_{C1})g_b R_2 - I_{SC} \qquad (1.63\text{q})$$

$$V_C = V_{CC} - V_{out}^+ + V_T \ln\left(\frac{I_{SC}}{I_{SD3}}\right) \qquad (1.63\text{r})$$

$$V_E = V_{CC} - V_{out}^- + V_T \ln\left(\frac{I_{SC}}{I_{SD4}}\right) \qquad (1.63\text{s})$$

The following expressions are functions of the following characteristic parameters of op.amp.:

C_2	:compensation capacitor
S_R^+	:positive going slew rate
S_R^-	:negative going slew rate
I_B	:average input bias current
I_{BOS}	:input current offset
V_{OS}	:offset voltage
$f_{0\,db}$	:0 db frequency of the fully compensated op. amp.
$\Delta\phi$	:excess phase at $f = f_{0\,db}$ due to non-dominant pole

$$\left(\frac{1}{2R_{C1}C_1}\right)$$

V_{CC}, V_{EE}	:bias sources
P_d	:power dissipation (in quiescent state)
CMRR	:common mode rejection ratio
R_{0-ac}	:AC output resistance
R_{out}	:DC output resistance
a_{VD}	:differential mode voltage gain
I_{SC}	:maximum current to the output
V_{out}^+	:positive voltage limiting
V_{out}^-	:negative voltage limiting

Exercises

Note: for the exercises to this chapter, the following data are used:

1. electron charge: $q = 1.6 \times 10^{-19}$ C,
2. Boltzmann constant: $k = 1.38 \times 10^{-23}$ J/K,
3. Temperature: $T = 300$ K.

Exercise 1.   Determine the $I-V$ characteristic of the circuit of Fig. E1.1, composed of two Boltzmann diodes (see equation (1.1)) connected in parallel and in opposition. Propose an application for this circuit. The two diodes are the same and are characterized by $I_S = 10^{-13}$ A, $n = 1$.

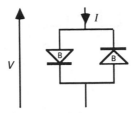

Figure E1.1

Exercise 2.   Using the Boltzmann diode characteristic, equation (1.1), plot the $I-V$ characteristic of two diodes connected back-to-back in series, as shown in Fig. E1.2. The two diodes are the same and are characterized by $I_S = 10^{-13}$ A, $n = 1$.

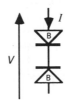

Figure E1.2

Exercise 3.   For the diode 1N5712 (the model given at the end of Exercises for this chapter), calculate the voltage V_{10} corresponding to a current of 10 mA (Boltzmann diode).

Exercise 4.   For a Boltzmann diode ($n = 1$), calculate the voltage ranges for which the following approximations are valid (with a maximum relative error $(I_{Da} - I_D)/I_D$ equal to 10^{-3}):

$$I_{Da} = I_S \exp\left\{\frac{V_D}{V_T}\right\} \text{ and } I_{Da} = -I_S$$

Calculate the voltage range for $(I_{Da} - I_D)/I_D$ equal to 10^{-2} and 10^{-1}.

Exercise 5.   For the diode 1N5712 (the model given at the end of Exercises for this chapter), using Boltzmann approximation, calculate the voltage range for which the static resistance $R_{Stat} = V_D/I_D$ varies from 1Ω to 1MΩ. What is the corresponding current range?

Exercise 6.   For the diode 1N5712 (the model given at the end of Exercises for this chapter), using Boltzmann approximation, calculate the voltage range for which the dynamic resistance $r_d = dv_D/dI_D$ varies from 1Ω to 1MΩ. What is the corresponding current range?
   Compare the values of currents and voltages with those of exercise 5.

Exercise 7.   For a Zener diode, the breakdown phenomenon is modelled by one diode (see equation (1.4)) with $I_S = 880.5 \times 10^{-18}$ A; $n = 1$; $I_{bd} = 16.748 \times 10^{-3}$ A, $V_{bd} = 4.3$ V and $n_{bd} = 1.7936$. Compare the value of each term of the equation as a function of voltage V.

Exercise 8.   For the diode 1N5712 (the model given at the end of Exercises for this chapter), we want to take into account the influence of R_S. Using the circuit shown in Fig. E1.3, compare the ideal (V_1) and real ($V_1 + V_2$) voltages across the diode.

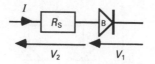

Figure E1.3

Exercise 9.   For the diode 1N5712 (the model given at the end of Exercises for this chapter), we want to take into account the influence of R_P. Using the simplified circuit shown in Fig. E1.4, for a given voltage, compare the ideal (I_1) and total ($I_1 + I_2$) current of the diode. $R_P = 10^{12}\ \Omega$.

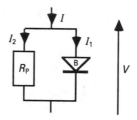

Figure E1.4

Exercise 10.   For the diode 1N5712 (the model given at the end of Exercises for this chapter), determine the voltage value V for which the following equalities are satisfied: $C_J = 2C_{J0}$, $C_J = 0.5C_{J0}$. Determine the C_J/C_{J0} value at the linearization point ($V = Fc\phi_0$). Calculate the voltage V for which diffusion capacitance C_D is equal to junction capacitance C_J. Give, for each case, the corresponding current.

Exercise 11.   The source of the circuit shown in Fig. E1.5(a) is described by the function of Fig. E1.5(b). Calculate the current $i(t)$ and the voltage $v(t)$ of the diode 1N5712 (the model given at the end of Exercises for this chapter) for the following three cases:

1.   the diode is equivalent to an ideal switch,
2.   the diode is represented by its $I-V$ characteristic (graphical approach) using the Boltzmann diode,
3.   the diode is represented by the model of Fig. E1.5(c) for direct bias and by the model of Fig. E1.5(d) for reverse bias.

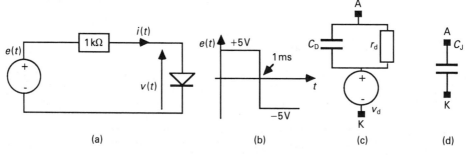

Figure E1.5

Exercise 12. Calculate by inspection the operating point of a bipolar transistor ($\beta_F = 100$, $\beta_R = 1$) biased as indicated in Fig. E1.6 ($R_1 = 10 \text{ k}\Omega$, $R_2 = 1 \text{ k}\Omega$). In this exercise, we assume that $V_{BE} = 0.6$ V when the base–emitter junction is forward biased. Produce calculations for the following values of E_1 and E_2:

1. $E_1 = -2$ V; $E_2 = +10$ V
2. $E_1 = +2$ V; $E_2 = +10$ V
3. $E_1 = +2$ V; $E_2 = +30$ V
4. $E_1 = -2$ V; $E_2 = -10$ V

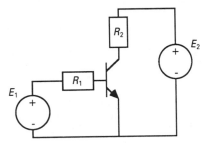

Figure E1.6

Exercise 13. For the parameters of the 2N2222 bipolar transistor (the model given at the end of Exercises for this chapter), build the small signal equivalent circuit considering the model of Fig. 1.16 (Early effect neglected) for the following operating point: $V_{CB}^* = 10$ V; $I_C^* = 5$ mA.

Exercise 14. Considering the following parameters for the 2N2222 bipolar transistor, describe qualitatively the Gummel–Poon small signal model (obtained from Fig. 1.21) corresponding to the following operating point: $V_{CB}^* = 10$ V; $I_C^* = 5$ mA. $I_S = 14.34 \times 10^{-15}$ A; $V_A = 74.03$ V; $\beta_F = 255.9$; $n_{EL} = 1.307$; $I_{SE} = 14.34 \times 10^{-15}$ A; $I_{KF} = 0.2847$ A; $\beta_R = 6.092$; $n_{CL} = 2$; $I_{SC} = 0$ A; $R_C = 1 \Omega$; $C_{JE0} = 22.01 \times 10^{-12}$ F; $M_{JE} = 0.377$; $\phi_E = 0.75$ V; $F_C = 0.5$; $C_{JC0} = 7.306 \times 10^{-12}$ F; $M_{JC} = 0.3416$; $\phi_C = 0.75$ V; $\tau_R = 46.91 \times 10^{-9}$ s; $\tau_F = 411.1 \times 10^{-12}$ s; $R_B = 10 \Omega$.

Exercise 15. For the parameters of an NMOS transistor (the model given at the end of Exercises for this chapter), plot the basic characteristics: $I_D = f(V_{DS})$ for V_{GS} constant and $I_D = f(V_{GS})$ for V_{DS} constant ($V_{BS}^* = -2$ V) (note: V_{Th0} is the value of V_{Th} for $V_{BS}^* = 0$ V).

Exercise 16. Using the basic model of the NMOS transistor (the model given at the end of Exercises for this chapter), take into account the influence of K_P' instead of K_P in the saturated region (see equation (1.50)). Compare the values of K_P' and K_P.

Exercise 17. For the parameters of the NMOS transistor (the model given at the end of Exercises for this chapter), calculate the dynamic conductances and transconductances of the small signal model at the following operating point: $V_{DS}^* = 8$ V; $V_{GS}^* = 5$ V; $V_{BS}^* = 0$ V.

Exercise 18. For the parameters of an NMOS transistor (the model given at the end of Exercises for this chapter), calculate the capacitances for the following operating points (see Fig. 1.30):

1. $V_{DS}^* = 8$ V; $V_{GS}^* = 5$ V; $V_{SB}^* = 0$ V.
2. $V_{DS}^* = 8$ V; $V_{GS}^* = 5$ V; $V_{SB}^* = -2$ V.
3. $V_{DS}^* = 1$ V; $V_{GS}^* = 5$ V; $V_{SB}^* = 0$ V.
4. $V_{DS}^* = 8$ V; $V_{GS}^* = 1$ V; $V_{SB}^* = 0$ V.

Exercise 19.   In the pinch-off region of a JFET transistor, we often use the following relation: $I_D = I_{DSS} \{1 - V_{GS}/V_P\}^2$; for the parameters of the 2N2608 JFET (the model given at the end of Exercises for this chapter), calculate the value of I_{DSS}.

Exercise 20.   For the parameters of the 2N2608 JFET (the model given at the end of Exercises for this chapter), construct the small signal equivalent circuit for the following operating point: $V_{GS}^* = -1$ V; $V_{DS}^* = 10$ V.

Exercise 21.   The slew rate of an operational amplifier is equal to 1 V/μs. This circuit is powered by DC voltage sources of 15 V and -15 V and used in an op.amp. low pass filter in order to have a gain of 10 at 3 kHz. What maximum input signal amplitude can be applied without distortion? What is the maximum input signal amplitude at 1 MHz?

Exercise 22.   Using simplified calculations, explain how the macromodel of the operational amplifier proposed in the PSPICE simulator takes into account the basic properties of the op.amp.

Models used for semiconductor devices.

Commutation diode: 1N5712.

$I_S = 432.8 \times 10^{-12}$ A; $n = 1$; $R_S = 12.15\ \Omega$;
$C_{J0} = 1.2 \times 10^{-12}$ F; $M = 0.333$; $\phi_0 = 0.75$ V; $F_C = 0.5$; $V_{bd} = 20$ V;
$I_{bd} = 10 \times 10^{-6}$ A; $\tau = 144.3 \times 10^{-12}$ s.

Standard transistor: 2N2222.

$I_S = 14.34 \times 10^{-15}$ A; $V_A = 74.03$ V; $\beta_F = 255.9$; $\beta_R = 6.092$;
$R_C = 1\ \Omega$; $C_{JE0} = 22.01 \times 10^{-12}$ F; $M_{JE} = 0.377$; $\phi_E = 0.75$ V; $F_C = 0.5$;
$C_{JC0} = 7.306 \times 10^{-12}$ F; $M_{JC} = 0.3416$; $\phi_C = 0.75$ V;
$\tau_R = 46.91 \times 10^{-9}$ s; $\tau_F = 411.1 \times 10^{-12}$ s; $R_B = 10\ \Omega$.

NMOS transistor.

$\gamma = 0.5$ V$^{1/2}$; $t_{ox} = 100 \times 10^{-9}$ m; $\mu_0 = 600$ cm^2 V^{-1}s^{-1};
$\phi_s = 0.6$ V; $K_P = 20.8 \times 10^{-6}$ AV^{-2}; $L = 2 \times 10^{-6}$ m;
$W = 20 \times 10^{-6}$ m; $V_{Th0} = 3.412$ V; $\lambda = 0.01$ V^{-1}; $R_S = 74.91$ mΩ;
$R_D = 26.95$ mΩ; $\varepsilon_S = 10^{-10}$F/m; $C_{GB0} = 1.115 \times 10^{-9}$ F/m;
$C_{GS0} = 679.6 \times 10^{-12}$ F/m; $C_{GD0} = 197.4 \times 10^{-12}$ F/m;
$C_{SB0} = C_{DB0} = 12.17 \times 10^{-15}$ F; $R_G = 5.286\ \Omega$; $\varepsilon_S = 10^{-10}$ F/m;
$\varepsilon_{ox} = 0.347\ 10^{-10}$ F/m.

FET transistor: 2N2608.

$\beta = 423.6 \times 10^{-6}$ AV^2; $\lambda = 15 \times 10^{-3}$ V^{-1}; $V_P = -2.45$ V;
$C_{JD0} = 6.7 \times 10^{-12}$ F; $C_{JS0} = 8.37 \times 10^{-12}$ F.

2 Some Network Transformations and Simple Numerical Calculations

2.1 Introduction

In this chapter some network transformations are presented which enable us to reduce relatively complex electronic circuits to simpler formats and therefore render the analysis simpler. Examples of simple numerical calculations of voltages and currents in dynamic elements are also given, that is, numerical integration methods are presented.

2.2 Some network transformations

2.2.1 Thevenin and Norton transformations

Ideal voltages and current sources used widely in circuit theory do not exist in practice. Actual non-ideal sources can, however, be modelled by ideal ones, connected with their internal impedance, Z_S, or admittance, Y_S, as is shown in Fig. 2.1.

To show the relationship between the voltage E and the current J, we can write

$$V = E \frac{Z_L}{Z_L + Z_S} \tag{2.1}$$

and

$$J = V(Y_S + Y_L) = V \frac{Z_S + Z_L}{Z_S Z_L} \tag{2.2}$$

Substituting (2.1) in (2.2) we obtain

$$J = \frac{E}{Z_S} \quad \text{or} \quad E = \frac{J}{Y_S}, \quad \text{where} \quad Z_S = \frac{1}{Y_S} \tag{2.3}$$

Thus, when the values of sources are related by (2.3) they can be transformed from a series connection of voltage source E and internal impedance Z_S into a parallel connection of current source J and internal admittance $Y_S = 1/Z_S$.

Note that the source voltage E (Fig. 2.1(a)) is equal to an open-circuit voltage

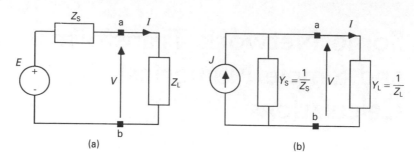

Figure 2.1   Actual sources loaded by impedances. (a) Series representation. (b) Parallel representation

$V_{OC} = V$ at terminals a–b when load is removed or $Z_L = \infty$, whereas the source current J (Fig. 2.1(b)) is equal to $I_{SC} = I$ when the load is short-circuited, that is $Y_L = \infty$ or $Z_L = 0$.

When more complicated networks are reduced to the formats shown in Fig. 2.1, then the part to the left of terminals a–b is called the Thevenin equivalent network for series representation (Fig. 2.1(a)) and the Norton equivalent network for parallel representation (Fig. 2.1(b)).

To reduce the network to Thevenin or Norton equivalent form, the following rules can be applied:

1. Select one element or part of the network between terminals a–b as a load.
2. Replace the load by an open circuit and calculate the open-circuit voltage V_{OC} at terminals a–b, then $V_{OC} = E$ of the Thevenin equivalent; or short circuit the a–b terminals and calculate the short-circuit current flowing into the short circuit, then $I_{SC} = J$ of the Norton equivalent.
3. To calculate the equivalent source impedance $Z_S = 1/Y_S$ short circuit all independent voltage sources and open circuit all independent current sources (to the left of terminals a–b) leaving dependent sources unchanged. Then, apply a voltage source V_a at the terminals a–b and calculate the current I supplied by this voltage source or apply a current source I_a to the terminals a–b and calculate the voltage V across the current source.
    The value of Z_S will be

$$Z_S = \frac{V_a}{I} \text{ or } Y_S = \frac{1}{Z_S} = \frac{I_a}{V} \tag{2.4}$$

2.2.2 Network reductions using Thevenin and Norton equivalents

In many practical cases we are not interested in knowledge of voltages and currents in all branches of the network but only in one selected branch, denoted as load impedance Z_L. In special cases, when the network does not contain controlled sources, the step-by-step reduction using Thevenin and Norton transformations will lead to the final equivalent Thevenin and Norton circuits from which we can easily find the interesting values of V_L and I_L for selected branches Z_L. An example of the step-by-step transformations for ladder network is shown in Fig. 2.2. Finally we find

$$V_L = 6\,\frac{R_L}{6 + R_L}\ [\text{V}], \quad I_L = 1 \times \frac{\dfrac{1}{R_L}}{\dfrac{1}{6} + \dfrac{1}{R_L}} = \frac{1 \times 6}{R_L + 6}\ [\text{A}],$$

$$V_L = R_L I_L = 6\,\frac{R_L}{6 + R_L}\ [\text{V}]$$

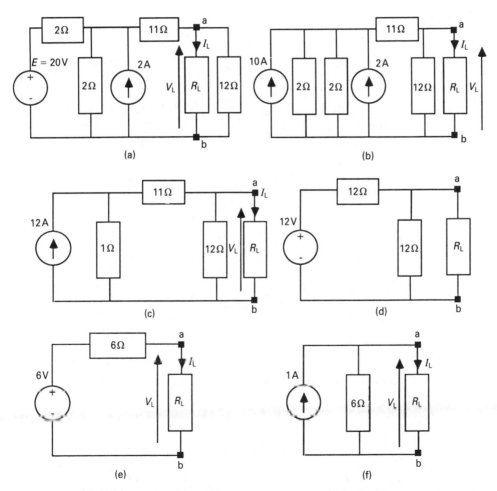

Figure 2.2   Step-by-step reduction of the circuits. (a) Initial circuit. (b) First transformation.
(c, d) Equivalent forms. (e) Thevenin. (f) Norton

When controlled sources exist in the initial (original) network, then the rules given in Section 2.2.1 can be applied. To simplify network reduction we can first find the Thevenin or Norton equivalent for the subcircuit containing the controlled sources and then follow the step-by-step reduction shown in Fig. 2.2. For the circuit shown in Fig. 2.3, we can first calculate the Thevenin equivalent for the subcircuit to the left of terminals a′−b′ (the right-hand side of the circuit is disconnected).

The node Kirchhoff equations for this circuit are

$$\left(\frac{1}{R_1}+\frac{1}{R_2}\right)V_1 - \frac{1}{R_2}V_2 = J_S$$

$$\left(\frac{-1}{R_2}+g_m\right)V_1 + \frac{1}{R_2}V_2 = 0$$

Solving for V_2, which is V_{eq} for Thevenin source, we obtain

$$V_{eq} = J_S\,\frac{\left(\dfrac{1}{R_2}-g_m\right)R_2}{\dfrac{1}{R_1}+g_m} = 2\times\frac{1.5\times0.5}{0.5+0.5} = 1.5\ \text{V}$$

The output resistance of the circuit R_{eq} can be calculated after disconnection of the independent current source J_S and application of the current source I_a between terminals a′−b′. Then we obtain the nodal Kirchhoff equations

$$\left(\frac{1}{R_1}+\frac{1}{R_2}\right)V_1 - \frac{1}{R_2}V_2 = 0$$

$$\left(\frac{-1}{R_2}+g_m\right)V_1 + \frac{1}{R_2}V_2 = I_a$$

Solving V_2 and calculating $R_{eq} = V_2/I_a$ we obtain

$$R_{eq} = \frac{\left(\dfrac{1}{R_2}+\dfrac{1}{R_1}\right)R_2}{\dfrac{1}{R_1}+g_m} = \frac{(0.5+2)\times0.5}{0.5+0.5} = 1.25\ \Omega$$

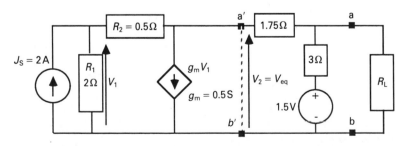

Figure 2.3   Circuit with independent and controlled sources

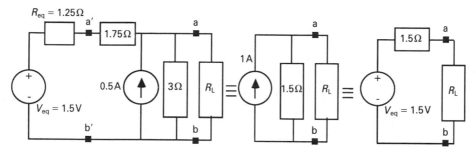

Figure 2.4 Intermediate and final transformation of circuit of Fig. 2.3

Further circuit reduction, shown in Fig. 2.4, is similar to that shown in Fig. 2.2.

2.2.3 Independent and controlled source transformations

In theoretical models some circuit branches can contain only ideal voltage sources without series impedances and/or ideal current sources without parallel admittances. In such cases these branches cannot be transformed from voltage to current sources and vice versa. To overcome this inconvenience we can add small resistances (much smaller than other resistances in the circuit) in series with a voltage source and/or small conductances (large resistances much larger than other resistances in the circuit) in parallel to a current source. After doing this, we can apply Thevenin and Norton transformations to independent and controlled sources.

In a nodal analysis the most convenient controlled source is VCCS (voltage controlled current source). Thus, when in models of active devices (transistors, operational amplifiers) other types of controlled sources are used, it is then useful to convert those sources into VCCSs. Such transformations are shown in Fig. 2.5.

2.2.4 Ideal voltage source and current source shifts

Insertion of small resistances in series with ideal voltage sources and attaching large resistances parallel to ideal current sources is not necessary when we apply the source shift method. The ideal voltage source shifts are illustrated in Fig. 2.6(a, b, c).

Note that in all cases the relation $V_3 = V_2 + V_a$ is valid, and the mesh equations are unchanged, showing that the behaviour of the circuit is not changed.

After shifting, the voltage sources are always in series with resistors and can be transformed into Norton equivalents if necessary. As an example of such manipulations let us consider the typical transformations of the bias circuit of a bipolar transistor shown in Fig. 2.7(a, b, c, d).

The ideal current source shifts are illustrated in Fig. 2.8(a, b).

Note that in both cases the Kirchhoff nodal equations remain unchanged.

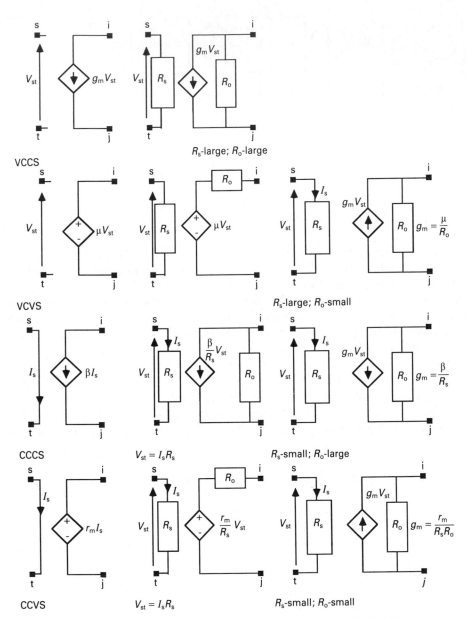

VCCS

R_s-large; R_o-large

VCVS

R_s-large; R_o-small

CCCS

$V_{st} = I_s R_s$

R_s-small; R_o-large

CCVS

$V_{st} = I_s R_s$

R_s-small; R_o-small

Figure 2.5 Transformations of controlled sources into VCCS form

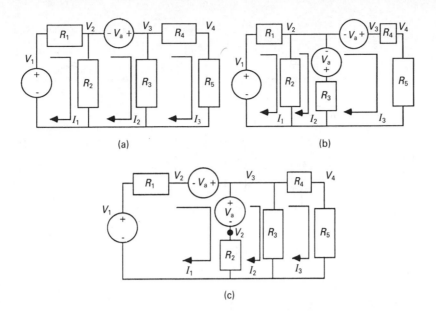

(a) (b)

(c)

Figure 2.6 (a) Circuit with an ideal voltage source V_a in mesh 2. (b) Circuit after shifting the V_a into the third mesh. (c) Circuit after shifting the V_a to the first mesh

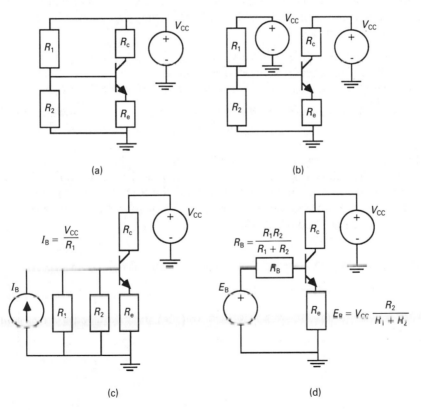

(a) (b)

(c) (d)

Figure 2.7 (a) Transistor bias circuit. (b) After shifting the V_{CC} source. (c) After Norton transformation. (d) After Thevenin transformation

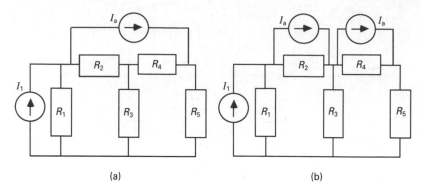

(a)   (b)

Figure 2.8   (a) Circuit with an ideal current source I_a. (b) After shifting

2.2.5 Substitution theorem

The substitution theorem states that if for a given circuit all branch voltages and currents are known and if the ith branch current I_i and voltage V_i are associated with the ith branch, then this branch can be replaced by any one of the following without altering the circuit properties:

1. An independent voltage source with value V_i.
2. An independent current source with value I_i.
3. A resistor of value $R_i = V_i/I_i$ or a conductor of value $G_i = I_i/V_i$.

To prove this theorem let us consider the simple resistive circuit shown in Fig. 2.9(a). Application of the first and second statements of the theorem is shown in Fig. 2.9(b, c). It can easily be proved that the behaviour of the circuit is not changed.
    To show the application of the third statement of the theorem, let us consider the MOS source follower operating at the low frequencies shown in Fig. 2.10(a). The equivalent model of the circuit, and its transformations, leading to replacement of a current branch composed of the controlled current source $g_m V_{gs}$ by a conductance g_m, is shown in Fig. 2.10(b, c, d, e).

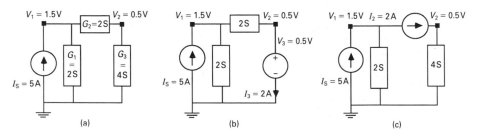

(a)   (b)   (c)

Figure 2.9   (a) Original resistive circuit. (b) Replacement of the third branch by voltage source. (c) Replacement of the second branch by current source

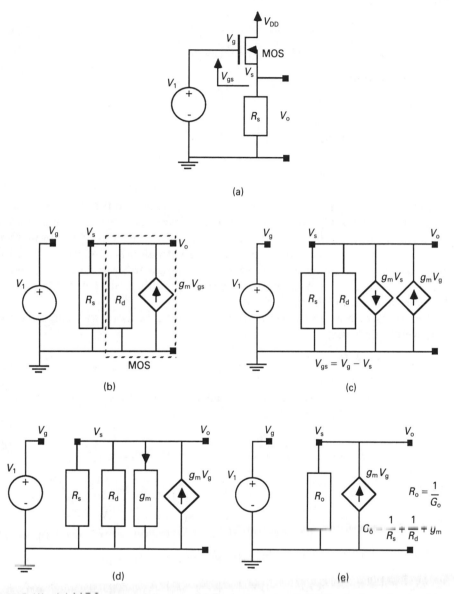

Figure 2.10 (a) MOS source-follower. (b) Equivalent circuit. (c) Equivalent circuit with split controlled sources. (d) Replacement of the current branch $I_i = g_m V_s$ by the conductance

$$g_m = \frac{I_i}{V_s}.$$ (e) Simplified equivalent circuit

Based on Fig. 2.10(e), we can easily calculate the voltage gain A_V of this follower:

$$A_V = \frac{V_o}{V_1} = \frac{V_s}{V_g} = \frac{g_m}{G_o} = \frac{g_m}{\dfrac{1}{R_s} + \dfrac{1}{R_d} + g_m} \tag{2.5}$$

and output resistance

$$R_o = \frac{1}{G_o} = \frac{1}{\dfrac{R_s + R_d}{R_s R_d} + g_m} \tag{2.6}$$

2.2.6 Superposition principle

The simplified or circuit-oriented version of the superposition principle says that the branch current or voltage of any one-port in a linear network containing several independent voltage and/or current sources is equal to the sum of the branch currents or voltages due to each source acting alone with all other independent sources set to zero. Setting a voltage source to zero corresponds to short circuiting, and setting a current source to zero corresponds to open circuiting. Simple examples shown in Fig. 2.11(a, b, c) explain this principle.

After transforming the current source of Fig. 2.11(a) to the Thevenin equivalent, we obtain directly

$$I_2 = \frac{V_2 - I_1 R_1}{R_2 + R_1} \tag{2.7}$$

Then the other current and voltage can be found directly. For example, the value of V_1 is

$$V_1 = V_2 \frac{R_1}{R_2 + R_1} + I_1 \frac{R_1 R_2}{R_2 + R_1} \tag{2.8}$$

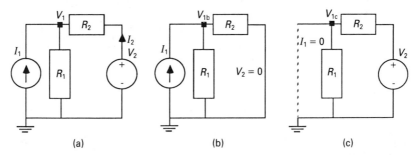

Figure 2.11   (a) Circuit with two independent sources. (b) Circuit with V_2 set to zero. (c) Circuit with I_1 set to zero

This result can be obtained from partial results, that is, V_{1b} of Fig. 2.11(b) and V_{1c} of Fig. 2.11(c) and, summing, $V_1 = V_{1b} + V_{1c}$. From Fig. 2.11(b), we obtain

$$V_{1b} = I_1 \frac{R_1 R_2}{R_2 + R_1} \tag{2.9}$$

and from Fig. 2.11(c), we obtain

$$V_{1c} = V_2 \frac{R_1}{R_2 + R_1} \tag{2.10}$$

Thus, as a result we obtain relation (2.8).

2.2.7 The Millmann theorem

The Millmann theorem allows us to find the voltage of several Thevenin equivalent circuits connected in parallel. This is shown in Fig. 2.12. From Fig. 2.12(a or b),

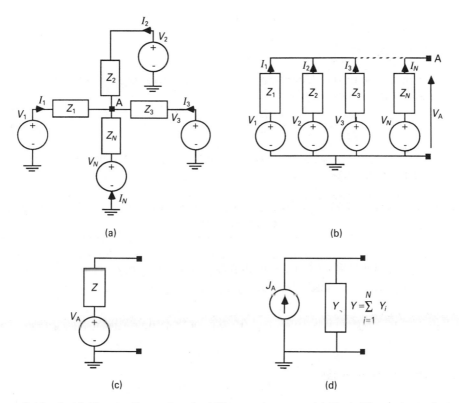

Figure 2.12   (a, b) Circuits illustrating the Millmann theorem. (c) Single Thevenin equivalent circuit. (d) Single Norton equivalent circuit

we obtain

$$I_1 + I_2 + \ldots + I_N = 0 \tag{2.11}$$

$$I_1 = (V_1 - V_A)Y_1$$

$$I_2 = (V_2 - V_A)Y_2 \tag{2.12}$$

$$\vdots$$

$$I_N = (V_N - V_A)Y_N$$

After rearranging, we obtain the formula for V_A:

$$V_A = \frac{\displaystyle\sum_{i=1}^{N} Y_i V_i}{\displaystyle\sum_{i=1}^{N} Y_i}; \tag{2.13}$$

and

$$Z = \frac{1}{Y}$$

$$Z = \frac{1}{\displaystyle\sum_{i=1}^{N} Y_i} \tag{2.14}$$

After transforming the Thevenin into the Norton equivalent circuit as in Fig. 2.12(d), we obtain

$$J_A = \frac{V_A}{Z} = \sum_{i=1}^{N} Y_i V_i = \sum_{i=1}^{N} \frac{V_i}{Z_i} \tag{2.15}$$

A simple computer program 'MILLMANN' used to calculate the voltage V_A is shown in Program 2.1.

See the listing of the program MILLMANN Directory of Chapter 2 on attached diskette.

As an application of this program, let us calculate V_A for the voltage divider of Fig. 2.13(a) and V_A for the more complicated circuit of Fig. 2.13(b).

After entering the data, the program gives

```
1. for Fig. 2.13(a): "The voltage VA=2.500 00 volts"
2. for Fig. 2.13(b): "The voltage VA=5.065 089 volts"
```

2.2.8 The Miller principle

The Miller principle is used to simplify the analysis of feedback circuits containing high gain inverting amplifiers. The feedback circuit, which includes an inverting

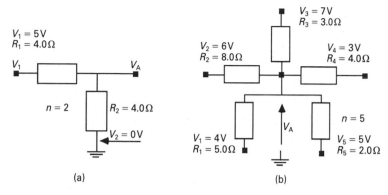

Figure 2.13 Illustration of the Millmann theorem. (a) Voltage divider. (b) More complicated voltage divider

voltage amplifier, and its equivalent circuit are shown in Fig. 2.14(a, b), respectively. The relations between output voltage V_o and input voltage V_i for the amplifier are

$$V_o = -A_V V_i \text{ or } V_i = \frac{-V_o}{A_V} \qquad A_V > 0 \tag{2.16}$$

The current I_f flowing into the admittance Y_f from the node V_i can be found directly from Ohm's law and equation (2.16):

$$I_f = Y_f(V_i - V_o) = V_i Y_f(1 + A_V) = V_i Y_M \tag{2.17}$$

$$Y_M = Y_f(1 + A_V) \tag{2.18}$$

where Y_M is called the Miller admittance.

Similarly, the current $I_f' = -I_f$ flowing into the admittance Y_f from the node

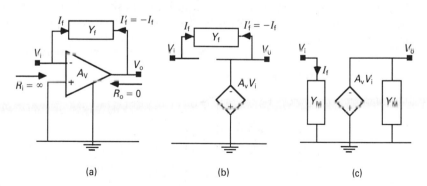

Figure 2.14 (a) Feedback circuit with inverting voltage amplifier. (b) Its equivalent circuit. (c) Equivalent circuit modified by the Miller principle

V_o can be written

$$I_\text{f}' = Y_\text{f}(V_\text{o} - V_\text{i}) = \left(V_\text{o} + \frac{V_\text{o}}{A_\text{V}}\right)Y_\text{f} = Y_\text{f}V_\text{o}\left(1 + \frac{1}{A_\text{V}}\right) = V_\text{o}Y_\text{M}' \tag{2.19}$$

$$Y_\text{M}' = Y_\text{f}\left(1 + \frac{1}{A_\text{V}}\right) \cong Y_\text{f} \text{ for } A_\text{V} \gg 1 \tag{2.20}$$

Therefore, the input admittance and output admittance of the considered circuit are

$$Y_\text{in} = \frac{I_\text{f}}{V_\text{i}} = Y_\text{M}, \qquad Y_\text{out} = \frac{I_\text{f}'}{V_\text{o}} = Y_\text{M}' \tag{2.21}$$

Thus, the properties of the circuit will not be changed if we redraw the equivalent circuit of Fig. 2.14(b) in the format shown in Fig. 2.14(c). After this modification, feedback disappears from the schematic diagram, but it is included in the Miller admittance at the input of the amplifier. For high gain $A_\text{V} \gg 1$, the output admittance is almost equal to the value of Y_f. In the most frequent practical cases, Y_f is capacitive: $Y_\text{f} = -j\omega C_\text{f}$, and this gives the Miller capacitance, C_M:

$$C_\text{M} = C_\text{f}(1 + A_\text{V}) \tag{2.22}$$

and

$$C_\text{M}' = C_\text{f}\left(1 + \frac{1}{A_\text{V}}\right) \tag{2.23}$$

It is necessary to note that the expressions derived in this section correspond to the ideal voltage amplifier or ideal voltage controlled voltage source, VCVS. Relations for the Miller admittance become more complicated when we assume that the output resistance, R_o, of the VCVS is finite.

2.2.9 Calculation of circuits containing ideal operational amplifiers

An ideal operational amplifier, with differential input as shown in Fig. 2.15, has a voltage gain A equal to infinity. Also, it is also assumed that its input impedance Z_i equals infinity, its output impedance Z_o equals zero and its DC properties are

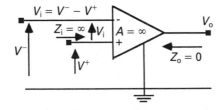

Figure 2.15   Ideal operational amplifier with differential input

such that on setting its input terminals at zero volts (grounded input) the output voltage gives zero.

Because of infinite gain $A = \infty$ we cannot draw the equivalent circuit using known circuit elements. Properties of circuits with operational amplifiers can easily be calculated, however, when each ideal operational amplifier has a feedback resistor connected between its output and inverting terminals (negative feedback). Two basic circuits with ideal operational amplifiers are shown in Fig. 2.16.

For an ideal operational amplifier, we can always say that with infinite voltage gain, for finite output voltage, V_o, the input voltage $V_i = 0$; but the current between the input terminals also equals zero because the input impedance Z_i equals infinity. Therefore, we say that between the input terminals there is a 'virtual short', that is, $V_i = 0$ and at the same time $I_i = 0$. When the non-inverting input is grounded, as it is for the inverting amplifier in Fig. 2.16(a), then we talk about 'virtual ground'. Therefore, for the inverting amplifier we can simply write

$$V_1 = Z_1 I_1$$

$$I_1 = I_f \tag{2.24}$$

$$- I_f Z_2 = V_o$$

Combining these equations, we obtain

$$\frac{V_1}{Z_1} = -\frac{V_o}{Z_2} \tag{2.25a}$$

or

$$V_o = -\frac{Z_2}{Z_1} V_1 \tag{2.25b}$$

When both impedances Z_2 and Z_1 are resistive, then we have the 'multiplying'

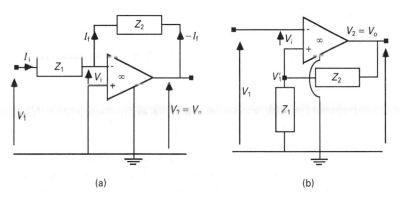

(a)   (b)

Figure 2.16   Basic circuits with an operational amplifier. (a) An inverting. (b) Non-inverting or 'scaling' circuit

$$V_o = -\frac{R_2}{R_1} V_1 \qquad (2.26a)$$

When $Z_1 = R_1$ and $Z_2 = 1/j\omega C_2$, then we have the 'integrating' circuit

$$V_o = -\frac{1}{j\omega C_2 R_1} V_1 \qquad (2.26b)$$

since in the time domain $v_o(t)$ is an integral of $v_1(t)$.

When, $Z_1 = 1/j\omega C_1$ and $Z_2 = R_2$, then we obtain the 'differentiating' circuit

$$V_o = -j\omega R_2 C_1 V_1 \qquad (2.26c)$$

since in the time domain $v_o(t)$ is a derivative of $v_1(t)$.

For the non-inverting circuit we can observe that because of the 'virtual short', $V_1' = V_1$. Therefore, we can write directly

$$V_1' = V_1 = \frac{Z_1}{Z_1 + Z_2} V_o \qquad (2.27a)$$

or

$$V_o = \left(1 + \frac{Z_2}{Z_1}\right) V_1 \qquad (2.27b)$$

The non-inverting circuit (Fig. 2.17), known as the non-inverting integrator, can also be analyzed directly by inspection:

$$V_a = \frac{R_1}{R_1 + \dfrac{1}{j\omega C_2}} V_o \qquad (2.28a)$$

$$V_a' = V_a \qquad (2.28b)$$

$$V_a' = \frac{\dfrac{1}{j\omega C_2}}{R_1 + \dfrac{1}{j\omega C_2}} V_1 \qquad (2.28c)$$

or

$$V_o = \frac{1}{j\omega R_1 C_2} V_1 \qquad (2.28d)$$

The more complicated circuit shown in Fig. 2.18(a), known as the 'lossy inverting integrator', can also be analyzed by inspection using the properties of the ideal operational amplifier and the Miller principle.

Two-step transformation leads to the equivalent circuits shown in Fig. 2.18(b, c). In the first step the inverting VCVS is created, with input resistance equal to R_1 (because of virtual ground at the input of the operational amplifier), and the coefficient $\mu = R_2/R_1$. In the second step, the application of the Miller

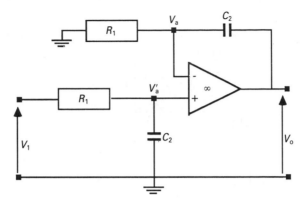

Figure 2.17 Non-inverting integrator

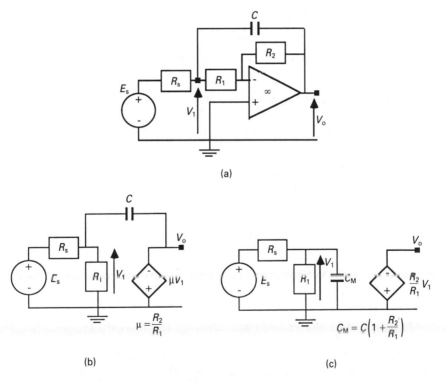

(a)

(b)   (c)

Figure 2.18 (a) Lossy non-inverting integrator. (b) Equivalent circuit with inverting voltage controlled voltage source. (c) Equivalent circuit after application of the Miller principle

principle allows us to write

$$V_o = -\frac{R_2}{R_1} V_1 \tag{2.29a}$$

$$V_1 = \frac{\dfrac{R_1}{j\omega C_M}}{\left[R_1 + \dfrac{1}{j\omega C_M}\right]\left[R_s + \dfrac{\dfrac{R_1}{j\omega C_M}}{R_1 + \dfrac{1}{j\omega C_M}}\right]} E_s \tag{2.29b}$$

$$C_M = C\left(1 + \frac{R_2}{R_1}\right) \tag{2.29c}$$

$$\frac{V_o}{E_s} = \frac{\dfrac{-R_2}{R_1 + R_s}}{1 + j\omega R_s C \dfrac{R_1 + R_2}{R_1 + R_s}} \tag{2.29d}$$

2.3 Calculation of voltage and current in dynamic elements: numerical calculations of integrals

2.3.1 Calculation of capacitance voltage and inductance current

The relationship between the branch current, electric charge and voltage for a capacitor is

$$i(t) = \frac{dq(t)}{dt} = \frac{dq(v(t))}{dv(t)} \frac{dv(t)}{dt} = C(v) \frac{dv(t)}{dt} \tag{2.30}$$

and for a linear capacitor

$$i(t) = C \frac{dv(t)}{dt} \tag{2.31}$$

Hence, the value of the voltage across the linear capacitor C, charged by the current $i(t)$ in the time period t_a to t_b, can be expressed as

$$v_C(t_b) = \frac{1}{C} \int_{t_a}^{t_b} i(t)\, dt + v_C(t_a) \tag{2.32}$$

where $v_C(t_a)$ is the voltage across the capacitor at the beginning of the given time period (initial condition). Similarly, the relationship between the branch voltage, magnetic flux and current flowing through the inductance is

$$v(t) = \frac{d\phi(t)}{dt} = \frac{d\phi(i(t))}{di(t)} \frac{di(t)}{dt} = L(i) \frac{di(t)}{dt} \tag{2.33}$$

and for linear inductance

$$v(t) = L \frac{di(t)}{dt} \tag{2.34}$$

Thus, the value of the current flowing through the linear inductance L, excited from the voltage source $v(t)$ in the time period $[t_a, t_b]$ is

$$i_L(t_b) = \frac{1}{L} \int_{t_a}^{t_b} v(t) \, dt + i_L(t_a) \tag{2.35}$$

where $i_L(t_a)$ is the current flowing through the inductance at the beginning of the given time period (initial condition). Hence, the problem of calculation of the voltage $v_C(t_b)$ or the current $i_L(t_b)$ at the end of the given period $[t_a, t_b]$ is simply the determination of the value of the definite integral of type $\int_{t_a}^{t_b} f(t) \, dt$, with given limits of integration and knowledge of the initial conditions $v_C(t_a)$ and $i_L(t_a)$, respectively.

2.3.2 Numerical calculations of the definite integral

The value of the definite integral of the function $\int_{t_a}^{t_b} f(t) \, dt$ in the limits t_a and t_b is equal to the area under this function in the range $[t_a, t_b]$. Therefore, numerical determination of the definite integral is just numerical calculation of this area. The most popular methods for numerical calculation of the area under the function $y = f(x)$ in the given range $[a, b]$ are

1.   Rectangular method.
2.   Trapezoidal method (linear interpolation).
3.   Simpson's method (second-degree parabolic interpolation).

Here the function must be given in an analytic form. These three methods are illustrated graphically in Fig. 2.19(a, b, c). In each case, the area is divided into N

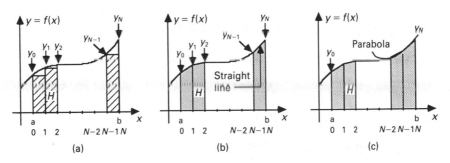

Figure 2.19   Graphical calculation of the area under the function $y = f(x)$ using the method: (a) rectangular, (b) trapezoidal, (c) Simpson (parabolic)

subareas (sections) of width H, called the integration step. Hence, the total area is the sum of the elementary areas in the range $[a, b]$, and the width H of the elementary area is

$$H = \frac{b - a}{N} \tag{2.36}$$

Rectangular method. It can be seen directly from Fig. 2.19 that for the rectangular method the value of the definite integral is

$$RECTINT = y_0 H + y_1 H + \ldots + y_{N-1} H = H \sum_{i=0}^{N-1} y_i \tag{2.37}$$

The error ε_R of the calculations by this method is

$$\varepsilon_R = \frac{(b-a)^2}{2N} f'(\zeta) = \frac{NH^2}{2} f'(\zeta), \qquad \zeta \in [a, b] \tag{2.38}$$

where $f'(\zeta)$ denotes the first derivative of $f(x)$ at $x = \zeta$.

From this equation we see that the accuracy of the calculation depends largely on the value of the integration step; for smaller H we have smaller error.

A simple example for definite integral calculation using the rectangular method is given in the program 'RECTINT' shown in Program 2.2.

See the listing of the program RECTINT Directory of Chapter 2 on attached diskette.

In this example program, the function $y = 2x^3$ is taken into account. For the limits of integration $a = 0$, $b = 1$ and number of sections (subareas) $N = 40$, the calculated value is 0.475 31, whereas the exact value is 0.5. Increasing the number N to 100, the calculated value increases to 0.490 05.

Trapezoidal method. Also based on graphical interpretation of the area calculation shown in Fig. 2.19(b) we can directly write the formula for the value of the definite integral using the trapezoidal method:

$$TRAPINT = H \frac{y_0 + y_1}{2} + H \frac{y_1 + y_2}{2} + \ldots + H \frac{y_{N-1} + y_N}{2} \tag{2.39}$$

or

$$TRAPINT = H \sum_{i=0}^{N} y_i - H \frac{y_0 + y_N}{2} = H \sum_{i=1}^{N-1} y_i + H \frac{y_0 + y_N}{2} \tag{2.40}$$

The error ε_T of calculations in this method is

$$\varepsilon_T = -\frac{(b-a)^3}{12N^2} f''(\zeta) = -\frac{NH^3}{12} f''(\zeta), \qquad \zeta \in [a, b] \tag{2.41}$$

where $f''(\zeta)$ denotes the second derivative of $f(x)$ at $x = \zeta$.

For small integration step H, the error is much smaller than for the rectangular method and is of opposite sign. An example of numerical calculation for the definite integral using the trapezoidal method is given in the program 'TRAPINT' shown in Program 2.3.

See the listing of the program TRAPINT Directory of Chapter 2 on attached diskette.

In this example program the function $y = 2x^3$ is again used for the calculations. For the limits of integration $a = 0$, $b = 1$ and number of sections $N = 40$, the calculated area value is 0.500 31; the exact value is 0.5. Increasing the number N to 100, the integral value is 0.500 05.

Simpson's method. The derivation of the formula for Simpson's method of area calculation is more complicated since the adjacent points y_i and y_{i+1} are connected by a second-degree parabola, as illustrated in Fig. 2.19(c). The resulting formula for the value of the definite integral using Simpson's method is

$$SIMPINT = \frac{H}{3}(y_0 + 4y_1 + y_2 + y_2 + 4y_3 + y_4 + \ldots + y_{N-2} + 4y_{N-1} + y_N) \quad (2.42)$$

Here, the summation concerns the triplets $y_i + 4y_{i+1} + y_{i+2}$ in the index i range from 0 to $N-2$ with a step of 2. Note that in this method the number N must be even! The error ε_S of calculation is equal to

$$\varepsilon_S = -\frac{(b-a)^5}{180N^4} f^{IV}(\zeta) = -\frac{NH^5}{180} f^{IV}(\zeta), \qquad \zeta[a, b] \quad (2.43)$$

where $f^{IV}(\zeta)$ is the fourth derivative of $f(x)$ at $x = \zeta$.

Simpson's method is the most accurate of those described here. An example program named 'SIMPINT' for numerical calculations is shown in Program 2.4.

See the listing of the program SIMPINT Directory of Chapter 2 on attached diskette.

In this example program the function $y = 2x^3$ is again used for the calculations. In this case for any limits of integration and for any even number $N \geqslant 2$, the result of the calculations is correct, since the fourth derivative of the function $y = 2x^3$ is always equal to zero.

Trapezoidal method for tabular representation of the function. In many practical cases, the relationship between the variables y and x is given as a set of discrete pairs of points, as it is, for example, during measurements. Also, the relationship $y = f(x)$ can be given in graphical form on the screen of an oscilloscope or as an output from a plotter. In such cases, instead of seeking analytical representation of $y = f(x)$ to integrate it in a given range $x \in [a, b]$, we can directly take the corresponding pairs of variables x_i and y_i as shown in Fig. 2.20 and find the area under the function

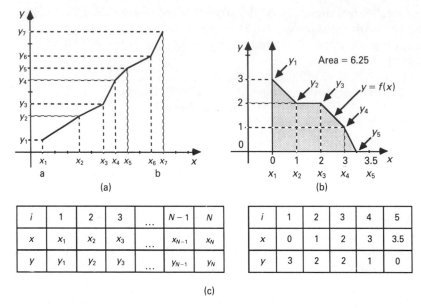

Figure 2.20   (a) Graphical representation of the function $y = f(x)$. (b) Example of graphical representation of the trapezoidal method. (c) Its tabular representation

using, for example, the trapezoidal method. Based on Fig. 2.20(b), we can write a formula for the area under the curve $y = f(x)$:

$$AREA = \frac{y_1 + y_2}{2}(x_2 - x_1) + \frac{y_2 + y_3}{2}(x_3 - x_2)$$

$$+ \frac{y_{N-2} + y_{N-1}}{2}(x_{N-1} - x_{N-2}) + \frac{y_{N-1} + y_N}{2}(x_N - x_{N-1}) \qquad (2.44)$$

or, after rearrangement,

$$AREA = \frac{1}{2}\left[y_1(x_2 - x_1) + y_N(x_N - x_{N-1}) + \sum_{i=2}^{N-1} y_i(x_{i+1} - x_{i-1}) \right] \qquad (2.45)$$

The example program named 'TABINT', for the trapezoidal method of integration when the relations between the variables y and x are given in tabular form, is shown in Program 2.5.

See the listing of the program **TABINT** Directory of Chapter 2 on attached diskette.

In this example program, we can enter the data points (x_i, y_i) from Fig. 2.20(b) to verify that the definite integral value in this case is equal to 6.25 (units squared). We note that the trapezoidal method gives sufficient accuracy for these cases where the data points are the results of measurements or graphical representation of the function $y = f(x)$.

2.3.3 Normalization of physical quantities

To calculate the voltage across a capacitor or the current flowing through an inductor according to equations (2.32) and (2.35), respectively, it is necessary to enter into the program, besides the function $i(t)$ or $v(t)$, the value of the capacitor C or inductor L, and the initial value $v_C(0) = v_C(t_a)$ or $i_L(0) = i_L(t_a)$. If we use non-normalized units in our calculations, then we should express the current in amperes, voltage in volts, time in seconds, capacitance in farads and inductances in henries. In such cases, we have to manipulate large and very small quantities since in practical electronic circuits we have voltages of several volts, currents of order of milliamperes (10^{-3} A) or microamperes (10^{-6} A), time of order of microseconds (10^{-6} s), capacitances of order of nanofarads (10^{-9} F), picofarads (10^{-12} F), or femtofarads (10^{-15} F) and inductances of order of millihenries (10^{-3} H) or microhenries (10^{-6} H). To overcome this inconvenience in a large spread of values leading to the decrease of accuracy in numerical calculations, we can normalize our physical quantities.

For example, to calculate the voltage on the capacitor charged from the current source $i(t)$ in the time range $[t_a, t_b]$ and with the initial voltage $v_C(0) = v_C(t_a)$, as is shown in Fig. 2.21(a), and to obtain the voltage in volts, we can enter the quantities in units as is shown in Fig. 2.21(b), since according to equation (2.32) we obtain

$$1V = \frac{1}{1F} \, 1A \, 1s = \frac{1}{1nF} \, 1mA \, 1\mu s = \frac{1}{1\mu F} \, 1A \, 1\mu s = \frac{1}{1pF} \, 1\mu A \, 1\mu s$$

A modified version of Program 2.4 for calculating the voltage on capacitance using Simpson's method, named 'VOLTSIMP', is shown in Program 2.6.

See the listing of the program VOLTSIMP Directory of Chapter 2 on attached diskette.

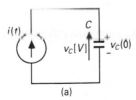

$v_C, v_C(0)$	i	t	C
V	A	s	F
V	mA	µs	nF
V	A	µs	µF
V	µA	µs	pF

(a)   (b)

Figure 2.21   (a) Capacitor C charged from the current source $i(t)$. (b) Table showing normalized quantities to obtain v_C in volts

Entering, for example, $t_a = 0$, $t_b = 5$ μs, $n = 8$, $C = 2$ nF and $V_0 = 3$ V, we obtain the result: 'The voltage V_C is 5.500 00 volts'.

In a similar way, the quantities for the current flowing through the inductance L excited from the voltage source can be normalized to obtain the current i_L in milliamperes, as is shown in Fig. 2.22(a), since according to equation (2.35) we obtain

$$1\text{mA} = 10^{-3}\text{A} = \frac{1}{1\text{H}}\ 1\text{V}\ 1\text{ms} = \frac{1}{1\text{mH}}\ 1\text{V}\ 1\mu\text{s} = \frac{1}{1\mu\text{H}}\ 1\text{V}\ 1\text{ns} = \frac{1}{\text{nH}}\ 1\text{mV}\ 1\text{ns}$$

Program 2.6 can easily be adapted to calculate the current i_L flowing through the inductance.

2.3.4 Numerical indefinite integral calculations

In many practical cases we are interested not only in the capacitance voltage at the end of a given time period $v_C(t_b)$, but in the variations of the voltage $v_C(t)$ in the given time period $t \in [t_0, t_b]$. In such cases, the upper limit of the integral becomes the time variable t:

$$v_C(t) = \frac{1}{C} \int_{t_0}^{t} i(\tau)\,d\tau + v_C(t_0), \qquad t > t_0 \tag{2.46}$$

and the right-hand side of (2.46) corresponds to an indefinite integral. To calculate this integral it is necessary to split the time period of interest into small subperiods, Δt, as is shown in Fig. 2.23, calculate the definite integrals in these subperiods, add the initial value $v_C(0)$ and sum up these values starting from t_0 up to the end of the time period t_b, since the integral in given limits is equal to the sum of integrals in sublimits:

$$\int_{1}^{4} f(t)\,dt = \int_{1}^{2} f(t)\,dt + \int_{2}^{3} f(t)\,dt + \int_{3}^{4} f(t)\,dt \tag{2.47}$$

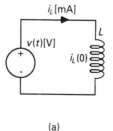

i_L, $i_L(0)$	v	t	L
A	V	s	H
mA	V	ms	H
mA	V	μs	mH
mA	V	ns	μH
mA	mV	ns	nH

(a)   (b)

Figure 2.22   (a) Inductor L excited from the voltage source $v(t)$. (b) Table showing normalized quantities to calculate i_L in mA

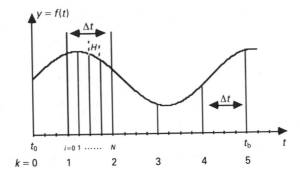

Figure 2.23 Illustration for numerical calculation of indefinite integral

Printing the values of definite integrals after summation gives us the values of $v_C(k\,\Delta t)$ at discrete time points, thus giving the approximate plot of the indefinite integral of the function $i(t)$.

In this case the program contains two summing loops, one for definite integral calculation in the time ranges Δt, and the other for summing up the values of the definite integrals. An example program called 'INDEFINT' for the calculation and printing of the indefinite integral is shown in Program 2.7.

See the listing of the program INDEFINT Directory of Chapter 2 on attached diskette.

This program can easily be adapted for the calculation of $v(t)$ or $i(t)$. In this particular program the function $f(x) = x^4$, whose indefinite integral (primary function), $f_p(x) = 1/5x^5$, is used and printed to show the accuracy of the calculations when definite integrals in the range Δt are calculated using Simpson's method.

Exercises

Exercise 1. Find the Thevenin equivalent circuit of the networks shown in Fig. E2.1.

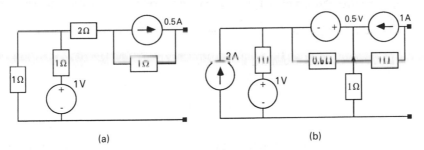

(a)   (b)

Figure E2.1

Exercise 2.   Find the Thevenin and Norton equivalent circuits of the networks shown in Fig. E2.2.

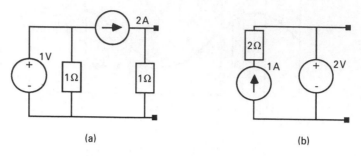

(a)   (b)

Figure E2.2

Exercise 3.   Find the Thevenin equivalent circuit of the networks shown in Fig. E2.3.

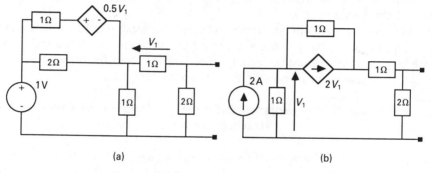

(a)   (b)

Figure E2.3

Exercise 4.   For the two networks shown in Fig. E2.4, use the shift of the source current J in order to find the Thevenin equivalent circuit between points A and B.

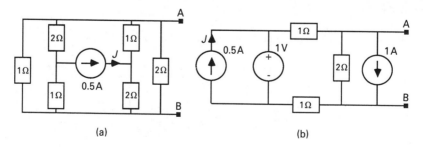

(a)   (b)

Figure E2.4

Exercise 5.   Use the principle of superposition in order to determine the voltage V of the circuit shown in Fig. E2.5.

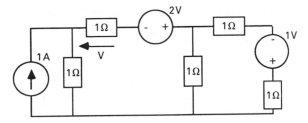

Figure E2.5

Exercise 6.   The networks of exercises 1 and 2 are loaded with a 1 Ω resistance; applying the principle of superposition, calculate the current flowing into the load.

Exercise 7.   Consider the MOSFET differential circuit shown in Fig. E2.6(a); Fig. E2.6(b) represents the corresponding MOSFET small signal model. We can assume that the transistors are identical. Draw the small signal equivalent circuit. Using the substitution theorem, remove all controlled sources except the VCVS source controlled by the input voltage. Calculate the analytical expression for the current flowing through the impedance Z. Give the value of the current when Z tends to zero.

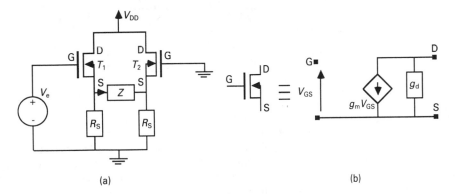

(a)   (b)

Figure E2.6

Exercise 8.   Using the program 'MILLMANN', calculate the voltage V in the circuit of Fig. E2.7 (the values of resistances are in ohms; the voltage sources are in volts).

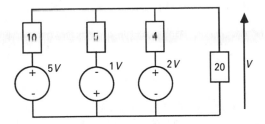

Figure E2.7

Exercise 9. Modify the program 'MILLMANN' in order to use complex variables (the impedances are defined in the form real part and imaginary part, and the voltage sources in the form magnitude and argument in degrees). Using this program, calculate the voltage V in the circuit shown in Fig. E2.8.

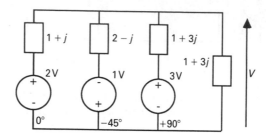

Figure E2.8

Exercise 10. For the circuit shown in Fig. E2.9, calculate the input admittance Y_{in} and the output voltage V_o. Consider that R_o tends to zero. Compare the results using Miller's theorem.

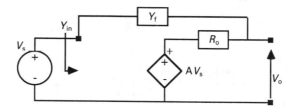

Figure E2.9

Exercise 11. Consider the circuit represented in Fig. E2.10. Apply Kirchhoff's current law at nodes 1 and 2. For node 2, the current flowing from node 2 to node 1 can be neglected. However, for node 1, the current flowing from node 1 to node 2 cannot be neglected. Using this approximation ('one side' approximation), calculate the voltage gain V_o/V_s. Find the equivalent Miller capacitance.

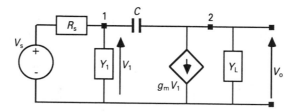

Figure E2.10

Exercise 12. The circuit shown in Fig. E2.11 is used to realize a compensated integrator. The gain K of the voltage amplifier under consideration is $K = \omega_t/s$. Using Miller's theorem, calculate the voltage gain V_o/V_s and show what value of R_c can be chosen in order to obtain an ideal integrator.

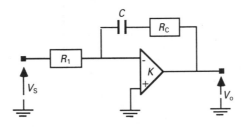

Figure E2.11

Exercise 13. Using the rectangular method, evaluate the integral between 0 and 1 of the function $\exp(x)$. The exact value is $\exp(1) - 1$. Modify the program 'RECTINT' in order to study the error variation as a function of the number of integration steps.

Exercise 14. Evaluate, using the trapezoidal method, the integral between 0 and 1 of the function $\sin(x)$. The exact value is $1 - \cos(1)$. Modify the program 'TRAPINT' in order to study the error variation as a function of the number of integration steps.

Exercise 15. When the expression of the derivative of the function $f(x)$ is known analytically, a corrective term can be added to the trapezoidal method:

$$\int_a^b f(x)\,dx = H \left[\frac{1}{2} y_0 + y_1 + \ldots + y_{N-1} + \frac{1}{2} y_N \right] - \frac{H^2}{12} \{f'(b) - f'(a)\}$$

1. Modify the program 'TRAPINT' enclosing this corrective term.
2. Apply this new program to the computation of the integral $\int_1^2 dx/x$.

Exercise 16. Using Simpson's method, evaluate the integral between 0 and 1 of the function $1/(1 + x^2)$. The exact value is $\pi/4$. Modify the program 'SIMPINT' in order to study the error variation as a function of the number of integration steps.

Exercise 17. When the expression of the third derivative of the function $f(x)$ is known analytically, a corrective term can be added to Simpson's method:

$$\int_a^b f(x)\,dx = \frac{H}{3} [y_0 + 4y_1 + 2y_2 + 4y_3 + \ldots + 4y_{N-1} + y_N] + \frac{H^4}{180}\{f'''(b) - f'''(a)\}$$

1. Modify the program 'SIMPINT' enclosing this corrective term.
2. Apply this new program to the computation of the integral $\int_1^2 dx/x$.
3. Compare with the results obtained in exercise E2.15.

Exercise 18. The current flowing through a device is presented in Fig. E2.12(a). The corresponding voltage is presented in Fig. E2.12(b). Modify the program 'TABINT' in order to compute the energy stored during one period.

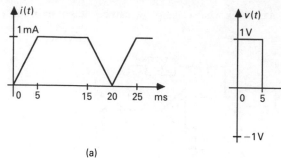

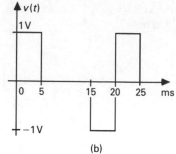

(a)   (b)

Figure E2.12   (a) Current flowing through a device. (b) Corresponding voltage

Exercise 19.   Modify the program 'VOLTSIMP' in order to calculate the current i_L flowing through an inductor biased by a voltage source $v(t)$, with initial condition $i_L(0) \neq 0$.

Exercise 20.   Modify the program 'INDEFINT' in order to determine the voltage on a 1 F capacitor initially discharged and biased by a current source $i(t) = \{e^{-t} - e^{-2t}\}1(t)$, where $1(t)$ is the unit pulse.

3 Automatic Immittance Matrix Generation for Linear Circuits

3.1 Introduction

This chapter describes some methods of automatic immittance (the word 'immittance' stands for impedance or admittance) matrix generation for linear passive and active circuits. For active circuit analysis, the nodal admittance matrix $\mathbf{Y}$ is usually used. In some cases, however, the application of a mesh impedance matrix to circuits with less independent meshes than nodes will lead to more economic computations. For example, Fig. 3.1 shows a circuit with three independent meshes (loops) and four independent voltage nodes.

This circuit also shows a grounded 'reference' node $V_0 = 0$, and 'reference' mesh I_0. The reference node/mesh is important for indefinite matrix creation, that is, a matrix in which reference node/mesh is included.

Note that the automatic creation of mesh impedance matrices is only possible for planar circuits having planar graphs as shown in Fig. 3.2(a). In such circuits only

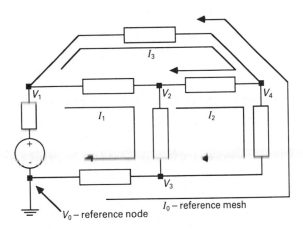

Figure 3.1 Example of a circuit having three independent meshes and four independent nodes

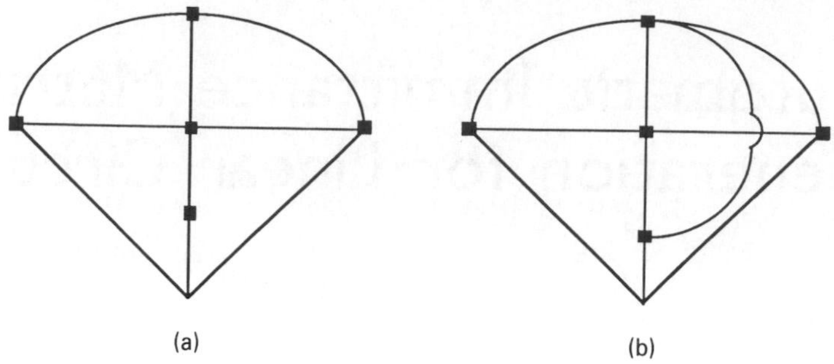

Figure 3.2   Examples of: (a) planar graph, (b) non-planar graph

two mesh currents, including the reference mesh current, can flow through each element, usually in opposite directions.

When working with node admittance matrices, the voltage controlled current sources (VCCSs) are used as the dependent sources since their coefficients g_m are mutual admittances. In mesh analysis, however, the current controlled voltage sources (CCVSs) are used. Other types of dependent source in models of electronic devices can always be transposed to a VCCS with g_m as the 'active' coefficient.

The creation methods of the impedance and admittance matrices are identical. The impedance matrix is based on Kirchhoff's voltage law (KVL) applied to a mesh description of the circuit. The admittance matrix is based on Kirchhoff's current law (KCL) applied to a nodal description. For example, consider the dual 'T' and 'Π' resistive circuits shown in Fig. 3.3(a, b).

Applying KVL to the circuit shown in Fig. 3.3(a):

$$(R_1 + R_2)I_1 - R_2I_2 = E_1$$

$$- R_2I_1 + r_mI_1 + (R_L + R_2 + R_3)I_2 = 0 \qquad (3.1)$$

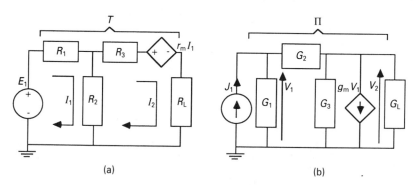

Figure 3.3   Dual resistive active circuits: (a) 'T'-type. (b) 'Π' type

This can be expressed in matrix form **R**

$$\mathbf{R} = \begin{bmatrix} R_1 + R_2 & -R_2 \\ -R_2 + r_m & R_2 + R_3 + R_L \end{bmatrix} = \begin{bmatrix} r_{11} & r_{12} \\ r_{21} & r_{22} \end{bmatrix} \tag{3.2}$$

Applying KCL to Fig. 3.3(b):

$$(G_1 + G_2)V_1 - G_2V_2 = J_1 \tag{3.3}$$

$$-G_2V_1 + g_mV_1 + (G_L + G_2 + G_3)V_2 = 0$$

or in matrix form **G**

$$\mathbf{G} = \begin{bmatrix} G_1 + G_2 & -G_2 \\ -G_2 + g_m & G_2 + G_3 + G_L \end{bmatrix} = \begin{bmatrix} g_{11} & g_{12} \\ g_{21} & g_{22} \end{bmatrix} \tag{3.4}$$

It can be seen from equations (3.1)–(3.4) that the form of the equations is identical and that the matrix elements r_{ij} and g_{ij} are created in a similar way. For example the 'mesh' resistance $r_{11} = R_1 + R_2$ is composed of all the resistances in mesh 1, whilst the 'node' conductance $g_{11} = G_1 + G_2$ is composed of all the conductances connected to node 1. The 'intermesh' resistance $r_{12} = -R_2$ corresponds to the resistance, with a negative sign, between meshes 1 and 2, whilst the 'internode' conductance $g_{12} = -G_2$ corresponds to the conductance, with a negative sign, between nodes 1 and 2, etc. Conductances and admittances shall, furthermore, be considered in matrix form.

These matrices, stored in the memory of a computer or on its printouts, can be either of definite or indefinite type. To recall the properties of the indefinite matrix, consider the simple circuit with grounded reference node ($V_0 = 0$) shown in Fig. 3.4(a), and the same circuit without a grounded reference node ($V_0 \neq 0$) as shown in Fig. 3.4(b). The conductance definite matrix for the circuit with grounded

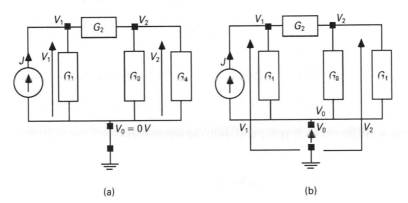

(a)   (b)

Figure 3.4 Simple circuit. (a) Grounded reference node ($V_0 = 0$). (b) Not grounded reference node ($V_0 \neq 0$)

reference node can be written by inspection:

$$\mathbf{G} = \begin{bmatrix} G_1 + G_2 & -G_2 \\ -G_2 & G_2 + G_3 + G_4 \end{bmatrix} \qquad (3.5)$$

The matrix for the same circuit without the reference grounded, called the indefinite matrix, can be obtained by starting with Kirchhoff's nodal equations

$$\text{(node 0)} \quad G_1(V_0 - V_1) + (G_3 + G_4)(V_0 - V_2) + J = 0$$

$$\text{(node 1)} \quad G_1(V_1 - V_0) + G_2(V_1 - V_2) - J = 0 \qquad (3.6)$$

$$\text{(node 2)} \quad G_2(V_2 - V_1) + (G_3 + G_4)(V_2 - V_0) = 0$$

which can be rearranged to obtain

$$\left. \begin{aligned} (G_1 + G_3 + G_4)V_0 - \quad G_1 V_1 \quad - \quad (G_3 + G_4)V_2 \quad &= -J \\ -G_1 V_0 \quad + (G_1 + G_2)V_1 - \quad G_2 V_2 \quad &= J \\ -(G_3 + G_4)V_0 \quad - \quad G_2 V_1 \quad + (G_2 + G_3 + G_4)V_2 &= 0 \end{aligned} \right\} \qquad (3.7)$$

which gives the indefinite conductance matrix shown in Fig. 3.5 (equation (3.8)).

$$\mathbf{G}_{\text{in}} = \begin{bmatrix} (G_1 + G_3 + G_4) & -G_1 & -(G_3 + G_4) \\ -G_1 & G_1 + G_2 & -G_2 \\ -(G_3 + G_4) & -G_2 & G_2 + G_3 + G_4 \end{bmatrix} \longleftarrow \text{Definite part} \qquad (3.8)$$

Figure 3.5

The number of rows and columns in the matrix is equal to the number of nodes, including the reference node, in the circuit. The important features of indefinite matrices can be seen directly from (3.8):

1. The sums of all the elements in each row are equal to zero.
2. The sums of all the elements in each column are equal to zero.

These properties can be used to check the validity of the circuit description.

3.2 Automatic admittance matrix generation

Automatic admittance matrix generation in computer programs is based on the properties of Kirchhoff's equations and can be obtained by inspection (or by the acquisition of knowledge from a circuit theory expert).

3.2.1 Matrix generation for passive resistive circuits

Some of the properties of the matrices mentioned in Section 3.1 can be utilized in computer programs for matrix generation for passive resistive circuits. For example, to create a node matrix element g_{ii}, for a node i with k conductances attached to it, we can write

$$g_{ii} = \sum_{i=1}^{k} g_i \qquad\qquad (3.9)$$

or, using Pascal syntax,

$$g_{ii} := 0;$$
$$g_{ii} := g_{ii} + g_i, \qquad i = 1, 2, ..., k \qquad (3.10)$$

Similarly, for internode matrix element g_{ij}, between the nodes i and j, between which k conductances are connected, we can write

$$g_{ij} = -\sum_{i=1}^{k} g_i \qquad\qquad (3.11)$$

or, using Pascal syntax,

$$g_{ij} := 0;$$
$$g_{ij} := g_{ij} - g_i, \qquad i = 1, 2, ..., k \qquad (3.12)$$

Knowing that the matrix of passive circuits is symmetrical along the main diagonal, then

$$g_{ij} := g_{ji} \qquad\qquad (3.13)$$

The procedure for passive resistive matrix generation can be written (in pseudo-Pascal) using the above properties:

```
readln(numberelements)
for k=1 to numberelements do begin
          readln(i,j,conductancevalue);
          g[i,i]:=g[i,i]+conductancevalue;
          g[i,j]:=g[i,j]-conductancevalue;
          g[j,i]:=g[i,j]
end;
```

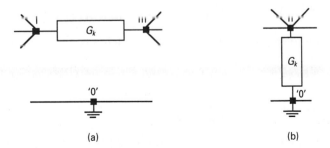

(a)   (b)

Figure 3.6 Location of the passive element G_k. (a) Between the nodes i and m. (b) Between the node n and the reference node '0'

To visualize the locations of passive elements in a matrix, including the indefinite matrix, consider the single element connected between two nodes in a circuit as shown in Fig. 3.6(a) and in Fig. 3.6(b). In each case the element G_k will appear in four places:

1. For Fig. 3.6(a),

$$
\begin{array}{cc}
 & \begin{array}{cc} \text{i} & \qquad \text{m} \end{array} \\
\begin{array}{c} \\ \text{i} \\ \\ \\ \text{m} \\ \\ \end{array} &
\begin{bmatrix}
. & . & . & . & . & . & . & . & . \\
. & . & +G_k & . & . & -G_k & . & . \\
. & . & . & . & . & . & . & . \\
. & . & . & . & . & . & . & . \\
. & . & -G_k & . & . & +G_k & . & . \\
. & . & . & . & . & . & . & . \\
. & . & . & . & . & . & . & . \\
\end{bmatrix}
\end{array}
\tag{3.14}
$$

2. For Fig. 3.6(b),

$$
\begin{array}{cc}
 & \begin{array}{cc} 0 & \qquad \text{n} \end{array} \\
\begin{array}{c} 0 \\ \\ \\ \text{n} \\ \\ \\ \end{array} &
\begin{bmatrix}
+G_k & . & . & -G_k & . & . \\
. & . & . & . & . & . \\
. & . & . & . & . & . \\
-G_k & . & . & +G_k & . & . \\
. & . & . & . & . & . \\
. & . & . & . & . & . \\
\end{bmatrix}
\end{array}
\tag{3.15}
$$

A convention similar to that of SPICE is usually used to enter the values of conductance; that is, entering the value of 28 mS for the conductance G between nodes 3 and 5 we can write

$$G \quad 3 \quad 5 \quad 28\,e-3$$

3.2.2 Matrix generation for passive capacitive and inductive circuits

The contribution of capacitive and inductive branches to the admittance matrix, $\mathbf{Y}$, is identical to the case of conductive branches, except that the admittances are imaginary:

$$Y_C = j\omega C \text{ and } Y_L = \frac{1}{j\omega L} = j\,\frac{-1}{\omega L} \tag{3.16}$$

Most computers are unable to handle complex numbers directly so these imaginary parts are stored and treated separately. The admittance of a complex branch composed of the parallel connection of G_k, C_k and L_k will therefore be stored in two submatrices: real $\mathbf{Y_R}$ and imaginary $\mathbf{Y_I}$:

$$\mathbf{Y} = \mathbf{Y_R} + j\mathbf{Y_I} = \mathbf{G}_k + j\left[\omega \mathbf{C}_k + \frac{-1}{\omega \mathbf{L}_k}\right] \tag{3.17}$$

Capacitive and inductive susceptances can be stored directly as in (3.17) or in separate 'capacitive' and 'inductive' submatrices:

$$\mathbf{Y} = \mathbf{G}_k + j\omega \mathbf{C}_k + \frac{j}{\omega}\frac{-1}{\mathbf{L}_k} \tag{3.18}$$

The latter method, however, requires more memory. The angular frequency $\omega = 2\pi f$ must be given prior to storing the imaginary components of $\mathbf{Y}$.

For example, the indefinite real and imaginary matrices for the simple RLC circuit shown in Fig. 3.7 are

$$\mathbf{Y} = \begin{bmatrix} G + \dfrac{1}{j\omega L} & -G & -\dfrac{1}{j\omega L} \\[2mm] -G & G + j\omega C & -j\omega C \\[2mm] -\dfrac{1}{j\omega L} & -j\omega C & j\omega C + \dfrac{1}{j\omega L} \end{bmatrix} = \mathbf{Y_R} + j\mathbf{Y_I} \tag{3.19}$$

$$\mathbf{Y_R} = \begin{bmatrix} G & -G & 0 \\ -G & G & 0 \\ 0 & 0 & 0 \end{bmatrix} = \begin{bmatrix} 0.005 & -0.005 & 0 \\ -0.005 & 0.005 & 0 \\ 0 & 0 & 0 \end{bmatrix} \tag{3.20}$$

$$\mathbf{Y_I} = \begin{bmatrix} \dfrac{-1}{\omega L} & 0 & \dfrac{1}{\omega L} \\[2mm] 0 & \omega C & -\omega C \\[2mm] \dfrac{1}{\omega L} & -\omega C & \omega C - \dfrac{1}{\omega L} \end{bmatrix} = \begin{bmatrix} -0.01 & 0 & 0.01 \\ 0 & 0.002 & -0.002 \\ 0.01 & -0.002 & -0.008 \end{bmatrix} \tag{3.21}$$

After entering the angular frequency ω [rad/s] in the program, the values of C and L are entered, in a similar manner as SPICE, by writing the type of element, the nodes between which the element C or L is located followed by its value.

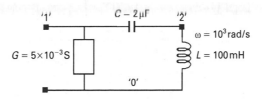

Figure 3.7 Simple RLC circuit

For example:

$$
\begin{array}{cccc}
C & 3 & 4 & 2e-6 \\
L & 5 & 8 & 3e-3
\end{array}
$$

3.2.3 Active circuit admittance matrix generation

The most popular dependent sources used for bipolar and MOS transistor modelling in the active circuits, are the voltage controlled current sources (VCCSs). Consequently, their use is considered in detail. In active circuits containing operational amplifiers, the voltage controlled voltage sources (VCVSs), which are usually used in operational amplifier models, can be transposed into the VCCSs as detailed in Section 2.2.3. To see where the mutual conductances g_m are placed in the conductance matrix, consider the simplified model of the transistor shown in Fig. 3.8.

Using KCL:

$$\text{(node 0)} \quad G_1(V_0 - V_1) + G_3(V_0 - V_2) - g_m(V_1 - V_0) = 0$$

$$\text{(node 1)} \quad G_1(V_1 - V_0) + G_2(V_1 - V_2) = 0 \tag{3.22}$$

$$\text{(node 2)} \quad G_2(V_2 - V_1) + G_3(V_2 - V_0) + g_m(V_1 - V_0) = 0$$

which after rearranging becomes

$$(G_1 + G_3 + g_m)V_0 - (G_1 + g_m)V_1 - G_3 V_2 = 0$$

$$-G_1 V_0 + (G_1 + G_2)V_1 - G_2 V_2 = 0 \tag{3.23}$$

$$-(G_3 + g_m)V_0 + (-G_2 + g_m)V_1 + (G_2 + G_3)V_2 = 0$$

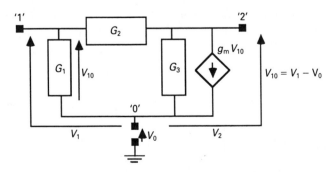

Figure 3.8 Simplified model of a transistor

Thus, the indefinite matrix $\mathbf{G}_{in}$ is

$$\mathbf{G}_{in} = \begin{array}{c} I_0 \\ I_1 \\ I_2 \end{array}\begin{bmatrix} G_1 + G_3 + g_m & -G_1 - g_m & -G_3 \\ -G_1 & G_1 + G_2 & -G_2 \\ -G_3 - g_m & -G_2 + g_m & G_2 + G_3 \end{bmatrix} = \begin{bmatrix} g_{00} & g_{01} & g_{02} \\ g_{10} & g_{11} & g_{12} \\ g_{20} & g_{21} & g_{22} \end{bmatrix} \qquad (3.24)$$
$$\qquad\qquad\qquad V_0 \qquad\qquad V_1 \qquad\qquad V_2$$

It can be seen from this example that the VCCS is located between nodes 2 and 0 (positive current flows from node 2 to 0) and is controlled by the voltage V_{10} (between nodes 1 and 0). Thus, the mutual conductance g_m appears four times in the indefinite matrix elements: g_{21}, g_{20}, g_{01} and g_{00} with positive sign in g_{21} and g_{00} and negative sign in g_{20} and g_{01}.

In general, when the VCCS is located between two nodes i and j (positive current flowing from i to j) and is controlled by the voltage V_{st} (positive at the s node), as shown in Fig. 3.9, then the following rules apply:

1. The coefficient g_m appears in the matrix elements g_{is}, g_{it}, g_{js} and g_{jt}.
2. The sign of g_m is positive when both indices are the upper (i.e. i and s) or lower (i.e. j and t).
3. The sign of g_m is negative when the indices are mixed (i.e. i, t and j, s).

When the VCCS is 'floating' (not connected to the reference node), g_m will be located in the four places shown:

$$\begin{array}{c} \\ \\ s \\ \\ \\ t \\ \\ \\ i \\ \\ \\ j \\ \\ \\ \end{array}\begin{array}{cccc} s & t & i & j \\ \end{array}\begin{bmatrix} \cdot & \cdot & \cdot & \cdot & \cdot & \cdot & \cdot & \cdot & \cdot & \cdot & \cdot & \cdot & \cdot \\ \cdot & \cdot & \cdot & \cdot & \cdot & \cdot & \cdot & \cdot & \cdot & \cdot & \cdot & \cdot & \cdot \\ \cdot & \cdot & \cdot & \cdot & \cdot & \cdot & \cdot & \cdot & \cdot & \cdot & \cdot & \cdot & \cdot \\ \cdot & \cdot & \cdot & \cdot & \cdot & \cdot & \cdot & \cdot & \cdot & \cdot & \cdot & \cdot & \cdot \\ \cdot & \cdot & \cdot & \cdot & \cdot & \cdot & \cdot & \cdot & \cdot & \cdot & \cdot & \cdot & \cdot \\ \cdot & \cdot & \cdot & \cdot & \cdot & \cdot & \cdot & \cdot & \cdot & \cdot & \cdot & \cdot & \cdot \\ \cdot & +g_m & \cdot & -g_m & \cdot & \cdot & \cdot & \cdot & \cdot & \cdot & \cdot & \cdot & \cdot \\ \cdot & \cdot & \cdot & \cdot & \cdot & \cdot & \cdot & \cdot & \cdot & \cdot & \cdot & \cdot & \cdot \\ \cdot & -g_m & \cdot & +g_m & \cdot & \cdot & \cdot & \cdot & \cdot & \cdot & \cdot & \cdot & \cdot \\ \cdot & \cdot & \cdot & \cdot & \cdot & \cdot & \cdot & \cdot & \cdot & \cdot & \cdot & \cdot & \cdot \end{bmatrix} \qquad (3.25)$$

These rules can easily be applied in a computer program for active matrix creation.

Figure 3.9   Location of the VCCS between nodes i and j, controlled by the voltage V_{st}

The formulae in Pascal syntax are similar to (3.10) and (3.12):

$$g_{is} := g_{is} + g_m$$

$$g_{jt} := g_{jt} + g_m$$

$$g_{it} := g_{it} - g_m \qquad (3.26)$$

$$g_{js} := g_{js} - g_m$$

To enter the value of the mutual conductance g_m into the computer program we have to enter the type of element, that is VCCS, the nodes i and j where the element is located (positive current flows from the node i to j), the value of g_m, and the nodes s and t to indicate the controlling voltage (positive at node s).

VCCS	i	j	g_m	s	t
VCCS	8	6	$2e-3$	2	5

3.3 Examples of programs for admittance matrix generation

Two simple programs for automatic nodal admittance matrix generation of active circuits with VCCS sources are presented in order to illustrate small approach differences. In the program 'ACTMATRX' shown in Program 3.1, the resistances R of the circuit are first entered and then converted to conductances. The VCCS sources are denoted by the letter s. The Pascal instruction 'case ... of' selects the different types of elements. In this program, it is assumed that there are three independent nodes ($n = 3$). To change this, the constant n must be amended in the source code.

See the listing of the program ACTMATRX Directory of Chapter 3 on attached diskette.

The program creates an indefinite matrix and then prints its real and imaginary part, $\mathbf{Y_R}$ and $\mathbf{Y_I}$, respectively, using the procedure 'printmat'. The main procedure 'readelem' reads the element locations and values and stores this information in memory in the arrays $[r_R]$ and $[r_I]$.

Consider the creation of the admittance matrix for the circuit having eight elements, shown in Fig. 3.10, assuming that $\omega = 2$ rad/s.

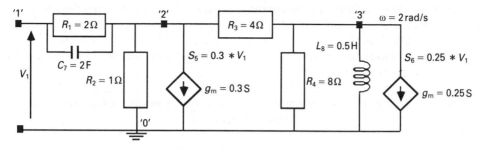

Figure 3.10 Example of an active circuit having eight elements

After running the program, the screen displays

```
Enter the number of elements : 8
Enter omega : 2
Enter the kind of element.1 : 'r', 's', 'c', 'l' : r
            for resistance 1 enter : i, j, res : 1 2 2
Enter the kind of element.2 : 'r', 's', 'c', 'l' : r
            for resistance 2 enter : i, j, res : 2 0 1
· · · ·

Enter the kind of element.5 : s
            for source 5 enter : i, j, gₘ, controlled
                                by s, t : 2 0 0.3  1 0
· · · ·

Enter the kind of element.7 : c
            for capacitor 7 enter : i, j, cap : 1 2 2
Enter the kind of element.8 : l
            for inductor 8 enter : i, j, ind : 3 0 0.5
```

The real part Y_R of indefinite matrix Y:

```
    1.675   -0.550   -1.000   -0.125
    0.000    0.500   -0.500    0.000
   -1.300   -0.200    1.750   -0.250
   -0.375    0.250   -0.250    0.375
```

Press a Key to continue!

The imaginary part Y_1 of indefinite matrix Y:

```
   -1.000    0.000    0.000    1.000
    0.000    4.000   -4.000    0.000
    0.000   -4.000    4.000    0.000
    1.000    0.000    0.000   -1.000
```

Another approach to the automatic creation of admittance matrices is shown in Program 3.2 called 'AUTOMATI'. Here, separate resistances R, mutual conductances g_m, capacitances C and inductances L are entered and stored in the memory. Printing procedures can be called from any place within the program showing the effects of entering an element.

> **See the listing of the program AUTOMATI Directory of Chapter 3 on attached diskette.**

Again, considering the example shown Fig. 3.10:

```
number of resistors    =4
number of VCCS         =2
number of capacitors   =1
number of inductors    =1
```

The resistors give the **G** definite matrix:

```
 0.500   -0.500    0.000
-0.500    1.750   -0.250
 0.000   -0.250    0.375
```

After entering the g_m values we obtain:

```
 0.500   -0.500    0.000
-0.200    1.750   -0.250
 0.250   -0.250    0.375
```

After entering, $\omega = 2$ and the value of the capacitor, the imaginary part of the matrix $\mathbf{Y_I}$, becomes

```
 4.000   -4.000    0.000
-4.000    4.000    0.000
 0.000    0.000    0.000
```

and after entering the value of the inductance:

```
 4.000   -4.000    0.000
-4.000    4.000    0.000
 0.000    0.000   -1.000
```

3.4 Normalization of quantities in the frequency domain

When the component values are entered into the program, resistances, R, are in ohms, capacitances, C, are in farads, inductances, L, are in henries, mutual conductances, g_m, are in siemens, and angular frequency ω is in radians per second. Thus, the real and imaginary parts of the calculated admittances are also in siemens, or reciprocals of ohms (sometimes denoted as mhos). In practice, however, the values of resistances are usually of the order of 10^2–10^9 ohms, capacitances 10^{-15}–10^{-6} farads, angular frequency 10^2–10^9 radians/second and inductances 10^{-9}–10^{-3} henries. The spread of these quantities can therefore be as high as 10^{20}, which can have an adverse effect on computer calculations and lead to overflows. To avoid this, normalization of circuit elements and frequency can be employed.

If impedances are denoted in kiloohms instead of ohms, the normalizing factor k, being $k = 10^3$, gives

$$R_n(\mathrm{k\Omega}) = \frac{R(\Omega)}{k} \tag{3.27}$$

or for $R = 5000\ \Omega$, $R_n = 5000/1000 = 5(\text{in k}\Omega)$.

Similarly, the normalization of conductance is (for $k = 10^3$)

$$G_n(\mathrm{mS}) = kG(\mathrm{S}) \tag{3.28}$$

For example, for a bipolar transistor having a transconductance value $g_m \cong 40 \times 10^{-3}$ S (corresponding to $I_E = 1$ mA) , the normalized value is $g_m = 40$ (in mS).

The frequency dependent impedances and admittances may be normalized either with or without frequency normalization. When the angular frequency ω is not normalized then

1. For capacitance:

$$Z_{Cn} = \frac{1}{kj\omega C} = \frac{1}{j\omega Ck} = \frac{1}{j\omega C_n} = \frac{1}{Y_{Cn}} \tag{3.29}$$

or

$$C_n = kC \tag{3.30}$$

that is, for $k = 10^{+3}$, C_n will be in millifarads; for example if $C = 10^{-12}$ F then the normalized value $C_n = 10^{-9}$ (mF) is obtained and Z_{Cn} will be in kiloohms.

2. For inductance:

$$Z_{Ln} = \frac{j\omega L}{k} = j\omega\,\frac{L}{k} = j\omega L_n = \frac{1}{Y_{Ln}} \tag{3.31}$$

or

$$L_n = \frac{L}{k} \tag{3.32}$$

that is, for $k = 10^{+3}$, L_n will be in kilohenries.

This is also inconvenient, for example $L_n = 5 \times 10^{-9}$ (in kH) is obtained for $L = 5\ \mu\mathrm{H} = 5 \times 10^{-6}$ H.

After normalizing the angular frequency using the formula

$$\omega_n = \frac{\omega}{n} \tag{3.33}$$

the frequency ω_n will be in kilorad/s for $n = 10^{+3}$, megarad/s for $n = 10^{+6}$, etc.

For impedance normalization, the following is obtained:

1. For capacitors:

$$Z_{Cn} = \frac{1}{kj\omega C} = \frac{1}{j \dfrac{\omega}{n} nC_n} = \frac{1}{j\omega_n C_n'} \tag{3.34}$$

or after combining impedance and frequency normalization

$$C_n' = nC_n = nkC \tag{3.35}$$

For example, for $\omega = 100$ rad/s, $C = 10^{-12}$ F, $k = 10^{+3}$ and $n = 10^{+3}$ then

$$Z_C = \frac{1}{j} 10^{+10}(\Omega)$$

$$C_n = 10^{-9}(\text{nF}), \quad \omega_n = 0.1(\text{krad/s}), \quad C_n' = 10^{-6}$$

Note that C_n' must be used together with ω_n.

2. For inductors:

$$Z_{Ln} = \frac{j\omega L}{k} = j \frac{\omega n L}{nk} = j\omega_n L_n'$$

or

$$L_n' = \frac{nL}{k} \tag{3.36}$$

For example, with $\omega = 100$ rad/s, $L = 5\ \mu\text{H}$, $k = 10^3$ and $n = 10^3$, then

$$Z_L = j \times 10^2 \times 5 \times 10^{-6} = j5 \times 10^{-4}(\Omega)$$

$$L_n = 5 \times 10^{-9}(\text{kH}), \quad \omega_n = 10^{-1}(\text{krad/s}), \quad L_n' = 5 \times 10^{-6}$$

$$Z_{Ln} = j \times 10^{-1} \times 5 \times 10^{-6} = j5 \times 10^{-7}(\text{k}\Omega)$$

Note that L_n' must be used together with ω_n. Consider the circuit shown in Fig. 3.11(a) with its equivalent circuit being shown in Fig. 3.11(b). After entering the normalized data for Fig. 3.11(b), the program 'AUTOMATI' of Program 3.2 will calculate the resulting real and imaginary parts of the admittance matrix, in millisiemens (the impedance normalization factor $k = 10^3$):

$$\mathbf{G} = \mathbf{Y_R} = \begin{bmatrix} 1.001 & -0.001 & 0.000 \\ -0.001 & 0.111 & -0.010 \\ -1000000.000 & 999999.990 & 10.010 \end{bmatrix} (\text{mS})$$

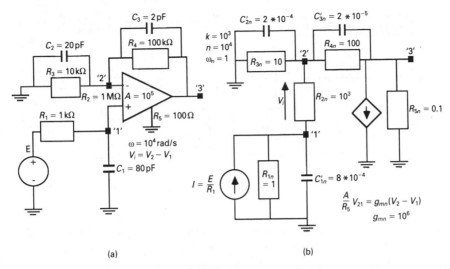

Figure 3.11   Active dynamic circuit. (a) Schematic diagram. (b) Simplified equivalent circuit with normalized element values

$$\mathbf{Y}_i = \begin{bmatrix} 0.0008 & 0 & 0 \\ 0 & 0.00022 & -0.00002 \\ 0 & -0.00002 & 0.00002 \end{bmatrix} \text{(mS)}$$

Exercises

The small signal model of bipolar transistors used in the exercises for this chapter is shown in Fig. E3.1(a). The small signal model used for MOS transistors is shown in Fig. E3.1(b).

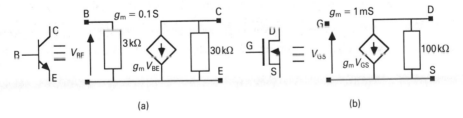

Figure E3.1

Exercise 1.   Write the indefinite admittance matrix for the two circuits shown in Fig. E3.2(a) and E3.2(b).

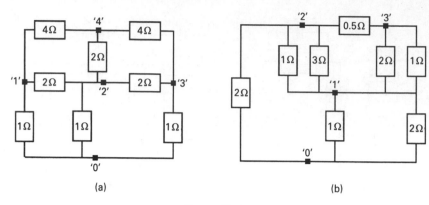

Figure E3.2

Exercise 2.   For the two circuits of exercise 1, use the program 'ACTMATRX' to find admittance matrices and compare the results with those of exercise 1.

Exercise 3.   Write the indefinite admittance matrix for the two circuits represented in Fig. E3.3(a) and E3.3(b) for $\omega = 1$ rad/s.

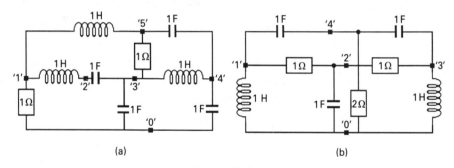

Figure E3.3

Exercise 4.   For the two circuits of exercise 3, use the program 'ACTMATRX' to find admittance matrices and compare the results obtained with those of exercise 3.

Exercise 5.   Using the transistor small signal model, draw the equivalent circuit in the medium frequency range for the two circuits shown in Fig. E3.4(a, b) (all the capacitors are replaced by short circuits at the considered frequency). Find the indefinite matrices for these circuits using the program 'ACTMATRX'. The models of the bipolar transistor and the MOS transistor are shown in Fig. E3.1. For Fig. E3.4(a), choose $k = 10^{+3}$; for Fig. E3.4(b), choose $k = 10^{+4}$.

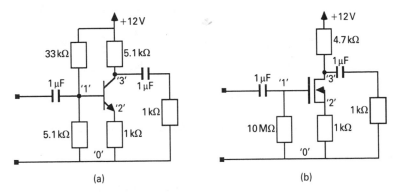

(a)   (b)

Figure E3.4

Exercise 6. Analyze the circuit shown in Fig. 3.7 with the program 'ACTMATRX' ; the results given by relations (3.20) and (3.21) will be used as references.

Exercise 7. In Chapter 3, the programs 'AUTOMATI' and 'ACTMATRX' are used to construct the matrix of the example shown in Fig. 3.10. It seems that the results are different. Explain why.

Exercise 8. Analyze the circuit shown in Fig. 3.7 with the program 'AUTOMATI' ; the results given by relations (3.20) and (3.21) will be used as references.

Exercise 9. For the two circuits of exercise 5, use the program 'AUTOMATI' to find definite admittance matrices and compare results obtained with those of exercise 5.

Exercise 10. The circuit shown in Fig. E3.5 represents (in normalized units) a third-order low pass filter. Determine the real value of each element for a cut-off frequency $\omega_0 = 2\pi \times 10^3$ rad/s and a characteristic resistance $R_0 = 50\ \Omega$.

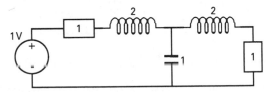

Figure E3.5

Exercise 11. Figure E3.6 represents a band pass Rauch structure. The voltage gain of the operational amplifier is equal to 10^{+5}, its output resistance is equal to $100\ \Omega$ and its input resistance (between $+$ and $-$) is equal to 1 MΩ. Choose $n = 10^{+4}$ and $k = 10^{+4}$ for the normalization coefficients for resistance and frequency. Draw the simplified equivalent circuit with the normalized values for each element. Using the program 'AUTOMATI', determine the real and imaginary parts of the admittance matrix ($\omega_n = 1$).

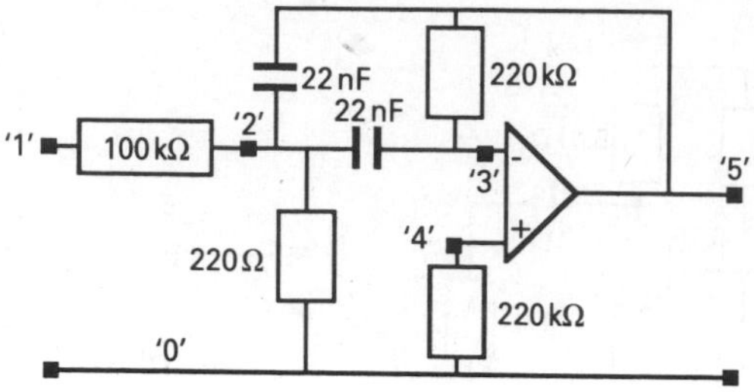

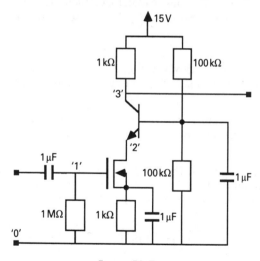

Figure E3.6

Exercise 12.   Analyze the circuit of exercise 11 with the same numerical values using the program 'ACTMATRX'.

Exercise 13.   Draw the small signal equivalent circuit in the medium frequency range of the cascode circuit shown in Fig. E3.7 (the capacitors are replaced by short circuits). Use the program 'ACTMATRX' to determine the admittance matrix of this circuit. The models of the bipolar transistor and the MOS transistor are shown in Fig. E3.1 ($k = 10^{+5}$).

Figure E3.7

Exercise 14.   Analyze the circuit of exercise 13 with the same numerical values utilizing the program 'AUTOMATI'.

Exercise 15.   Analyze the circuit of Fig. 3.11 with the same numerical values utilizing the program 'ACTMATRX'.

Exercise 16.   Draw the small signal equivalent circuit in the medium frequency range of the common base Darlington circuit shown in Fig. E3.8 (the capacitors are replaced by short circuits). Use the program 'ACTMATRX' to determine the admittance matrix of this circuit. The models of the two bipolar transistors are shown in Fig. E3.1(a) ($k = 10^{+3}$).

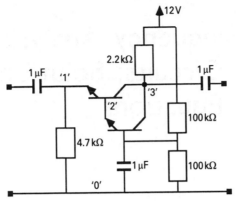

Figure E3.8

Exercise 17.   Analyze the circuit of exercise 16 with the same numerical values using the program 'AUTOMATI'.

4 Low Frequency Analysis of Linear Circuits; Solutions of Linear Equations

4.1 Introduction

In the low frequency range we assume that the internal reactance elements of electronic devices and the parasitics of elements and their interconnections can be neglected. For simplicity we will not use capacitors and inductors in the circuits considered in this chapter. Therefore, we will deal with linear equations with real coefficients, which simplifies explanations and examples of calculations. Only in the last section of this chapter is a program presented for the solution of linear equations with complex coefficients in order to show some differences in programming technique. The admittance matrix will be used to describe the circuits considered. After writing the KCL equations for a given circuit, the main problem is to solve these equations, that is, to find nodal voltages for given current excitations. While teaching or learning circuit theory in a classroom we usually use the determinant (Cramer) method for solving two or three equations. This method is, however, useless for four or more equation solutions; in such cases numerical methods supported by computers are usually used, such as Gauss–Jordan, Gauss-elimination and LU (lower upper) decomposition.

The efficiency of numerical methods for the solution of equations can be expressed in the number of long operations, that is, multiplications and divisions; the lower this number, the more efficient the method.

For the set of nodal equations describing a given linear circuit

$$\mathbf{YV} = \mathbf{I} \tag{4.1}$$

the vector of unknown node voltages $\mathbf{V}$ can be found through inversion of the admittance $\mathbf{Y}$ matrix and multiplication by the excitation current $\mathbf{I}$ vector:

$$\mathbf{V} = \mathbf{Y}^{-1}\mathbf{I} \tag{4.2}$$

It is known, however, that the number of long operations required to invert an $N \times N$ matrix is equal to N^3 and the number of long operations required to calculate the product $\mathbf{Y}^{-1}\mathbf{I}$ is equal to N^2. Therefore, the total number of long operations during the computation of (4.2) is equal to $N^3 + N^2$. The number of long operations can be reduced significantly by the use of the Gauss–Jordan, Gauss-elimination and LU decomposition methods.

Let us compare approximate numbers of long operations for different methods, for the number of equations $N = 10$:

1. Determinant (Cramer) method: $\cong (N-1)(N+1)! \cong 3.6 \times 10^{+8}$

2. Matrix inversion method: $\cong N^3 + N^2$ $= 1100$

3. Gauss–Jordan method: $\cong \dfrac{N^3 - N}{2}$ $= 450$

4. Gauss-elimination method: $\cong \dfrac{N^3}{3} + N^2$ $= 430$

5. LU decomposition method: $\cong \dfrac{N^3}{3} + N^2$ $= 430$

It can be seen from this comparison that the Cramer method is totally unsuitable. The best efficiency is obtained for the Gauss-elimination and LU decomposition methods. The Gauss–Jordan method is slightly less efficient but it is simple to code in the computer program, so it is also considered in this chapter.

4.2 Solution of linear equations using the Gauss–Jordan method

The set of nodal equations of type (4.1), for a circuit having three independent nodes, is:

$$y_{11}V_1 + y_{12}V_2 + y_{13}V_3 = I_1$$
$$y_{21}V_1 + y_{22}V_2 + y_{23}V_3 = I_2 \qquad\qquad (4.3)$$
$$y_{31}V_1 + y_{32}V_2 + y_{33}V_3 = I_3$$

We know from mathematics that the vector of unknowns [V] will not be changed when we apply the allowed linear operations to the set of equations, that is, multiplication of both sides of the equation by a constant and the addition of equations. Therefore, we can manipulate equations (4.3) to obtain the format

$$1 \times V_1 + 0 \times V_2 + 0 \times V_3 = E_1$$
$$0 \times V_1 + 1 \times V_2 + 0 \times V_3 = E_2 \qquad\qquad (4.4a)$$
$$0 \times V_1 + 0 \times V_2 + 1 \times V_3 = E_3$$

or, in a matrix form,

$$\begin{bmatrix} 1 & 0 & 0 \\ 0 & 1 & 0 \\ 0 & 0 & 1 \end{bmatrix} \begin{bmatrix} V_1 \\ V_2 \\ V_3 \end{bmatrix} = \mathbf{1V} = \mathbf{E} \qquad\qquad (4.4b)$$

where $\mathbf{1}$ is the unity matrix. After such a transformation we obtain the required solution directly:

$$V_1 = E_1, \quad V_2 = E_2, \quad V_3 = E_3 \tag{4.5}$$

To obtain the coefficient equal to 1 on the main diagonal of the matrix for the ith equation it is necessary to divide this equation by y_{ii}.

For example for the first equation

$$1 \times V_1 + \frac{y_{12}}{y_{11}} V_2 + \frac{y_{13}}{y_{11}} V_3 = \frac{I_1}{y_{11}} \tag{4.6}$$

To obtain zero in the first term of the second equation, we can multiply the modified first equation (4.6) by y_{21} and subtract from the second equation, that is,

$$[y_{21} - y_{21} \times 1] V_1 + \left[y_{22} - y_{21} \frac{y_{12}}{y_{11}}\right] V_2 + \left[y_{23} - y_{21} \frac{y_{13}}{y_{11}}\right] V_3 = I_2 - y_{21} \frac{I_1}{y_{11}} \tag{4.7}$$

These operations can be performed iteratively leading to the format of (4.4). The manipulations are performed on matrix $\mathbf{Y}$ coefficients and excitation vector $\mathbf{I}$ coefficients. Therefore, in the computer program, we are creating a so-called augmented admittance $N \times (N+1)$ matrix $\mathbf{Y}_a$ composed of the original $\mathbf{Y}$ matrix and excitation $\mathbf{I}$ vector:

$$\mathbf{Y}_a = \mathbf{Y} \,|\, \mathbf{I} = \begin{bmatrix} y_{11} & y_{12} & y_{13} & | & I_1 \\ y_{21} & y_{22} & y_{23} & | & I_2 \\ y_{31} & y_{32} & y_{33} & | & I_3 \end{bmatrix} \tag{4.8}$$

After manipulations this matrix is transformed into the form

$$\begin{bmatrix} 1 & 0 & 0 & | & E_1 \\ 0 & 1 & 0 & | & E_2 \\ 0 & 0 & 1 & | & E_3 \end{bmatrix} = \begin{bmatrix} 1 & 0 & 0 & | & V_1 \\ 0 & 1 & 0 & | & V_2 \\ 0 & 0 & 1 & | & V_3 \end{bmatrix} \tag{4.9}$$

Thus, the last column of the transformed augmented matrix is the vector of unknowns $\mathbf{V}$. To show the details of manipulation, let us find the values of voltages V_1 and V_2 for the circuit shown in Fig. 4.1.

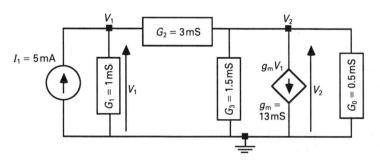

Figure 4.1   Active circuit for V_1 and V_2 calculation

Since voltages are equal to the ratio of I/Y, then for computations we can use conductances in [mS] and current in [mA] to obtain voltage in volts. Thus, the augmented matrix $\mathbf{Y}_a$ for the circuit is

$$\mathbf{Y}_a = \begin{bmatrix} 4 & -3 & | & 5 \\ 10 & 5 & | & 0 \end{bmatrix}$$

The first step is the normalization (division) of the first row by y_{11}:

$$\mathbf{Y}_{a1} = \begin{bmatrix} 1 & \dfrac{-3}{4} & | & \dfrac{5}{4} \\ 10 & 5 & | & 0 \end{bmatrix}$$

In the second step we need to have 0 as the first element in the second row. To obtain this we multiply the first row by $y_{21} = 10$ and subtract the multiplied first row from the second row (element by element):

$$\mathbf{Y}_{a2} = \begin{bmatrix} 1 & \dfrac{-3}{4} & | & \dfrac{5}{4} \\ 10-(10\times 1) & 5 - \left[10\times\dfrac{-3}{4}\right] & | & 0 - \left[10\times\dfrac{5}{4}\right] \end{bmatrix} = \begin{bmatrix} 1 & \dfrac{-3}{4} & | & \dfrac{5}{4} \\ 0 & \dfrac{50}{4} & | & -\dfrac{50}{4} \end{bmatrix}$$

The third step is the normalization (division) of the second row by modified y'_{22}:

$$\mathbf{Y}_{a3} = \begin{bmatrix} 1 & \dfrac{-3}{4} & | & \dfrac{5}{4} \\ 0 & 1 & | & -1 \end{bmatrix}$$

In the fourth step we need to have 0 as the second element in the first row, which is obtained by multiplying the second row by the modified y'_{12} element and subtracting from the first row (element by element):

$$\mathbf{Y}_{a4} = \begin{bmatrix} 1 & \dfrac{-3}{4}-\left[\dfrac{-3}{4}\times 1\right] & | & \dfrac{5}{4}-\left[\dfrac{-3}{4}\times(-1)\right] \\ 0 & 1 & | & -1 \end{bmatrix} = \begin{bmatrix} 1 & 0 & | & 0.5 \\ 0 & 1 & | & -1 \end{bmatrix}$$

The process of manipulation is finished since we have obtained the unity matrix **1** in the left-hand part of $\mathbf{Y}_a$. Therefore, the last column is the desired vector **V**, or

$$V_1 = 0.5 \text{ volt}, \quad V_2 = -1 \text{ volt}$$

In the computer program for the Gauss–Jordan method it is necessary to create four loops. In the first one (with index i) normalizing coefficient alpha $= y_{ii}$ is defined and in the second (inner loop) the ith row is normalized (divided by alpha). In the third (inner loop) (with index k) the coefficient beta $= y_{ki}$ is defined and in the fourth (inner loop) (with index j) zeros are created except for the elements on the main diagonal and the last column. The algorithm for the Gauss–Jordan method can be seen in Fig. 4.2.

The full example program for solutions of linear equations using the Gauss–Jordan method is shown in Program 4.1.

> **See the listing of the program GAUSSJOR Directory of Chapter 4 on attached diskette.**

In this program it is assumed for simplicity that the matrix $\mathbf{Y}$ for the given circuit is known and y_{ij} elements are entered directly into the program. However, this program can be combined with the program for the creation of the admittance matrix allowing the elements of the circuit considered to be automatically entered into the program. In the program a printing procedure can be inserted inside the loops to show modification of coefficients during the matrix transformation.

The number of equations to be considered is denoted by the constant $n = \ldots$, and the number of columns $n1$ in the augmented matrix is denoted by $n1 = n + 1$. As an example, let us compute the node voltages for the circuit of Fig. 4.3(a) with the simplified transistor model shown in Fig. 4.3(b).

The matrix $\mathbf{Y_R} = \mathbf{G}$ of the equivalent circuit of Fig. 4.3(c) can be found by inspection or by using the program in Section 3:

$$\mathbf{G} = \begin{bmatrix} 0.425 & -0.2 & 0 & -0.025 & 0 \\ -20.2 & 22.2 & 0 & 0 & 0 \\ 20.0 & -20.0 & 0.5 & -0.4 & 0 \\ -0.025 & 0 & -40.4 & 40.925 & 0 \\ 0 & 0 & 40.0 & -40.0 & 0.125 \end{bmatrix}$$

After entering the augmented matrix $\mathbf{G_a}$ composed of matrix $\mathbf{G}$ and vector $[\mathbf{I}]^t = [0.2 \ 0 \ 0 \ 0 \ 0]$ as the last column into the program 'GAUSSJOR', shown in Program 4.1, we obtain the results

$V_1 = 0.3003$
$V_2 = 0.2733$
$V_3 = -5.1467$
$V_4 = -5.0805$
$V_5 = 21.1862$

```
for i=1 to n
  alpha=yii
  for j=i to n+1
    yij=yij/alpha
      for k=1 to n, except k=i
        beta=yki
          for j=i to n+1
            ykj=ykj-beta*yij
```

Figure 4.2 The algorithm for the Gauss–Jordan method

Having used the input voltage $E_1 = 1$ V, we have obtained voltages with values corresponding to the transfer functions between the given node and the input; for example,

$$T_{51} = \frac{V_5}{E_1} = 21.1862 \quad \text{and} \quad T_{41} = \frac{V_4}{E_1} = -5.0805$$

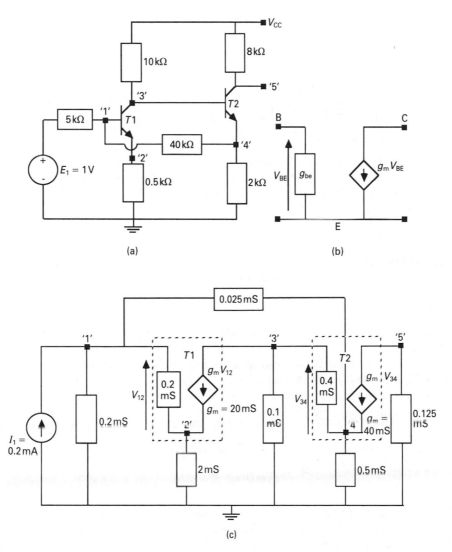

Figure 4.3   (a) Transistor circuit. (b) Simplified model of transistor. (c) Equivalent circuit

4.3 Solution of linear equations using the Gauss-elimination method

As it was mentioned in Section 4.1, the Gauss-elimination method requires slightly shorter operations than does the Gauss–Jordan method. This is because the set of equations, as for example (4.3), is manipulated to obtain the upper triangular format

$$1 \times V_1 + y'_{12} \times V_2 + y'_{13} \times V_3 = I'_1$$

$$0 \times V_1 + 1 \times V_2 + y''_{23} \times V_3 = I''_2 \qquad\qquad (4.10a)$$

$$0 \times V_1 + 0 \times V_2 + 1 \times V_3 = I'''_3$$

or, in matrix form,

$$\begin{bmatrix} 1 & y'_{12} & y'_{13} \\ 0 & 1 & y''_{23} \\ 0 & 0 & 1 \end{bmatrix} \begin{bmatrix} V_1 \\ V_2 \\ V_3 \end{bmatrix} = \begin{bmatrix} I'_1 \\ I''_2 \\ I'''_3 \end{bmatrix} \qquad\qquad (4.10b)$$

where the coefficients y'_{ij} and I'_i have been modified once, while the coefficient I'''_3 has been modified three times. This upper triangularization is named the 'forward reduction' and is similar to the procedure in the Gauss–Jordan method. From (4.10) we obtain directly

$$V_3 = I'''_3 \qquad\qquad (4.11)$$

Then to obtain the rest of the unknowns, that is, V_2 and V_1, it is necessary to substitute V_3 back into the second equation:

$$V_2 = I''_2 - y''_{23}\, I'''_3 \qquad\qquad (4.12)$$

and then V_2 and V_3 into the first equation to find V_1. This step is called 'back substitution'.

Similarly, as in the Gauss–Jordan method, here we are creating the augmented matrix $\mathbf{Y}_a$ with the last column composed of the excitation vector $[\mathbf{I}]$. An algorithm for the Gauss-elimination method is shown in Fig. 4.4.

The example program 'GAUSSELIM' for linear equation solution using the Gauss-elimination method is shown in Program 4.2.

See the listing of the program GAUSSELIM Directory of Chapter 4 on attached diskette.

The procedure 'Readmat' allows the augmented matrix $\mathbf{Y}_a$ to be read in for the circuit to be analyzed. After modification of the matrix, the last column becomes the required vector of unknown node voltages, $[\mathbf{V}]$, which is printed out as the solution. After entering the data corresponding to the circuit of Fig. 4.3 we obtain the same results as were computed earlier.

FORWARD REDUCTION

```
for i = 1 to n
   alpha = yᵢᵢ
   for k = 1 to n + 1
      yᵢₖ = yᵢₖ/alpha
   if i < n then
      begin
      for j = i + 1 to n
         yⱼ,ₙ₊₁ = yⱼ,ₙ₊₁ - yᵢ,ₙ₊₁ *yⱼᵢ
         for k = n downto 1
            yⱼₖ = yⱼₖ - yᵢₖ*yⱼᵢ
```

BACKWARD SUBSTITUTION

```
for k = n - 1 downto 1
   sum = 0
   for j = k + 1 to n
      sum = sum + yₖⱼ*yⱼ,ₙ₊₁
   yₖ,ₙ₊₁ = yₖ,ₙ₊₁ - sum
```

Figure 4.4   The algorithm for the Gauss-elimination method

4.4 Solution of linear equations using the *LU* decomposition method

When multiple solutions of a set of algebraic equations of the type

$$\mathbf{AX} = \mathbf{B} \tag{4.13}$$

are required for the same matrix $\mathbf{A}$ but different excitation vectors $\mathbf{B}$, the Gauss–Jordan and Gauss-elimination methods are not efficient. This is because in both methods the manipulations are performed on matrix $\mathbf{A}$ and vector $\mathbf{B}$; nevertheless the matrix $\mathbf{A}$ is the same in all the computations. In such cases the efficiency of the calculations can be improved by decomposition of the matrix $\mathbf{A}$ into the product of two triangular matrices, $\mathbf{L}$ – the lower triangular and $\mathbf{U}$ – the upper triangular:

$$\mathbf{A} = \mathbf{L} \times \mathbf{U} \tag{4.14}$$

where the $\mathbf{L}$ and $\mathbf{U}$ matrices have the format

$$\mathbf{L} = \begin{bmatrix} l_{11} & 0 & 0 & . & . & . & 0 \\ l_{21} & l_{22} & 0 & . & . & . & 0 \\ l_{31} & l_{32} & l_{33} & . & . & . & 0 \\ . & . & . & . & . & . & . \\ . & . & . & . & . & . & . \\ . & . & . & . & . & . & . \\ l_{n1} & l_{n2} & l_{n3} & . & . & . & l_{nn} \end{bmatrix} \tag{4.15}$$

$$U = \begin{bmatrix} 1 & u_{12} & u_{13} & . & . & . & u_{1n} \\ 0 & 1 & u_{23} & . & . & . & u_{2n} \\ 0 & 0 & 1 & . & . & . & u_{3n} \\ . & . & . & . & . & . & . \\ . & . & . & . & . & . & . \\ . & . & . & . & . & . & . \\ 0 & 0 & 0 & . & . & . & 1 \end{bmatrix} \tag{4.16}$$

To find the vector of unknowns [X] it is necessary to make two substitutions: forward and backward. Substituting (4.14) into (4.13) we obtain

$$LUX = B \tag{4.17}$$

or

$$LD = B \tag{4.18}$$

where

$$D = UX \tag{4.19}$$

The auxiliary vector D can easily be found in the forward substitution (L and B are known):

$$
\begin{aligned}
& l_{11}d_1 + 0 = b_1 \\
& l_{21}d_1 + l_{22}d_2 + 0 = b_2 \\
& \qquad . \qquad . \\
& \qquad . \qquad . \\
& \qquad . \qquad . \\
& l_{n1}d_1 + l_{n2}d_2 + l_{n3}d_3 + \ldots + l_{nn}d_n = b_n
\end{aligned}
\tag{4.20}
$$

which leads to

$$d_1 = \frac{b_1}{l_{11}}$$

$$d_2 = \frac{b_2 - l_{21}d_1}{l_{22}} \tag{4.21}$$

and so on, or generally,

$$d_k = \frac{b_k - \displaystyle\sum_{j=1}^{k-1} l_{kj}\, d_j}{l_{kk}} \qquad k = 1, 2, \ldots, n \tag{4.22}$$

After having found the auxiliary vector D, we can find the unknown vector X from

(4.19) using backward substitution:

$$1 \times x_1 + u_{12} \times x_2 + u_{13} \times x_3 + \ldots + u_{1n} \times x_n = d_1$$

$$1 \times x_2 + u_{23} \times x_3 + \ldots + u_{2n} \times x_n = d_2$$

$$\vdots$$

$$1 \times x_{n-1} + u_{n-1,n} \times x_n = d_{n-1}$$

$$1 \times x_n = d_n$$

(4.23)

from which we have elements of the unknown vector **X**:

$$x_n = d_n$$

$$x_{n-1} = d_{n-1} - u_{n-1,n}x_n$$

(4.24)

or, generally,

$$x_k = d_k - \sum_{j=k+1}^{n} u_{kj}x_j \quad k = n, n-1, \, n-2, \ldots, 2, 1$$

(4.25)

Therefore, the *LU* decomposition method consists of the following three steps:

1. The **LU** decomposition of the matrix **A**.
2. Forward substitution to find the auxiliary vector **D** from knowledge of **L** and excitation vector **B**.
3. Backward substitution to find the unknown vector **X** from **U** and **D**.

In multiple solutions of a given set of equations, after changing the excitations **B**, only forward and backward substitutions have to be performed to find **D** and **X**, while *LU* decomposition is performed only once at the beginning of the calculations. This saves on computer time in comparison to the Gauss-elimination method. It is necessary to note that during the *LU* decomposition, the matrix **A** is overwritten by the **L** and **U** matrices and therefore the **A** matrix is destroyed. However, this saves the computer memory since extra space for two matrices is not required. The format of **LU** matrices is as follows (for $n = 3$):

$$\begin{bmatrix} l_{11} & u_{12} & u_{13} \\ l_{21} & l_{22} & u_{23} \\ l_{31} & l_{32} & l_{33} \end{bmatrix}$$

(4.26)

Note that there are units on the main diagonal of the **U** matrix. There are several different types of *LU* decompositions. One of them is based on Gauss elimination.

An algorithm for this decomposition, together with forward and backward substitutions for the equation solution, is shown in Fig. 4.5.

According to this algorithm, elements of the matrix **A** are manipulated column after column starting from the second; the first column of the **L** matrix is the same as for that of the **A** matrix. To visualize this, let us consider the circuit of Fig. 4.6

```
                    LU DECOMPOSITION OF [A]

          for k = 1 to n - 1
            for j = k+1 to n
            akj = akj/akk
            for i= k+1 to n
              aij = aij - akj *aik


          FORWARD SUBSTITUTION ; [A]-matrix, [B]-vector

          b1 = b1/a11
          for i = 2 to n
            for j = 1 to i - 1
            bi = bi - aij*bj
          bi = bi/aii                         (bi ≡ di-auxilary)

                    BACKWARD SUBSTITUTION

          for k = 1 to n -1
            i = n-k
            for j = i+1 to n
            bi = bi - aij*bj
```

Figure 4.5 Algorithm for *LU* decomposition of matrix **A**, forward and backward
substitutions with **B** as excitation vector

for which the **Y** ≡ **A** matrix is

$$\mathbf{A} \equiv \mathbf{Y} = \begin{bmatrix} 5 & -1 & -1 \\ -1 & 4 & -1 \\ 4 & -1 & 2 \end{bmatrix}$$

In the first pass, for $k = 1$, $j = 2$ and $i = 3$, we obtain the modified second
column (Fig. 4.7).

For $k = 1$, $j = 3$, $i = 2$ and $i = 3$, we obtain the initial modification of the third
column (Fig. 4.8).

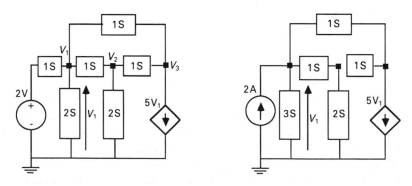

Figure 4.6 Example of active circuit

k = 1
j = 2
i = 2
i = 3
5 −0.2 −1
−1 3.8 −1
4 −0.2 2

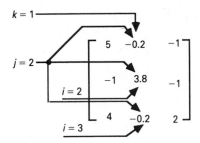

Figure 4.7 First pass; modification of the second column

k = 1
j = 3
i = 2
i = 3
5 −0.2 −0.2
−1 3.8 −1.2
4 −0.2 2.8

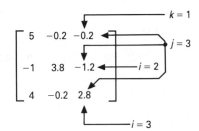

Figure 4.8 Initial modification of the third column

For $k=2$, $j=3$ and $i=3$, we obtain the final result of the decomposition (Fig. 4.9).

Using the forward substitution after entering the excitation vector $\mathbf{B}^t = [2\ 0\ 0]$ we obtain the following in the first pass:

1. For $i=2$ and $j=1$

$$\begin{matrix} d_1 = \dfrac{b_1}{l_{11}} \rightarrow \\ i=2,\ j=1 \rightarrow \\ \ \end{matrix} \begin{bmatrix} 0.4 \\ 0.1053 \\ 0 \end{bmatrix}$$

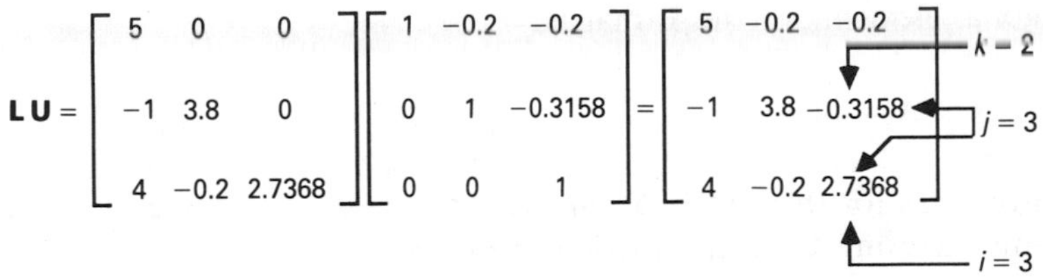

LU =
5 0 0
−1 3.8 0
4 −0.2 2.7368
1 −0.2 −0.2
0 1 −0.3158
0 0 1
=
5 −0.2 −0.2
−1 3.8 −0.3158
4 −0.2 2.7368
k = 2
j = 3
i = 3

Figure 4.9 Final result of the decomposition

2. And after the second pass, $i = 3$, $j = 1$

$$
\begin{array}{c}
d_1 \rightarrow \\
d_2 \rightarrow \\
d_3 \rightarrow
\end{array}
\left[
\begin{array}{c}
0.4 \\
0.1053 \\
-1.6
\end{array}
\right]
\leftarrow i = 3, j = 1
$$

3. And $i = 3$, $j = 2$

$$
i = 3, \ j = 2 \rightarrow
\left[
\begin{array}{c}
0.4 \\
0.1053 \\
-\dfrac{1.58}{2.7368}
\end{array}
\right]
$$

Thus we have obtained the auxiliary vector **D**. After backward substitution we obtain the results

$$
\mathbf{V} =
\begin{bmatrix}
V_1 \\
V_2 \\
V_3
\end{bmatrix}
=
\begin{bmatrix}
0.2692 \\
-0.0769 \\
-0.5769
\end{bmatrix}
$$

To repeat the calculations for a different excitation vector we need to enter the new vector **B** and involve only the 'forward–backward' procedure. For example, entering $\mathbf{B}^t = [4 \ 0 \ 0]$ we obtain the results

$$
[\mathbf{V}] =
\begin{bmatrix}
V_1 \\
V_2 \\
V_3
\end{bmatrix}
=
\begin{bmatrix}
0.5384 \\
-0.1538 \\
-1.1538
\end{bmatrix}
$$

The example program 'LUDECOMP' for the linear equation solution is shown in Program 4.3.

See the listing of the program LUDECOMP Directory of Chapter 4 on attached diskette.

It is necessary to note that after LU decomposition we can easily calculate the determinant Δ of the decomposed matrix, since its value is equal to the product of the main diagonal elements of the **L** matrix:

$$
\Delta = \det \mathbf{A} = \prod_{i=1}^{n} l_{ii} \tag{4.27}
$$

We can prove this for the matrix **Y** for the example circuit of Fig. 4.6. Calculating the determinant using Cramer's method we obtain

$$
\det \mathbf{Y} = 40 + 4 - 1 + 16 - 5 - 2 = 52
$$

and using formula (4.27):

$$\Delta = \det \mathbf{Y} = l_{11} l_{22} l_{33} = 5 \times 3.8 \times 2.7368 = 52$$

4.5 Matrix inversion using the programs for solution of linear equations

When we need to invert a matrix, for example to find the nodal admittance $\mathbf{Y}$ matrix for a given circuit when the mesh impedance $\mathbf{Z}$ matrix is given:

$$\mathbf{Y} = \mathbf{Z}^{-1} \tag{4.28}$$

then we can use the programs for linear equation solution. It is necessary to note that when the excitation vector $\mathbf{B}$ of the set of equations

$$\mathbf{AX} = \mathbf{B}$$

is a versor $\mathbf{VE}_1$, that is, a vector with unity at the first position and zeros elsewhere:

$$\mathbf{B} = \mathbf{VE}_1 = \begin{bmatrix} 1 \\ 0 \\ \vdots \\ 0 \end{bmatrix} \tag{4.29}$$

then after the solution of the set of equations, we obtain the vector $\mathbf{X}_1$, which is the first column of the inverted matrix. Similarly, when the set of equations is excited by the versor $\mathbf{VE}_2$ with unity at the second position, then after the solution we obtain as a result the vector $\mathbf{X}_2$, which is the second column of the inverted matrix, and so on. To prove this, let us solve the set of equations excited by the versor with unity at the first position:

$$y_{11} V_1 + y_{12} V_2 = 1$$

$$y_{21} V_1 + y_{22} V_2 = 0 \tag{4.30}$$

using the Cramer method

$$V_{11} = \frac{\begin{vmatrix} 1 & y_{12} \\ 0 & y_{22} \end{vmatrix}}{\Delta y} = \frac{1 \times y_{22}}{\Delta y} \tag{4.31}$$

where $\Delta y = y_{11} y_{22} - y_{12} y_{21}$

$$V_{21} = \frac{\begin{vmatrix} y_{11} & 1 \\ y_{21} & 0 \end{vmatrix}}{\Delta y} = 1 \times \frac{-y_{21}}{\Delta y} \tag{4.32}$$

And now with an excitation versor with unity at the second position:

$$y_{11}V_1 + y_{12}V_2 = 0$$

$$y_{21}V_1 + y_{22}V_2 = 1 \tag{4.33}$$

This gives

$$V_{12} = \frac{\begin{vmatrix} 0 & y_{12} \\ 1 & y_{22} \end{vmatrix}}{\Delta y} = 1 \times \frac{-y_{12}}{\Delta y} \tag{4.34}$$

and

$$V_{22} = \frac{\begin{vmatrix} y_{11} & 0 \\ y_{21} & 1 \end{vmatrix}}{\Delta y} = 1 \times \frac{y_{11}}{\Delta y} \tag{4.35}$$

After combining the results in matrix form, we obtain

$$\begin{bmatrix} V_{11} & V_{12} \\ V_{21} & V_{22} \end{bmatrix} = \begin{bmatrix} \dfrac{y_{22}}{\Delta y} & \dfrac{-y_{12}}{\Delta y} \\ \dfrac{-y_{21}}{\Delta y} & \dfrac{y_{11}}{\Delta y} \end{bmatrix} = \begin{bmatrix} z_{11} & z_{12} \\ z_{21} & z_{22} \end{bmatrix} = \mathbf{Z} \tag{4.36}$$

Thus, we have obtained the impedance matrix $\mathbf{Z}$, which is equal to the inverted admittance matrix $\mathbf{Y}^{-1}$.

In the sequence of solutions during matrix inversion only the excitation vector is changed. Therefore, using the Gauss–Jordan method for solutions the initial set of equations can be written

$$y_{11}V_1 + y_{12}V_2 + \cdots + y_{1n}V_n = 1 \times I_1 + 0 \times I_1 + \cdots + 0 \times I_1$$

$$y_{21}V_1 + y_{22}V_2 + \ldots + y_{2n}V_n = 0 \times I_2 + 1 \times I_2 + \cdots + 0 \times I_2$$

$$\vdots \tag{4.37}$$

$$y_{n1}V_1 + y_{n2}V_2 + \cdots + y_{nn}V_n = 0 \times I_n + 0 \times I_n + \cdots + 1 \times I_n$$

or, in matrix form,

$$\mathbf{YV} = \mathbf{1\,I} \tag{4.38}$$

where $\mathbf{1}$ is the unity matrix.

After transforming (4.37) according to the Gauss–Jordan method, we obtain

$$1 \times V_1 + 0 \times V_2 + \cdots + 0 \times V_n = z_{11}I_1 + z_{12}I_2 + \cdots + z_{1n}I_n$$

$$0 \times V_1 + 1 \times V_2 + \cdots + 0 \times V_n = z_{21}I_1 + z_{22}I_2 + \cdots + z_{2n}I_n$$

$$\vdots \tag{4.39}$$

$$0 \times V_1 + 0 \times V_2 + \cdots + 1 \times V_n = z_{n1}I_1 + z_{n2}I_2 + \cdots + z_{nn}I_n$$

or

$$1V = ZI = Y^{-1}I \tag{4.40}$$

The 'GAUSSJOR' program shown in Program 4.1 can easily be adapted for matrix inversion. The only difference is that the augmented matrix Y should have the form

$$Y_a = Y \,|\, 1 \tag{4.41}$$

or, for three equations,

$$Y_a = \begin{bmatrix} y_{11} & y_{12} & y_{13} & | & 1 & 0 & 0 \\ y_{21} & y_{22} & y_{23} & | & 0 & 1 & 0 \\ y_{31} & y_{32} & y_{33} & | & 0 & 0 & 1 \end{bmatrix} \tag{4.42}$$

and after manipulations will be transformed into the form

$$\begin{bmatrix} 1 & 0 & 0 & | & z_{11} & z_{12} & z_{13} \\ 0 & 1 & 0 & | & z_{21} & z_{22} & z_{23} \\ 0 & 0 & 1 & | & z_{31} & z_{32} & z_{33} \end{bmatrix} \tag{4.43}$$

Inversion of the admittance matrix Y into the impedance matrix Z for a two-port network corresponds to the transformation of the 'Π'-type circuit into the 'T'-type circuit. As an example, let us consider the 'Π'-type circuit of Fig. 4.10(a), for which the Y matrix is

$$Y = \begin{bmatrix} 4 & -3 \\ 10 & 5 \end{bmatrix} \times 10^{-3} \text{ S}$$

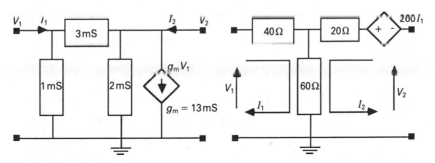

Figure 4.10 Example of active circuit. (a) 'Π'-type. (b) 'T'-type

Calculating the z_{ij} parameters directly using (4.36) we obtain

$$z_{11} = \frac{y_{22}}{\Delta y} = \frac{5 \times 10^{-3}}{50 \times 10^{-6}} = 100 \ \Omega, \quad z_{12} = \frac{-y_{12}}{\Delta y} = \frac{3 \times 10^{-3}}{50 \times 10^{-6}} = 60 \ \Omega$$

$$z_{21} = \frac{-y_{21}}{\Delta y} = \frac{-10 \times 10^{-3}}{50 \times 10^{-6}} = -200 \ \Omega, \quad z_{22} = \frac{y_{11}}{\Delta y} = \frac{4 \times 10^{-3}}{50 \times 10^{-6}} = 80 \ \Omega$$

The equivalent circuit of the 'T'-type for these parameters is shown in Fig. 4.10(b).

For the Gauss–Jordan method we form the augmented matrix $\mathbf{Y}_a$ (in millisiemens):

$$\begin{bmatrix} 4 & -3 & | & 1 & 0 \\ 10 & 5 & | & 0 & 1 \end{bmatrix}$$

Then the successive steps of transformations are as follows:

1. First step:

$$\begin{bmatrix} 1 & \dfrac{-3}{4} & \dfrac{1}{4} & 0 \\[2mm] 10 & 5 & 0 & 1 \end{bmatrix}$$

2. Second step:

$$\begin{bmatrix} 1 & \dfrac{-3}{4} & \dfrac{1}{4} & 0 \\[3mm] 10 - 10 \times 1 & 5 - 10 \times \left[\dfrac{-3}{4}\right] & 0 - 10 \times \left[\dfrac{1}{4}\right] & 1 \end{bmatrix} = \begin{bmatrix} 1 & \dfrac{-3}{4} & \dfrac{1}{4} & 0 \\[3mm] 0 & \dfrac{50}{4} & -\dfrac{10}{4} & 1 \end{bmatrix}$$

3. Third step:

$$\begin{bmatrix} 1 & \dfrac{-3}{4} & \dfrac{1}{4} & 0 \\[3mm] 0 & \dfrac{50}{4} \times \dfrac{4}{50} & -\dfrac{10}{4} \times \dfrac{4}{50} & 1 \times \dfrac{4}{50} \end{bmatrix} = \begin{bmatrix} 1 & \dfrac{-3}{4} & \dfrac{1}{4} & 0 \\[2mm] 0 & 1 & -0.2 & 0.08 \end{bmatrix}$$

4. Fourth step:

$$\begin{bmatrix} 1 & \dfrac{-3}{4} - \left[\dfrac{-3}{4} \times 1\right] & \dfrac{1}{4} - \left[\dfrac{-3}{4}\right] \times \left[\dfrac{-1}{5}\right] & 0 - \left[\dfrac{-3}{4}\right] \times \left[\dfrac{4}{50}\right] \\[3mm] 0 & 1 & -0.2 & 0.08 \end{bmatrix}$$

$$= \begin{bmatrix} 1 & 0 & 0.1 & 0.06 \\ 0 & 1 & -0.2 & 0.08 \end{bmatrix}$$

Thus the impedance matrix $\mathbf{Z}$ (in kiloohms) (because of the division of each row by y_{ii} in millisiemens) is

$$\mathbf{Z} = \begin{bmatrix} z_{11} & z_{12} \\ z_{21} & z_{22} \end{bmatrix} = \begin{bmatrix} 0.1 & 0.06 \\ -0.2 & 0.08 \end{bmatrix}$$

whose elements are equal to the z_{ij} parameters calculated directly.

The matrix inversion is more efficient when the LU decomposition method is applied. In this case for the $N \times N$ matrix, only one LU decomposition is performed and then forward and backward substitution is performed N times for excitation versors with unity changing from the first to the nth position. For this purpose the 'LUDECOMP' program shown in Program 4.3 can easily be adapted by inserting the procedures 'Readvector' and 'Forw-Back-Subst' in a loop for an n-fold equation solution. Of course, instead of reading the vectors n times, we can clear the vector and insert the expression V [i] := 1 in a 'for i := 1 to n do' loop to obtain '1' at the ith position. The example program 'MATINVLU' for matrix inversion is shown in Program 4.4.

See the listing of the program MATINVLU Directory of Chapter 4 on attached diskette.

As an example, when we enter the matrix

$$\begin{bmatrix} 4 & -3 \\ 10 & 5 \end{bmatrix}$$

The LU decomposition is

$$\begin{bmatrix} 4 & -0.75 \\ 10 & 12.5 \end{bmatrix}$$

and the program results are

```
"The elements of the 1st...column are:"
      X(1) = 0.1
      X(2) = -0.2
"The elements of the 2nd...column are:"

      X(1) - 0.06
      X(2) = 0.08
```

4.6 Pivoting technique in the solution of linear equations

Using any of the above-mentioned numerical methods for the linear equation solution, we cannot solve the following set of equations:

$$-2X_2 = 1$$
$$-2X_1 + 5X_2 = -9 \tag{4.44}$$

since the element a_{11} of the matrix corresponding to equations (4.44), by which it is necessary to divide the rest of the coefficients in the first equation (in any method), is equal to zero. The element a_{11} laid on the main diagonal of the matrix of a given set of equations is named 'the pivot'. We see from different methods of equation solutions that the pivot must not be zero. Also, the accuracy of numerical calculations can be improved in some cases when the elements on the main diagonal (the pivots) of a matrix have the greatest values of all coefficient values in each row (equation). Therefore, in the pivoting technique the values of the elements in the first row of the matrix are checked initially. When there is an element, say a_{1k}, having the greatest absolute value, then the kth column is interchanged with the first one and after this the coefficient a_{1k} becomes a_{11}; thus we obtain the pivot with the greatest value. During the manipulations of matrix coefficients, second and subsequent rows are checked for the greatest absolute value and the coefficients with the greatest absolute value are put on the main diagonal of the matrix. To obtain the numerical solution of equations (4.44) we can interchange the columns:

$$-2X_2 + 0 = 1$$
$$+5X_2 - 2X_1 = -9$$

(4.45)

since now the pivot in the first row equals -2. It is necessary to note that after interchanging the columns the sequence of unknown values is also changed; for equations (4.45) the first value (the first element in the last column of the augmented matrix) will be X_2 and the second will be X_1. This should also be taken into account in printing out the results of the calculations.

Zero-valued coefficients can also occur during manipulations of the augmented matrix, as in the Gauss-elimination method. Let us consider, for example, the following set of equations and its corresponding augmented matrix:

$$\begin{bmatrix} 4 & 8 & -2 & | & 2 \\ 1 & 2 & 1 & | & 2 \\ 1 & -1 & 1 & | & 0 \end{bmatrix}$$

Initially, using the pivoting technique, the first and second columns are interchanged (giving the sequence of unknowns X_2, X_1, X_3):

$$\begin{bmatrix} 8 & 4 & -2 & | & 2 \\ 2 & 1 & 1 & | & 2 \\ -1 & 1 & 1 & | & 0 \end{bmatrix}$$

and after normalization of the first row:

$$\begin{bmatrix} 1 & 0.5 & -0.25 & | & 0.25 \\ 2 & 1 & 1 & | & 2 \\ -1 & 1 & 1 & | & 0 \end{bmatrix}$$

After the second pass, zero in element a_{21} is created, and at the same time we obtain the new value for a_{22}: $a_{22} = 1 - 2 \times 0.5 = 0$ yielding

$$\begin{bmatrix} 1 & 0.5 & -0.25 & | & 0.25 \\ 0 & 0 & 1.5 & | & 1.5 \\ -1 & 1 & 1 & | & 0 \end{bmatrix}$$

Now, again, it is necessary to interchange the second and third columns to be able to continue the solution procedure:

$$\begin{bmatrix} 1 & -0.25 & 0.5 & | & 0.25 \\ 0 & 1.5 & 0 & | & 1.5 \\ -1 & 1 & 1 & | & 0 \end{bmatrix}$$

This interchange again changes the sequence of the unknowns; now it will be X_2, X_3, X_1. After completing the manipulations, we obtain the printout of the results in the sequence

$$X_2 = 0.6667, \qquad X_3 = 1.0000, \qquad X_1 = -0.3333$$

The procedure 'pivot' is included in the Gauss-elimination program 'PIVOTGE' shown in Program 4.5.

See the listing of the program PIVOTGE Directory of Chapter 4 on attached diskette.

4.7 The *LU* decomposition method for solutions of equations with complex coefficients

When we analyze circuits containing reactive elements, that is, capacitors and inductors, then the coefficients of the nodal equations (elements of the admittance matrix), and the voltages and currents in the circuits, become complex numbers which are functions of the frequency ω. Therefore, the matrix elements should be stored in the computer in two real matrices containing the real part $\mathbf{Y_R}$ and the imaginary part $\mathbf{Y_I}$ of the $\mathbf{Y}$ matrix:

$$\mathbf{Y} = \mathbf{Y_R} + j\mathbf{Y_I} \qquad (4.46)$$

Similarly, the complex voltage and current vectors are stored in two separate real vectors:

$$\mathbf{V} - \mathbf{V_R} + j\mathbf{V_I} \qquad (4.47)$$

and

$$\mathbf{I} = \mathbf{I_R} + j\mathbf{I_I} \qquad (4.48)$$

At the beginning of the calculation leading to the solution of the equation it is necessary to enter the real and imaginary parts of the admittance matrix of the circuit to be analyzed and also the real and imaginary parts of the excitation current vector for the single fixed frequency ω. Then, using any of the linear equation solution methods, it is necessary to introduce into the program procedures for complex number multiplication:

$$c_r + jc_i = (a_r + ja_i)(b_r + jb_i) \tag{4.49}$$

or

$$c_r = a_r b_r - a_i b_i$$
$$\tag{4.50}$$
$$c_i = a_r b_i + a_i b_r$$

and division

$$c_r + jc_i = (a_r + ja_i)/(b_r + jb_i) \tag{4.51}$$

or

$$c_r = \frac{(a_r b_r + a_i b_i)}{(b_r b_r + b_i b_i)}$$
$$\tag{4.52}$$
$$c_i = \frac{(a_i b_r - a_r b_i)}{(b_r b_r + b_i b_i)}$$

and call these procedures in every place in the program where multiplication or division is required. An example program using the LU decomposition method for linear equation solution with complex numbers is shown in Program 4.6.

> **See the listing of the program COMPLXLU Directory of Chapter 4 on attached diskette.**

As an example, let us calculate the node voltages for the circuit of Fig. 4.11 at the frequency $\omega = 10^7$ rad/s.

The values of the coefficients of the real and imaginary parts of the admittance matrix in millisiemens are:

$$\mathbf{Y_R} = \begin{bmatrix} 60.1 & -50 & 0 & 0 \\ -50 & 50.4 & -0.4 & 0 \\ 0 & -40 & 45.4 & -0.002 \\ 0 & 40 & -40.002 & 0.502 \end{bmatrix}$$

$$\mathbf{Y_I} = \begin{bmatrix} 0 & 0 & 0 & 0 \\ 0 & 2.1 & -2 & -0.1 \\ 0 & -2 & 2 & 0 \\ 0 & -0.1 & 0 & 0.1 \end{bmatrix}$$

To obtain voltages in volts, it is necessary to represent the currents in milliamps.

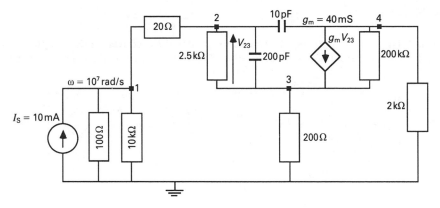

Figure 4.11 Example circuit to calculate complex node voltages

Thus, the transposed vectors for the real and imaginary parts of $\mathbf{I_S}$ are:

$$\mathbf{I_R}^t = [1\ \ 0\ \ 0\ \ 0]$$

$$\mathbf{I_I}^t = [0\ \ 0\ \ 0\ \ 0]$$

After entering these data in the program 'COMPLXLU' the resulting printouts are as follows.

1. The real part of *LU* decomposition:

$$\mathbf{LU_R} = \begin{bmatrix} 60.1 & -0.8319 & 0 & 0 \\ -50 & 8.8027 & -0.0943 & -0.0026 \\ 0 & -40 & 42.0383 & -0.004 \\ 0 & 40 & -36.2104 & 0.5061 \end{bmatrix}$$

2. The imaginary part of *LU* decomposition:

$$\mathbf{LU_I} = \begin{bmatrix} 0 & 0 & 0 & 0 \\ 0 & 2.1 & -0.2047 & -0.0107 \\ 0 & -2 & -6.3771 & -0.0104 \\ 0 & 0.1 & 8.1791 & 0.1560 \end{bmatrix}$$

and the final results of the forward and backward substitutions:

$$V(1) = 0.946 + j(-0.111)$$

$$V(2) = 0.937 + j(-0.133)$$

$$V(3) = 0.826 + j(-0.112)$$

$$V(4) = -8.116 + j(3.466)$$

Exercises

The small signal model of the bipolar transistors used in the exercises for this chapter is shown in Fig. E4.1(a). The small signal model used for the MOS transistors is shown in Fig. E4.1(b).

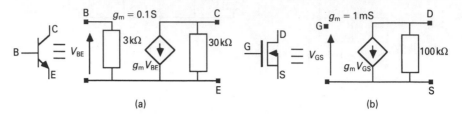

(a)   (b)

Figure E4.1

Exercise 1.   Consider the resistive ladder network shown in Fig. E4.2.

1.   Determine by inspection, the voltages at different nodes of the circuit.
2.   Write the admittance matrix and find the previous results using the program 'GAUSSJOR'. All resistances are in ohms.

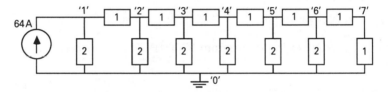

Figure E4.2

Exercise 2.   For the two circuits shown in Fig. E4.3, calculate the voltage V_{30} using the program 'GAUSSJOR'. These two circuits are the same as that in exercise 1 of Chapter 3. Only current sources have been added.

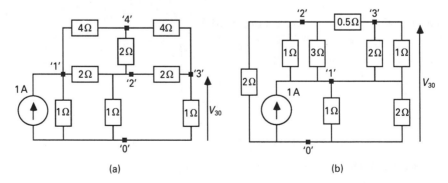

(a)   (b)

Figure E4.3

Exercise 3.   Apply the program 'GAUSSJOR' in order to determine the 'small signal' voltage gain V_o/V_i of the circuit shown in Fig. E4.4. This circuit is the same as that in exercise

5 of Chapter 3. Capacitors are equivalent to short circuits at the frequency considered $(k = 10^{+3})$

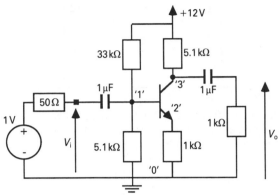

Figure E4.4

Exercise 4. Apply the program 'GAUSSELIM' to the circuit shown in Fig. 4.1 of Chapter 4 with the numerical values proposed in the text.

Exercise 5. Apply the program 'GAUSSELIM' in order to determine the 'small signal' voltage gain V_o/V_i of the common base circuit shown in Fig. E4.5. Capacitors are equivalent to short circuits at the frequency considered $(k = 10^{+3})$.

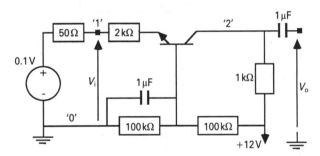

Figure E4.5

Exercise 6. Apply the program 'LUDECOMP' to the matrix represented by the Table E4.1 and note, for the successive steps, the substitution of terms equal to zero. Do the same for the matrix represented by Table E4.2 obtained from the previous matrix after permutation of the first and last rows and then the first and last columns.

Table E4.1

$$
\begin{bmatrix}
1 & 2 & 3 & 4 & 5 \\
6 & 7 & 0 & 0 & 0 \\
8 & 0 & 9 & 0 & 0 \\
10 & 0 & 0 & 11 & 0 \\
12 & 0 & 0 & 0 & 13
\end{bmatrix}
$$

Table E4.2

$$\begin{bmatrix} 13 & 0 & 0 & 0 & 12 \\ 0 & 7 & 0 & 0 & 6 \\ 0 & 0 & 9 & 0 & 8 \\ 0 & 0 & 0 & 11 & 10 \\ 5 & 2 & 3 & 4 & 1 \end{bmatrix}$$

Exercise 7. Apply the program 'LUDECOMP' in order to determine the 'small signal' voltage gain V_o/V_i of the common collector circuit shown in Fig. E4.6. Capacitors are equivalent to short circuits at the frequency considered ($k = 10^{+3}$).

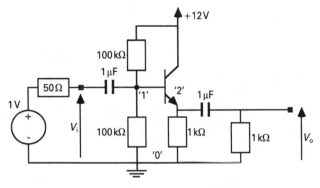

Figure E4.6

Exercise 8. Using a graphical interpretation, explain why the system

$$x - y = 0$$
$$x + y = 2$$

is named the well-conditioned system, and why the system

$$1.002x - 1.001y = 0.001$$
$$x \quad - \quad y \quad = \quad 0$$

is named the ill-conditioned system.

Exercise 9. Apply the three methods used in the programs 'GAUSSJOR', 'GAUSSELIM' and 'LUDECOMP' to the following ill-conditioned system (perform computations with 15 significant digits):

$$\begin{bmatrix} 5 & 7 & 6 & 5 \\ 7 & 10 & 8 & 7 \\ 6 & 8 & 10 & 9 \\ 5 & 7 & 9 & 10 \end{bmatrix} \begin{bmatrix} x \\ y \\ z \\ t \end{bmatrix} = \begin{bmatrix} 23 \\ 32 \\ 33 \\ 31 \end{bmatrix}$$

Exercise 10. Modify the program 'LUDECOMP' in order to calculate the determinant of a matrix. Apply this new program to the computation of the following Vandermonde determinant D (order 4):

$$\begin{vmatrix} 1 & x_1 & x_1^2 & x_1^3 \\ 1 & x_2 & x_2^2 & x_2^3 \\ 1 & x_3 & x_3^2 & x_3^3 \\ 1 & x_4 & x_4^2 & x_4^3 \end{vmatrix}$$

and verify that $D = \Pi_i (x_i - x_j)$ for $i > j$ for the following values: $x_1 = -1$; $x_2 = 2$; $x_3 = -3$; $x_4 = 4$.

Exercise 11.   Apply the program 'MATINVLU' in order to determine the inverse matrix of the following matrix:

$$\begin{bmatrix} 2 & -1 & 0 & 0 & 0 \\ -1 & 2 & -1 & 0 & 0 \\ 0 & -1 & 2 & -1 & 0 \\ 0 & 0 & -1 & 2 & -1 \\ 0 & 0 & 0 & -1 & 2 \end{bmatrix}$$

Verify that, generally, the inverse matrix of a band matrix is not a band matrix.

Exercise 12.   Using the program 'MATINVLU', explain why the following matrix is named the 'rotation' matrix:

$$\begin{bmatrix} \cos(\theta) & -\sin(\theta) & 0 \\ \sin(\theta) & \cos(\theta) & 0 \\ 0 & 0 & 1 \end{bmatrix}$$

Exercise 13.   We want to demonstrate that, for the matrix **A** shown in Table E4.3, the inverse matrix **B** is shown in Table E4.4 (even matrix order). Verify that the program 'MATINVLU' does not lead to a good result. Apply the program 'PIVOTGE' to matrix **A** using special second member vectors in order to verify the previous affirmation.

Table E4.3

$$A = \begin{bmatrix} 0 & 1 & 1 & 1 & 1 & \cdots \\ -1 & 0 & 1 & 1 & 1 & \cdots \\ -1 & -1 & 0 & 1 & 1 & \cdots \\ -1 & -1 & -1 & 0 & 1 & \cdots \\ -1 & -1 & -1 & -1 & 0 & \cdots \\ \cdots & \cdots & \cdots & \cdots & \cdots & \cdots \end{bmatrix}$$

Table E4.4

$$B = \begin{bmatrix} 0 & -1 & 1 & -1 & 1 & \cdots \\ 1 & 0 & -1 & 1 & -1 & \cdots \\ -1 & 1 & 0 & -1 & 1 & \cdots \\ 1 & -1 & 1 & 0 & -1 & \cdots \\ -1 & 1 & -1 & 1 & 0 & \cdots \\ \cdots & \cdots & \cdots & \cdots & \cdots & \cdots \end{bmatrix}$$

Exercise 14.   Develop a program to generalize the extension of circuit of exercise 1 to any circuit composed of n R-$2R$ cells. Analyze the form of the resulting admittance matrix. For the different steps of program 'GAUSSJOR', observe the transformation of matrix elements equal to zero into elements not equal to zero.

Exercise 15.   Using program 'COMPLXLU', verify that the input impedance R_i of the two-port shown in Fig. E4.7 is equal to unity for all values of ω with $R_0 = 1\ \Omega$. Explain why the program does not give any result for $\omega = 1$ rad/s.
   What is the input impedance for $R_0 = 2\ \Omega$ and $\omega = 0.5,\ 2.0$ rad/s?

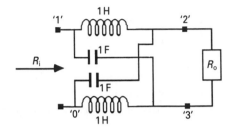

Figure E4.7

Exercise 16.   For the circuit shown in Fig. E4.8, choose $k = 10^{+4}$, $n = 10^{+5}$ for impedance and frequency normalization. Analyze the circuit using the program 'COMPLXLU' and verify that this circuit behaves as a rejector circuit (minimum magnitude of the output voltage V_o).

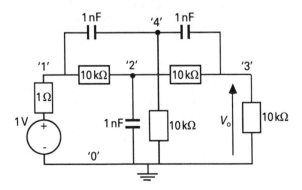

Figure E4.8

Exercise 17.   For the circuit composed of the three cells shown in Fig. E4.9, choose $k = 10^{+3}$, $n = 10^{+4}$ for impedance and frequency normalization. Modify the program 'COMPLXLU' in order to display the value V_o/V_i. Running this program, determine the normalized value of ω for which the output voltage V_o is in opposite phase to the input V_i.
   Determine the value of attenuation and input impedance using the previous value for ω.

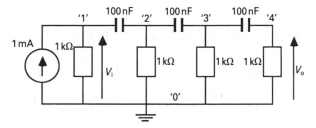

Figure E4.9

Exercise 18.   Analyze the frequency behaviour of the MOS circuit shown in Fig. E4.10; modify the output of the program 'COMPLXLU' in order to determine the magnitude and the phase of output voltage ($k = 10^{+3}$, $n = 10^{+2}$).

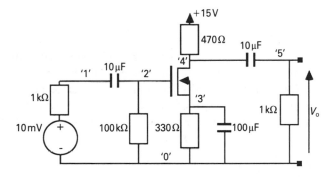

Figure E4.10

Exercise 19.   Consider the artificial line shown in Fig. E4.11. Apply the program 'COMPLXLU' in order to determine the voltages to the different nodes of the circuit for $Z = 1$, Z near to zero and Z near to infinite value ($\omega = 1$ rad/s). Inductances are in henries, and capacitances are in farads.

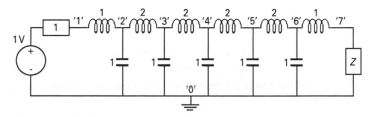

Figure E4.11

Exercise 20.   Modify the program 'COMPLXLU' in order to calculate a determinant with complex values.

Exercise 21.   Modify the program 'COMPLXLU' in order to determine the inverse matrix of a matrix of complex values.

5 Adjoint Network and its Application to Sensitivity and Thevenin Circuit Computation; Numerical Computation of Network Function Derivatives

5.1 Introduction

Derivative computation of the network function, $f(x)$ of one variable or $f(x_1, x_2, ..., x_n)$ of many variables, is frequently used in the analysis and design of electronic circuits. Examples are sensitivity analysis for performance evaluation of circuits or minimization of performance (or cost) function using gradient methods during circuit design process. However, numerical derivative computations using approximate differentiation formulae such as

$$\frac{\mathrm{d}f(x)}{\mathrm{d}x} \cong \frac{\Delta f(x)}{\Delta x}\bigg|_{x_0} = \frac{f(x_0 + \Delta x) - f(x)}{\Delta x} \tag{5.1a}$$

or

$$\frac{\partial f(x_1, x_2, ..., x_i, ..., x_n)}{\partial x_i}\bigg|_{\substack{x_{i0} \\ x_k = \text{const} \\ k \neq i}} \cong \frac{f(..., x_{i0} + \Delta x_i, ...) - f(..., x_{i0}, ...)}{\Delta x_i} \tag{5.1b}$$

are not convenient and are error prone. Another method is therefore needed that allows us to compute efficiently derivatives of a given network function $f(x_1, x_2, ..., x_n)$ with respect to many variables. This method is based on the Tellegen theorem and application of the so-called 'adjoint network'. The most popular application of derivative computation is for sensitivity analysis of electronic circuits.

5.2 Sensitivity of electronic circuits

Sensitivity is one of the measures of electronic circuit performance, since it reflects deviations in some performance function or characteristic of the circuit due to a change in the nominal value of one or more of the elements of the circuit.

There are several definitions of different sensitivities. In each one, however, the derivative of the performance function $P(x)$ with respect to element value x is included (for simplicity of notation $P(x)$ is treated as the function of a single variable x; however, x can be a vector).

1. Absolute sensitivity:

$$S_{ax}^P = \frac{dP}{dx} \tag{5.2}$$

2. Semirelative sensitivity:

$$S_{sx}^P = \frac{dP}{\frac{dx}{x}} = \frac{dP}{dx}\, x = x S_{ax}^P \tag{5.3}$$

or

$$S_x^{sP} = \frac{\frac{dP}{P}}{dx} = \frac{S_{ax}}{P} \tag{5.4}$$

3. Relative sensitivity:

$$S_x^P = \frac{\frac{dP}{P}}{\frac{dx}{x}} = \frac{d\ln P}{d\ln x} = \frac{x}{P}\, S_{ax}^P \tag{5.5}$$

In the case where variables form the vector $\mathbf{x}^t = [x_1, x_2, \ldots, x_n]$, then partial derivatives with respect to a given variable x_i are used in equations (5.2)–(5.5).

Also, a multi-parameter sensitivity can be used as an overall performance characteristic of the circuit when the performance characteristic P is a function of many parameters x_i, $i = 1, 2, \ldots, n$, and all the parameters change simultaneously. In such a case the change in P is expressed by the total differential

$$dP = \sum_{i=1}^{n} \frac{\partial P}{\partial x_i}\, dx_i \tag{5.6}$$

or

$$\frac{dP}{P} = \sum_{i=1}^{n} \frac{\partial P/P}{\partial x_i/x_i} \tag{5.7}$$

The circuit performance function $P(x)$ can be any appropriate characteristic, such as the transfer function, input impedance, pole or zero location on a complex frequency plane, or even the resistance R of a resistor; while x can represent any element value, such as resistance, capacitance, inductance, mutual conductance, temperature, etc. For example, semirelative sensitivity given by equation (5.4), applied to a resistance–temperature relation $R(t)$, where t stands for temperature, is known as a temperature coefficient of the resistance

$$S_t^{sR} = \frac{\dfrac{dR}{R}}{dt} = \frac{1}{R}\frac{dR}{dt} \tag{5.8}$$

while the relative sensitivities of the transfer function of simple voltage divider $T = V_2/V_1 = P(R_1, R_2)$ with

$$P(R_1, R_2) = \frac{V_2}{V_1} = V_2|_{V_1=1} = \frac{R_2}{R_1 + R_2} \tag{5.9}$$

computed using equation (5.5), are

$$s_{R_1}^P = \frac{-R_1}{R_1 + R_2} \tag{5.10a}$$

$$S_{R_2}^P = \frac{R_1}{R_1 + R_2} \tag{5.10b}$$

A positive sign in the value of sensitivity S_x^P indicates that, for an increase in x, the value of P also increases; while a negative value of S_x^P shows that, for an increase in x, the value of P decreases.

The relative sensitivities are especially useful in the evaluation of circuit properties, since, usually, a circuit is better when its sensitivities are smaller. Also, in the design of electronic circuits, the relative sensitivity indicates the size of the percentage deviation of a given performance function $P(x)$ for a 1% variation of element value x:

$$\frac{dP}{P}[\%] = 100\left(\frac{dx}{x}S_x^P\right) = S_x^P\bigg|_{dx/x=1\%} \tag{5.11}$$

Sensitivities can easily be computed directly for simple circuits. However, for large circuits, a numerical method based on the Tellegen theorem is used.

5.3 Application of the Tellegen theorem and the adjoint network

The Tellegen theorem in its simplest form states that the sum of the products of branch currents I_k and branch voltages V_k in an arbitrary circuit is equal to zero:

$$\sum_{k=1}^{n_b} V_k I_k = 0 \tag{5.12}$$

where n_b is the number of branches including excitation sources. This is one of the forms of power conservation statement, that is, the total power consumed by the network must equal the total power generated in the network. It is not very surprising.

However, the consequences of application of the Tellegen theorem to two different circuits having the same topology (structure) are very interesting. When speaking about the circuits of the same topology we mean circuits having the same incidence matrix, that is, the same number of branches and branch interconnections.

The Tellegen theorem, applied to two different circuits of the same topology and number of branches n_b, the first of which has branch voltages V_k and branch currents I_k and the second has branch voltages V_k' and branch currents I_k' is as follows:

$$\sum_{k=1}^{n_b} V_k I_k' = 0 \qquad\qquad (5.13a)$$

$$\sum_{k=1}^{n_b} V_k' I_k = 0 \qquad\qquad (5.13b)$$

To prove the validity of equations (5.13), consider two simple resistive circuits shown in Fig. 5.1, having the same topology, and the same notation for branch currents and voltages.

For the circuit of Fig. 5.1(a), we have branch voltages and currents:

$$V_1 = -9 \text{ V}, \ V_2 = 9 \text{ V}, \ V_3 = 4 \text{ V}, \ V_4 = 5 \text{ V}$$

$$I_1 = 1 \text{ A}, \ I_2 = 0.9 \text{ A}, \ I_3 = 0.1 \text{ A}, \ I_4 = 0.1 \text{ A}$$

and for the circuit of Fig. 5.1(b):

$$V_1' = -3.2 \text{ V}, \ V_2' = 3.2 \text{ V}, \ V_3' = -6.8 \text{ V}, \ V_4' = 10 \text{ V}$$

$$I_1' = -0.16 \text{ A}, \ I_2' = 0.04 \text{ A}, \ I_3' = -0.2 \text{ A}, \ I_4' = -0.2 \text{ A}$$

Inserting current and voltage values in equation (5.13a), we obtain

$$-9 \times (-0.16) + 9 \times (0.04) + 4 \times (-0.2) + 5 \times (0.2) = 0$$

Similarly, using equation (5.13b) we obtain

$$-3.2 \times (1) + 3.2 \times (0.9) - 6.8 \times (0.1) + 10 \times (0.1) = 0$$

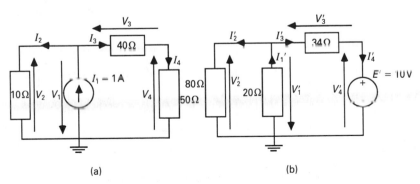

(a)   (b)

Figure 5.1   Two circuits having the same topology to illustrate the Tellegen theorem

Equations (5.13) are also valid for sinusoidal steady state analysis of circuits containing resistors R, conductors G, capacitors C, inductors L and VCC sources with mutual conductance g_m; in such a case the voltages V and currents I are the phasors (complex amplitudes).

To define the adjoint network, and its properties, consider two networks having the same topology and n internal branches excited by two current sources, as shown in Fig. 5.2. It is assumed that both networks are composed of n_r resistors, n_g conductors, n_c capacitors, n_l inductors and n_m voltage controlled current sources. The total number of internal branches is

$$n_r + n_g + n_c + n_l + n_m = n \tag{5.14}$$

Both networks have $n_b = n + 2$ branches. The zero-valued current sources $I_o = 0$ and $I'_s = 0$ are added to distinguish directly the branches where the voltages V_o and V'_s exist.

Now, let us calculate the absolute sensitivity of the transfer function $T = V_o / I_s$ with respect to network element changes. Since the excitation current $I_s = 1$, then the voltage V_o will represent this transmittance $T = V_o$. Thus we are interested in the relation of the increments ΔV_o in terms of element value increments ΔR, ΔG, ΔC, ΔL and Δg_m, which can easily be obtained from the Tellegen theorem.

Assuming, again, that the branch voltages and currents in the kth branch in the original network N and in the network N' are denoted by V_k, I_k, V'_k, I'_k, respectively, we can use equations (5.13a) and (5.13b); after subtracting we obtain

$$\sum_{k=1}^{n_b} (V_k I'_k - V'_k I_k) = 0 \tag{5.15}$$

If the elements in N change their values, that is, $R_k + \Delta R_k$, $C_k + \Delta C_k$, etc., then the voltages and currents also change their values and become $V_k + \Delta V_k$ and $I_k + \Delta I_k$, respectively. Thus (5.15) can be rewritten as

$$\sum_{k=1}^{n_b} [(V_k + \Delta V_k)I'_k - V'_k(I_k + \Delta I_k)] = 0 \tag{5.16}$$

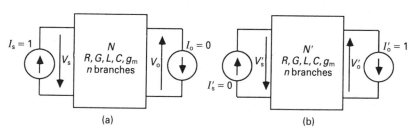

Figure 5.2   Two networks with the same topology. (a) Excited by $I_S = 1$. (b) Excited by $I'_o = 1$

yielding

$$\sum_{k=1}^{n_b} [\Delta V_k I_k' - V_k' \Delta I_k] = 0 \tag{5.17}$$

since according to (5.13) $\sum_{k=1}^{n_b} V_k I_k' = 0$ and $\sum_{k=1}^{n_b} V_k' I_k = 0$.

Extracting the external source branches, this summation can be expanded indicating the different types of elements as follows:

$$\Delta V_s I_s' - \Delta I_s V_s' + \Delta V_o I_o' - \Delta I_o V_o'$$

$$+ \sum^{n_r} (\Delta V_R I_R' - \Delta I_R V_R') + \sum^{n_g} (\Delta V_G I_G' - \Delta I_G V_G')$$

$$+ \sum^{n_c} (\Delta V_C I_C' - \Delta I_C V_C') + \sum^{n_l} (\Delta V_L I_L' - \Delta I_L V_L')$$

$$+ \sum^{n_m} (\Delta V_{mc} I_{mc}' - \Delta I_{mc} V_{mc}' + \Delta V_{gm} I_{gm}' - \Delta I_{gm} V_{gm}') = 0 \tag{5.18}$$

where the upper limits of the summations, n_r, n_g, n_c, n_l and n_m, fulfil relation (5.14); also, the voltages and currents for the controlled sources used in (5.18) are indicated in Fig. 5.3. Note that the subscripts used on V' and I' in the summation concerning the VCC sources indicate the correspondence between the branches in N and N'.

Now let us consider branch relationship in N when each element changes its value. The input current source $I_s = 1$, so that $\Delta I_s = 0$ independently of changes in the element values. Also, the output current source is described by $I_o = 0$, so that $\Delta I_o = 0$. Therefore, the second and fourth terms in (5.18) disappear. Resistance branches are described by $V_R = R I_R$; thus, after changes in element values, we obtain

$$V_R + \Delta V_R = (R + \Delta R)(I_R + \Delta I_R) = R I_R + \Delta R I_R + R \Delta I_R + \Delta R \Delta I_R \tag{5.19}$$

After subtraction, V_R and $R I_R = V_R$, and neglecting the second-order term $\Delta R \Delta I_R$ we have

$$\Delta V_R = \Delta R I_R + R \Delta I_R \tag{5.20}$$

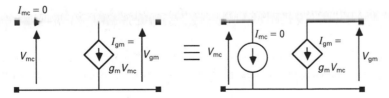

Figure 5.3 VCC source and its representation in the network N and equation (5.18)

In a similar way, we obtain the relations

$$\Delta I_G = \Delta G V_G + G \Delta V_G \tag{5.21}$$

for conductances branches,

$$\Delta I_C = j\omega \Delta C V_C + j\omega C \Delta V_C \tag{5.22}$$

for capacitor branches and

$$\Delta V_L = j\omega \Delta L I_L + j\omega L \Delta I_L \tag{5.23}$$

for inductor branches.

The controlling branch, according to the notation in Fig. 5.3, is characterized by

$$I_{mc} = 0, \text{ so that } \Delta I_{mc} = 0$$

and the controlled branch is characterized by

$$I_{gm} = g_m V_{mc} \tag{5.24}$$

so that similarly to the conductance branches, we obtain

$$\Delta I_{gm} = \Delta g_m V_{mc} + g_m \Delta V_{mc} \tag{5.25}$$

After substitution of these relations, equation (5.18) becomes (recall that $\Delta I_s = 0$, and $\Delta I_o = 0$)

$$\Delta V_s I_s' + \Delta V_o I_o' + \sum^{n_r} [(RI_R^k - V_R^k) \Delta I_R + I_R I_R^k \Delta R]$$

$$+ \sum^{n_g} [(I_G^k - G V_G^k) \Delta V_G - V_G V_G^k \Delta G]$$

$$+ \sum^{n_c} [(I_C^k - j\omega C V_C^k) \Delta V_C - V_C V_C^k j\omega \Delta C]$$

$$+ \sum^{n_l} [(j\omega L I_L^k - V_L^k) \Delta I_L + I_L I_L^k j\omega \Delta L]$$

$$+ \sum^{n_m} [I_{gm}' \Delta V_{gm} + (I_{mc}' - g_m V_{gm}') \Delta V_{mc} - V_{mc} V_{gm}' \Delta g_m] = 0 \tag{5.26}$$

We need, however, to find the relationship between the change in the output voltage ΔV_o and the changes in the parameters ΔR, ΔG, ΔC, ΔL and Δg_m. We can achieve this if we eliminate the first term in (5.26) by setting $I_s' = 0$, as is already shown in Fig. 5.2(b), and also the terms involving ΔI_R, ΔV_G, ΔV_C, ΔI_L, ΔV_{gm} and ΔV_{mc}.

For simplicity, we also set $I_o' = 1$ (see Fig. 5.2(b)). We can eliminate terms involving ΔI_R if resistive branches in N' are chosen as resistors with the same values as in N, so that

$$V_R^k = RI_R^k \text{ or } RI_R^k - V_R^k = 0 \tag{5.27}$$

Similarly, the terms involving ΔV_G, ΔV_C and ΔI_L can be eliminated if conductance, capacitance and inductance branches in N' are chosen to be conductors, capacitors and inductors with their values as in original network N. Therefore, we see that for passive reciprocal circuits (not containing controlled sources) the network N' should be exactly the same as N. Finally, to eliminate the terms involving ΔV_{gm} and ΔV_{mc} in (5.26) we can set in N':

$$I'_{gm} = 0 \quad \text{and} \quad I'_{mc} = g_m V'_{gm} \tag{5.28}$$

If we compare description (5.28) with (5.24) we can observe that it corresponds to a voltage controlled current source acting in the reverse direction, that is, the controlling branch is now controlled by the current $I'_{mc} = g_m V'_{gm}$ and the dependent branch $I'_{gm} = 0$ is now controlling with voltage V'_{gm}, as is illustrated in Fig. 5.4.

But, reversing the direction of operation of the VCCS corresponds to the transposition of a matrix $\mathbf{Y}$ describing the source. To show this let us assume that $V_{mc} \equiv V_{10}$ in Fig. 5.4. Then the matrix descriptions for both sources in Fig. 5.4 will be

$$\mathbf{Y} = \begin{array}{c} \\ 1 \\ 2 \end{array} \overset{\begin{array}{cc} 1 & 2 \end{array}}{\begin{bmatrix} 0 & 0 \\ g_m & 0 \end{bmatrix}} \qquad \mathbf{Y'} = \begin{array}{c} \\ 1 \\ 2 \end{array} \overset{\begin{array}{cc} 1 & 2 \end{array}}{\begin{bmatrix} 0 & g_m \\ 0 & 0 \end{bmatrix}} = \mathbf{Y^t} \tag{5.29}$$

The same can be proved for more complicated networks with many VCC sources.

Therefore, to obtain the properties that allow us to cancel some terms in equation (5.26) the network N' should have the $\mathbf{Y'}$ matrix, which is the transposed matrix $\mathbf{Y}$ of the original network N:

$$\mathbf{Y'} = \mathbf{Y^t} \tag{5.30}$$

Thus, the network N' of Fig. 5.2(b), having $\mathbf{Y'} = \mathbf{Y^t}$ and excited by $I_s' = 0$ and $I_o' = 1$, is called the transpose network or adjoint network with respect to N. Now, knowing the properties of the adjoint network, and assuming that $I_o = -1$. allowing positive I_o' to flow into upper pole of output port, expression (5.26) can

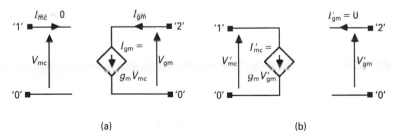

(a)   (b)

Figure 5.4   Explanation of operation on VCCS. (a) In forward direction in N. (b) In reverse direction in N

be rewritten as

$$\Delta V_o = \sum^{n_r} (I_R I_R') \, \Delta R + \sum^{n_g} (-V_G V_G') \, \Delta G + \sum^{n_c} (-j\omega V_C V_C') \, \Delta C$$

$$+ \sum^{n_l} (j\omega I_L I_L') \, \Delta L + \sum^{n_m} (-V_{mc} V_{gm}') \, \Delta g_m \qquad (5.31)$$

The change in V_o in a given network can be related to changes in individual parameters by the total differential

$$\Delta V_o = \sum^{n_r} \frac{\partial V_o}{\partial R} \Delta R + \sum^{n_g} \frac{\partial V_o}{\partial G} \Delta G + \sum^{n_c} \frac{\partial V_o}{\partial C} \Delta C + \sum^{n_l} \frac{\partial V_o}{\partial L} \Delta L + \sum^{n_m} \frac{\partial V_o}{\partial g_m} \Delta g_m \qquad (5.32)$$

After comparing expressions (5.31) and (5.32) we see that

$$\frac{\partial V_o}{\partial R} = I_R I_R' \qquad (5.33)$$

$$\frac{\partial V_o}{\partial G} = -V_G V_G' \qquad (5.34)$$

$$\frac{\partial V_o}{\partial C} = -j\omega V_C V_C' \qquad (5.35)$$

$$\frac{\partial V_o}{\partial L} = j\omega I_L I_L' \qquad (5.36)$$

$$\frac{\partial V_o}{\partial g_m} = -V_{mc} V_{gm}' \qquad (5.37)$$

The voltages V and V' and the currents I and I' correspond to individual elements in N and N'. For example, if we have a network composed only of resistors, then relation (5.31) becomes

$$\Delta V_o = \sum_{k=1}^{n_r} (I_{Rk} I_{Rk}') \, \Delta R_k \qquad (5.38)$$

If we assume additionally that only ith resistor changes its value by ΔR_i and the rest remain constant:

$$\Delta R_i \neq 0, \qquad \Delta R_k = 0 \qquad \text{for } k \neq i$$

then (5.38) becomes

$$\Delta V_o = I_{Ri} I_{Ri}' \, \Delta R_i \qquad (5.39)$$

and absolute sensitivity of V_o with respect to R_i is

$$S_{aR_i}^{V_o} = \frac{\Delta V_o}{\Delta R_i} = I_{Ri} I_{Ri}' \qquad (5.40)$$

The relative sensitivity is given by the expression

$$S_{R_i}^{V_o} = \frac{\Delta V_o}{\Delta R_i} \frac{R_i}{V_o} = I_{Ri} I'_{Ri} \frac{R_i}{V_o} \tag{5.41}$$

From relations (5.31)–(5.41), we see that the sensitivity of the output voltage V_o for a given network N, with respect to all elements in the network, can be obtained without direct differentiation by the following two analyses:

1. of the original network N excited by the current source $I_s = 1$ and computing certain branch voltages and currents at given frequency ω.
2. of the adjoint network N' excited at the output by the current source $I'_o = 1$ and computing certain branch voltages and currents at given frequency ω.

Finally, the absolute sensitivities are calculated as the weighted products of voltages and currents in the network N with the corresponding voltages and currents in the adjoint network N', as is given in (5.33)–(5.37). As an example of absolute sensitivity S_{ax}^P calculations to prove the validity of derived expressions, let us consider the circuit shown in Fig. 5.5(a) and the corresponding adjoint network shown in Fig. 5.5(b).

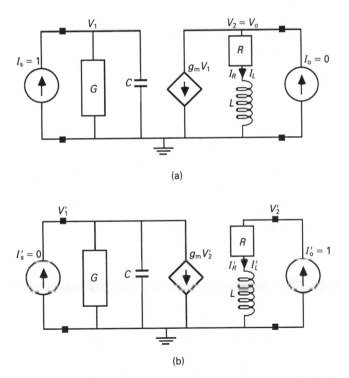

(a)

(b)

Figure 5.5 (a) Active circuit to illustrate sensitivity computation. (b) Adjoint network

For the circuit of Fig. 5.5(a) we obtain by inspection

$$V_1 = V_G = V_C = \frac{1}{G + j\omega C}$$

and

$$V_o = V_2 = -g_m V_1 (R + j\omega L) = \frac{-g_m (R + j\omega L)}{G + j\omega C}$$

The current through the resistor and inductor is

$$I_R = I_L = \frac{V_2}{R + j\omega L} = \frac{-g_m}{G + j\omega C}$$

The direct analysis of the adjoint network of Fig. 5.5(b) yields

$$V_2' = (R + j\omega L)$$

$$I_R' = I_L' = 1$$

$$V_1' = \frac{-g_m V_2'}{G + j\omega C} = \frac{-g_m (R + j\omega L)}{G + j\omega C}$$

$$V_C' = V_G' = V_1'$$

Now we can compute derivatives (absolute sensitivities) of V_o with respect to all parameters by direct differentiation:

$$\frac{\partial V_o}{\partial g_m} = \frac{(R + j\omega L)}{G + j\omega C}$$

$$\frac{\partial V_o}{\partial G} = \frac{g_m (R + j\omega L)}{(G + j\omega C)^2}$$

$$\frac{\partial V_o}{\partial R} = \frac{-g_m}{G + j\omega C}$$

$$\frac{\partial V_o}{\partial C} = \frac{j\omega g_m (R + j\omega L)}{(G + j\omega C)^2}$$

$$\frac{\partial V_o}{\partial L} = \frac{-j\omega g_m}{G + j\omega C}$$

and by application of expressions (5.33)–(5.37):

$$\frac{\partial V_o}{\partial g_m} = -V_1 V_2' = \frac{-1}{G + j\omega C} (R + j\omega L) = \frac{-(R + j\omega L)}{G + j\omega C}$$

$$\frac{\partial V_o}{\partial G} = -V_G V_G' = \frac{1}{G + j\omega C} \frac{g_m (R + j\omega L)}{(G + j\omega C)} = \frac{g_m (R + j\omega L)}{(G + j\omega C)^2}$$

$$\frac{\partial V_o}{\partial R} = I_R I_R' = \frac{-g_m \times 1}{G + j\omega C} = \frac{-g_m}{G + j\omega C}$$

$$\frac{\partial V_o}{\partial C} = -j\omega V_C V_C' = \frac{j\omega}{G + j\omega C} \frac{g_m(R + j\omega L)}{(G + j\omega C)} = \frac{j\omega g_m(R + j\omega L)}{(G + j\omega C)^2}$$

$$\frac{\partial V_o}{\partial L} = j\omega I_L I_L' = \frac{-j\omega g_m}{G + j\omega C} \times 1 = \frac{-j\omega g_m}{G + j\omega C}$$

We see that the results of both approaches are identical. Also, it can easily be shown that the $\mathbf{Y}'$ matrix of the adjoint network equals the transpose of the $\mathbf{Y}$ matrix of the original network; for the circuit of Fig. 5.5(a), we have

$$\mathbf{Y} = \begin{bmatrix} G + j\omega C & 0 \\ g_m & \dfrac{1}{R + j\omega L} \end{bmatrix}$$

and for the adjoint network of Fig. 5.5(b) we have

$$\mathbf{Y}' = \begin{bmatrix} G + j\omega C & g_m \\ 0 & \dfrac{1}{R + j\omega L} \end{bmatrix} = \mathbf{Y}^t$$

which equals the transpose of the $\mathbf{Y}$ of the original circuit.

Numerically, the sensitivities are obtained as complex numbers for given fixed frequency ω. If, as an example, we assume for the circuit of Fig. 5.5(a):

$$\omega = 2 \text{ rad/s}, \qquad G = 1 \text{ S}, \qquad C = 0.5 \text{ F}$$
$$R = 4 \text{ }\Omega, \qquad L = 2 \text{ H}, \qquad g_m = 5 \text{ S}$$

then we obtain

$$V_1 = 0.5 - j0.5 \text{ V}$$
$$V_o = V_2 = -20 \text{ V}$$
$$I_R = I_l = -2.5 + j2.5 \text{ A}$$
$$V_C = V_G = V_1 = 0.5 - j0.5 \text{ V}$$

and for the adjoint network of Fig. 5.5(b):

$$V_1' = -20 \text{ V}$$
$$V_C' = V_G' = V_1' = -20 \text{ V}$$
$$V_2' = 4 + j4 \text{ V}$$
$$I_R' = I_l' = 1 \text{ A}$$

The calculated values of absolute sensitivites are

$$S_{a g_m}^{V_o} = \frac{d V_o}{d g_m} = -4$$

$$S_{aG}^{V_o} = \frac{d V_o}{d G} = +10 - j10$$

$$S_{aC}^{V_o} = \frac{d V_o}{d C} = +20 + j20$$

$$S_{aR}^{V_o} = \frac{d V_o}{d R} = -2.5 + j2.5$$

$$S_{aL}^{V_o} = \frac{d V_o}{d L} = -5 \quad -j5$$

and the values of relative sensitivities:

$$S_{g_m}^{V_o} = +1, \qquad S_G^{V_o} = -0.5 + j0.5, \qquad S_C^{V_o} = -0.5 - j0.5$$
$$S_R^{V_o} = +0.5 - j0.5, \qquad S_L^{V_o} = +0.5 + j0.5,$$

since generally

$$S_x^P = S_{ax}^P \frac{x}{P}$$

5.4 Description of the procedure and program for calculation of sensitivity

As was explained in Section 5.3, to calculate the absolute sensitivity of the output voltage, that is, the transfer function V_o/I_s with unity excitation current $I_s = 1$, for a given circuit, we have to perform the following two analyses:

1. for the original network having the admittance matrix $\mathbf{Y}$ excited by the current vector:

    $$\mathbf{I}^t = [1 \quad 0 \quad 0 \ldots 0],$$

2. for the adjoint network having the admittance matrix $\mathbf{Y}' = \mathbf{Y}^t$ excited by the current vector:

    $$\mathbf{I}'^t = [0 \quad 0 \quad 0 \ldots 1],$$

find the nodal voltages in both networks and compute branch voltages for the conductances G, capacitances C, controlling voltages for VCCS and the currents through resistances R and inductances L in both networks.

The relationship between the admittance matrices of the original and adjoint networks, $\mathbf{Y}' = \mathbf{Y}^t$, facilitates the computations when we use the LU decomposition method for equation solution, since in this case, for the two analyses required, only one LU decomposition is needed. This is due to the fact that the LU decomposition of the transposed matrix equals the product of transposed $\mathbf{U}^t$ and $\mathbf{L}^t$ matrices, which can be explained as follows:

$$\mathbf{Y} = \mathbf{L} \times \mathbf{U} \tag{5.42a}$$

$$\mathbf{Y}' = \mathbf{Y}^t = (\mathbf{L} \times \mathbf{U})^t = \mathbf{U}^t \times \mathbf{L}^t \tag{5.42b}$$

or, in expanded form (for three equations),

$$\mathbf{Y} = \begin{bmatrix} y_{11} & y_{12} & y_{13} \\ y_{21} & y_{22} & y_{23} \\ y_{31} & y_{32} & y_{33} \end{bmatrix} = \begin{bmatrix} l_{11} & u_{12} & u_{13} \\ l_{21} & l_{22} & u_{23} \\ l_{31} & l_{32} & l_{33} \end{bmatrix} = \mathbf{L} \times \mathbf{U} \tag{5.43a}$$

$$\mathbf{Y}' = \mathbf{Y}^t = \begin{bmatrix} y_{11} & y_{21} & y_{31} \\ y_{12} & y_{22} & y_{32} \\ y_{13} & y_{23} & y_{33} \end{bmatrix} = \begin{bmatrix} l_{11} & l_{21} & l_{31} \\ u_{12} & l_{22} & l_{32} \\ u_{13} & u_{23} & l_{33} \end{bmatrix} = \mathbf{U}^t \times \mathbf{L}^t \tag{5.43b}$$

Thus, instead of performing the LU decomposition of the $\mathbf{Y}'$ matrix of the adjoint network in the second analysis, it is sufficient to reverse the indices in the $\mathbf{L}$ and $\mathbf{U}$ matrices so that the transposed matrix $\mathbf{U}^t$ becomes the lower triangular matrix and the transposed matrix $\mathbf{L}^t$ becomes the upper triangular matrix. After finding the nodal voltages in the original and adjoint networks, we can easily find the appropriate branch voltages and currents.

To summarize the procedure, let us list all the necessary steps:

1. Enter the real and imaginary parts of the $\mathbf{Y}$ matrix of the original network (or use the program for $\mathbf{Y}$ matrix generation described in Chapter 3),
2. Factorize $\mathbf{Y}$ into the lower $\mathbf{L}$ and the upper $\mathbf{U}$ triangular matrices so that $\mathbf{Y} = \mathbf{L} \times \mathbf{U}$,
3. Enter the excitation current vector $\mathbf{I}^t = [1 \quad 0 \quad 0 \dots 0]$,
4. Perform the forward substitution to find the auxiliary vector $\mathbf{D}$, so that $\mathbf{L} \times \mathbf{D} = \mathbf{I}$, where $\mathbf{D} = \mathbf{U} \times \mathbf{V}$,
5. Perform the backward substitution to find the node voltages $\mathbf{V}$ so that $\mathbf{U} \times \mathbf{V} = \mathbf{D}$,
6. Enter the excitation current vector $\mathbf{I}'^t = [0 \quad 0 \quad 0 \dots 1]$ for the adjoint network analysis,
7. Transpose the decomposed $(\mathbf{L} \times \mathbf{U})$ matrix so that $(\mathbf{L} \times \mathbf{U})^t = \mathbf{U}^t \times \mathbf{L}^t = \mathbf{L}' \times \mathbf{U}'$,
8. Perform the forward substitution to find the auxiliary vector $\mathbf{D}'$ so that $\mathbf{L}' \times \mathbf{D}' = \mathbf{U}^t \times \mathbf{D}' = \mathbf{I}'$,
9. Perform the backward substitution to find the node voltages $\mathbf{V}'$ in the adjoint network so that $\mathbf{U}' \times \mathbf{V}' = \mathbf{L}^t \times \mathbf{V}' = \mathbf{D}'$,
10. Find appropriate branch voltages and currents in the original and adjoint networks,
11. Calculate the weighted products of branch voltages and currents in the original

network, with the corresponding branch voltages and currents in the adjoint network, to find the absolute sensitivities,

12. Make use of the absolute sensitivities obtained to compute the relative sensitivities (if necessary).

However, steps 7, 8 and 9 need not be performed as separate procedures, since transposition of the matrix is just changing the indices in its elements. Also, because of the properties of the transposed upper triangular matrix and excitation current vector $\mathbf{I}'^{\,t}$, the matrix equation for the forward substitution in the step 8 is

$$
\begin{bmatrix}
1 & 0 & . & . & 0 \\
u_{12} & 1 & . & . & 0 \\
. & . & . & . & . \\
. & . & . & . & . \\
u_{1n} & u_{2n} & . & . & 1
\end{bmatrix}
\begin{bmatrix}
d_1' \\
d_2' \\
. \\
. \\
d_n'
\end{bmatrix}
=
\begin{bmatrix}
0 \\
0 \\
. \\
. \\
1
\end{bmatrix}
\tag{5.44}
$$

The result of this matrix equation gives

$$
d_n' = 1 \tag{5.45}
$$

Therefore, the forward substitution in step 8 can be omitted and result (5.45) can be entered directly into the backward substitution in step 9:

$$
\begin{bmatrix}
l_{11} & l_{21} & . & . & . & l_{n1} \\
0 & l_{22} & . & . & . & l_{n2} \\
. & . & . & . & . & . \\
. & . & . & . & . & . \\
0 & 0 & . & . & . & l_{nn}
\end{bmatrix}
\begin{bmatrix}
V_1' \\
V_2' \\
. \\
. \\
V_n'
\end{bmatrix}
=
\begin{bmatrix}
0 \\
0 \\
. \\
. \\
1
\end{bmatrix}
\tag{5.46}
$$

Thus, to find the node voltages in the adjoint network only backward substitution according to relation (5.46) is required. This backward substitution differs slightly from that in step 5 since in equation (5.46) we have the elements l_{ii} on the main diagonal. The program 'SENSITLU' for the original and adjoint networks analysis is shown in Program 5.1.

See the listing of the program SENSITLU Directory of Chapter 5 on attached diskette.

In this program the procedure 'forw-back-subst' is used for the original network analysis (with excitation $V_{r1} = I_1 = 1$) and the procedure 'back-subst' with excitation $V_{rn} = d_n' = 1$ is used for the adjoint network analysis.

To compute the absolute sensitivities of the output voltage to the changes in all the elements of the network considered, it is necessary to compute the appropriate products of branch voltages and branch currents in the original and adjoint networks. To use the program "SENSITLU" for sensitivity computation for the circuit of Fig. 5.5 with elements values:

$$G = 1\ \text{S},\ R = 4\ \Omega,\ g_m = 5\ \text{S},\ \omega = 2\ \text{rad/s},\ C = 0.5\ \text{F},\ L = 2\ \text{H}$$

we should enter the real and imaginary parts of the **Y** matrix:

$$\mathbf{Y}_r = \begin{bmatrix} 1 & 0 \\ 5 & 0.125 \end{bmatrix}, \quad \mathbf{Y}_i = \begin{bmatrix} 1 & 0 \\ 0 & -0.125 \end{bmatrix}$$

The *LU* decompositions of these matrices are exactly the same as the original since the matrices are triangular. The printout of the computer response is

```
The results of the first analysis are
     V(1)     =0.5     +j(-0.5)
     V(2)     =-20     +j(0.0)
The results of the second analysis are
     V'(1)    =-20.0 +j(0.0)
     V'(2)    =4.0     +j(4.0)
```

Thus, the results are identical as computed directly in the previous section.

When we modify slightly the circuit of Fig. 5.5 by inclusion of the additional conductance $G' = 0.001$ S and capacitance $C' = 0.0005$ F between nodes 1 and 2, then the real and imaginary parts of admittance matrix become

$$\mathbf{Y}_r = \begin{bmatrix} 1.001 & -0.001 \\ 4.999 & 0.126 \end{bmatrix}, \quad \mathbf{Y}_i = \begin{bmatrix} 1.001 & -0.001 \\ -0.001 & -0.124 \end{bmatrix}$$

After entering $\mathbf{Y}_r$ and $\mathbf{Y}_i$ into the program 'SENSITLU' we obtain the printouts

```
The final LU decomposition is
     1.001    -0.001
     4.999     0.131

     1.001     0.000
    -0.001    -0.124
The results of the first analysis are
     V(1)     =0.480        +j(-0.499)
     V(2)     =-19.5699 +j(0.5406)
The results of the second analysis are
     V'(1)    =-19.5699  +j(0.5406)
     V'(2)    =4.0262      +j(3.8112)
```

Having the node voltages we can easily compute branch voltages and currents and hence the sensitivities. For example, the absolute sensitivity of V_o to changes in G is

$$S_{aG}^{V_o} = 9.366 - j10.024$$

5.5 Sensitivities to parasitic circuit elements

A parasitic element in a circuit is one that has a zero nominal value in an idealized circuit. As long as the circuit is ideal it has no influence on the properties of the network function. However, small deviations from the ideal case render parasitic element values different from zero and then their influence may result in large changes in the network function of the circuit considered. Typical parasitic elements are stray capacitances of a layout (connections of elements), leakage conductances in capacitors, resistances of inductors, etc ... . To compute the absolute sensitivity of the transfer function $T = V_0/I_s$ of any circuit analytically it is first necessary to include the possible parasitic elements in the circuit, calculate the transfer function, calculate the derivative of the transfer function with respect to the parasitic element and then set the value of parasitic element to zero (since zero is the nominal value of the parasitic element).

As an example, consider the simple circuit of Fig. 5.6(a), in which capacitance C_2 is parasitic and calculate the transfer function $T = V_0/I_s$:

$$T = \frac{V_0}{I_s} = V_0 = \frac{G_2 + j\omega C_2}{(G_2 + j\omega C_2)(j\omega C_1 + j\omega C_3) + j\omega C_1 j\omega C_3} = \frac{N}{D} \tag{5.47}$$

The derivative

$$\frac{dT}{dC_2} = \frac{j\omega \times j\omega C_1 \times j\omega C_3}{D^2} \tag{5.48}$$

Inserting the nominal value $C_2 = 0$ yields

$$V_0\big|_{C_2=0} = T\big|_{C_2=0} = \frac{G_2}{G_2(j\omega C_1 + j\omega C_3) + j\omega C_1 j\omega C_3} \tag{5.49}$$

and

$$\frac{dV_0}{dC_2}\bigg|_{C_2=0} = \frac{dT}{dC_2}\bigg|_{C_2=0} = \frac{j\omega \times j\omega C_1 \times j\omega C_3}{(G_2(j\omega C_1 + j\omega C_3) + j\omega C_1 j\omega C_3)^2} \tag{5.50}$$

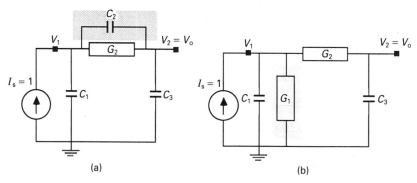

(a)   (b)

Figure 5.6   Simple circuit with (a) parasitic capacitance C_2, (b) parasitic conductance G_1

From the two last relations we see that at the nominal conditions, that is, $C_2 = 0$, the capacitance C_2 has no influence on the transfer function T but influences the sensitivity dT/dC_2. It is necessary to mention that we cannot use the relative sensitivity for parasitic element description since it is zero at nominal conditions. Instead, we can use the semirelative sensitivity as defined by relation (5.4).

Similarly, consider the circuit of Fig. 5.6(b) with parasitic conductance G_1 and calculate the transfer function T and the derivative dT/dG_1:

$$T = \frac{V_o}{I_s} = \frac{G_2}{G_2(G_1 + j\omega C_1 + j\omega C_3) + (G_1 + j\omega C_1)j\omega C_3} = \frac{N}{D} \tag{5.51}$$

$$\frac{dV_o}{dG_1} = \frac{dT}{dG_1} = \frac{-(G_2 + j\omega C_3)G_2}{D^2} \tag{5.52}$$

After inserting $G_1 = 0$, we obtain

$$V_o\big|_{G_1=0} = T\big|_{G_1=0} = \frac{G_2}{G_2(j\omega C_1 + j\omega C_3) + j\omega C_1 j\omega C_3} \tag{5.53}$$

$$\frac{dV_o}{dG_1}\bigg|_{G_1=0} = \frac{dT}{dG_1}\bigg|_{G_1=0} = \frac{-G_2(G_2 + j\omega C_3)}{(G_2(j\omega C_1 + j\omega C_3) + j\omega C_1 j\omega C_3)^2} \tag{5.54}$$

For larger circuits the computational effort will be much higher. Observe, however, when we calculate the absolute sensitivites of V_o with respect to G_2 or C_2 for the circuit of Fig. 5.6(a) and with respect to C_1 or G_1 for the circuit of Fig. 5.6(b), we obtain similar expressions differing only by the factor $j\omega$ once we compute $S_C^{V_o}$ and $S_G^{V_o}$. Or, in other words, if we compute the absolute sensitivities for different admittances which are in parallel, then the expressions are the same.

For Fig. 5.6(a):

$$\frac{dT}{dG_2} = \frac{dT}{dj\omega C_2} = \frac{j\omega C_1 j\omega C_3}{((G_2 + j\omega C_2)(j\omega C_1 + j\omega C_3) + j\omega C_1 j\omega C_3)^2} \tag{5.55}$$

For Fig. 5.6(b):

$$\frac{dT}{dG_1} = \frac{dT}{dj\omega C_1} = \frac{-G_2(G_2 + j\omega C_3)}{(G_2(G_1 + j\omega C_1 + j\omega C_3) + (G_1 + j\omega C_1)j\omega C_3)^2} \tag{5.56}$$

Therefore, for computation of absolute sensitivity, we can use the same computational technique as for normal elements, that is, the adjoint network method. Thus, the absolute sensitivity of V_o with respect to C_2 in Fig. 5.6(a) is

$$S_{aC_2}^{V_o} = \frac{dV_o}{dC_2} = -j\omega(V_1 - V_2)(V_1' - V_2') \tag{5.57}$$

and the absolute sensitivity of V_o with respect to G_1 in Fig. 5.6(b) is

$$S_{aG_1}^{V_o} = \frac{dV_o}{dG_1} = -V_1 V_1' \tag{5.58}$$

where the voltages V' correspond to the adjoint networks. To prove this, let us assume the element values for the circuits of Fig. 5.6:

$$C_1 = 1, \quad G_2 = 1, \quad C_3 = 1, \quad \omega = 1$$

Using the program 'SENSITLU' we obtain the voltages

$$V_1 = \quad 0.2 + j(-0.6)$$
$$V_2 = -0.2 + j(-0.4)$$
$$V_1' = -0.2 + j(-0.4)$$
$$V_2' = \quad 0.2 + j(-0.6)$$

Thus, for the circuit of Fig. 5.6(a):

$$V_{C2} = V_1 - V_2 = 0.4 - j(0.2)$$
$$V_{C2}' = V_1' - V_2' = -0.4 + j(0.2)$$

and

$$S_{aC_2}^{V_o} = -j(0.4 - j0.2)(-0.4 + j0.2) = 0.16 + j0.12$$

Inserting the element values into the expression for dV_o/dC_2 at $C_2 = 0$ yields

$$S_{aC_2}^{V_o}|_{C_2=0} = \frac{dV_o}{dC_2} = \frac{-j}{(j2-1)^2} = \frac{-j}{-3-j4} = 0.16 + j0.12$$

For the circuit of Fig. 5.6(b):

$$V_{G1} = V_1, \quad V_{G1}' = V_1'$$

and:

$$S_{aG_1}^{V_o} = -V_1 V_1' = -(0.2 - j0.6)(-0.2 - j0.4) = +0.28 - j0.04$$

Using directly the formula for dV_o/dG_1 at $G_1 = 0$, we obtain

$$S_{aG_1}^{V_o}|_{G_1=0} = \frac{dV_o}{dG_1} = \frac{-(1+j1)}{(j2-1)^2} = 0.28 - j0.04$$

Similarly, when we have the parasitic resistance R_p in series with inductance L then we can use the formula

$$S_{aR_p}^{V_o} = \frac{dV_o}{dR_p} = I_{Rp} I_{kp}' \tag{5.59}$$

where I_{Rp} and I_{kp}' are currents flowing through resistance R_p in the original and adjoint networks.

5.6 Application of the adjoint network for calculation of the equivalent Thevenin circuit

As it was described in Chapter 2, to find the elements of the Thevenin equivalent circuit for the representation of a given network, it is necessary to compute the open-circuit voltage V_{OC} at two selected nodes and to compute the output impedance seen from the two nodes selected, with independent sources set to zero or removed (i.e. voltage sources shorted and current sources open), as shown in Fig. 5.7.

The network of Fig. 5.7(a) can be described by its nodal admittance matrix $\mathbf{Y}$ and excitation vector $\mathbf{I}$, in which all independent sources are included. Thus, to find V_{OC}, we can use any program for linear equation solution, or procedures for LU decomposition and the forward–backward substitutions from the 'SENSITLU' program 5.1. Then to find the output impedance Z_{out}, as shown in Fig. 5.7(b), it is only necessary to change the excitation vector $\mathbf{I}$ by removing all previous elements and to insert only the value '1' at the position corresponding to the number of the output node (in our case node n).

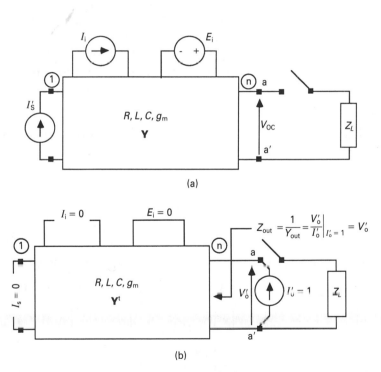

(a)

(b)

Figure 5.7 Network for illustration of Thevenin equivalent circuit computation; (a) E_i and I_i represent the internal independent voltage and current sources for open-circuit voltage V_{OC} computation at node n. (b) E_i and I_i represent internal independent voltage and current sources for output impedance Z_{out} computation

However, the output impedance Z_{out} or admittance, $Y_{out} = 1/Z_{out}$, for a given network described by the nodal admittance $\mathbf{Y}$, can be obtained using the transposed network described by the transposed admittance matrix $\mathbf{Y}' = \mathbf{Y}^t$. For example, for the simple original and transposed networks of Fig. 5.8 we obtain nodal matrices $\mathbf{Y}$ and $\mathbf{Y}' = \mathbf{Y}^t$:

$$\mathbf{Y} = \begin{bmatrix} y_1 + y_2 & -y_2 \\ -y_2 + g_m & y_2 + y_3 \end{bmatrix} \tag{5.60a}$$

$$\mathbf{Y}' = \begin{bmatrix} y_1 + y_2 & -y_2 + g_m \\ -y_2 & y_2 + y_3 \end{bmatrix} = \mathbf{Y}^t \tag{5.60b}$$

With excitation vector $\mathbf{I}^t = [0\ 1]$ we can compute the voltage V_o for original network:

$$V_o = \frac{\begin{vmatrix} y_1 + y_2 & 0 \\ -y_2 + g_m & 1 \end{vmatrix}}{\Delta} = \frac{1 \times (y_1 + y_2)}{(y_1 + y_2)(y_2 + y_3) - (-y_2 + g_m)(-y_2)} \tag{5.61a}$$

and for the transposed network:

$$V_o' = \frac{\begin{vmatrix} y_1 + y_2 & 0 \\ -y_2 & 1 \end{vmatrix}}{\Delta} = \frac{1 \times (y_1 + y_2)}{(y_1 + y_2)(y_2 + y_3) - (-y_2 + g_m)(-y_2)} \tag{5.61b}$$

The results are identical. Thus, the output impedances for both circuits are identical:

$$Z_{out} = \frac{V_o}{1} = V_o = \frac{y_1 + y_2}{\Delta}$$

Similarly, for a three-node circuit excited by the current source $I_o' = 1$, at the output node we obtain the output voltage V_o for the original circuit:

$$V_o = V_3 = \frac{\begin{vmatrix} y_{11} & y_{12} & 0 \\ y_{21} & y_{22} & 0 \\ y_{31} & y_{32} & 1 \end{vmatrix}}{\Delta} = \frac{1 \times (y_{11}y_{22} - y_{12}y_{21})}{\Delta} \tag{5.62a}$$

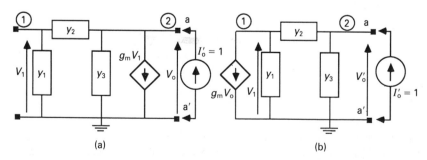

(a)   (b)

Figure 5.8   (a) Simple original circuit. (b) Its transposed circuit

and for the transposed circuit:

$$V_o' = V_3' = \frac{\begin{vmatrix} y_{11} & y_{21} & 0 \\ y_{12} & y_{22} & 0 \\ y_{13} & y_{23} & 1 \end{vmatrix}}{\Delta} = \frac{1 \times (y_{11}y_{22} - y_{12}y_{21})}{\Delta} = V_o \tag{5.62b}$$

But, the analysis of the transposed or adjoint network is performed in the second part of the sensitivity computation in the program 'SENSITLU'.

Thus, when we select the V_{OC} of the Thevenin equivalent circuit as the nth node voltage V_n, then the V_{OC} and Z_{out} of the Thevenin equivalent circuit are obtained by simply running the 'SENSITLU' program. The last voltage V_n of the first analysis corresponds to $V_{OC} = V_n$, and the last voltage V_n' of the second analysis is numerically equal to the output impedance $Z_{out} = V_n'$.

Consider as an example the circuit of Fig 5.9(a) with nodal admittance matrix **Y** (in siemens):

$$\mathbf{Y} = \begin{bmatrix} 3 & -1 & 0 \\ -0.5 & 4 & -1 \\ 0.4 & -1.4 & 3 \end{bmatrix}$$

Entering this matrix into the 'SENSITLU' program we obtain the following results:

$$V_3 = -0.0293 \text{ V}$$

$$V_3' = 0.3746 \text{ V}$$

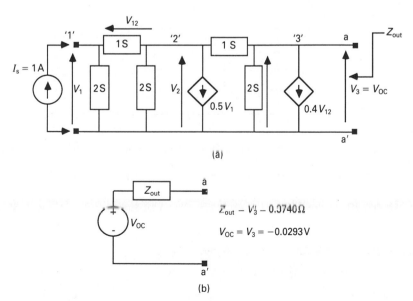

(a)

$$Z_{out} - V_3' - 0.0740\Omega$$

$$V_{OC} = V_3 = -0.0293\text{V}$$

(b)

Figure 5.9 (a) Example active circuit. (b) Its Thevenin equivalent

Thus, $V_{\text{OC}} = -0.0293$ V and $Z_{\text{out}} = V_3^i = 0.3746\ \Omega$, as is shown in the equivalent circuit in Fig. 5.9(b).

Exercises

Exercise 1. Using the definition of relative sensitivity, give an expression as a function of S_x^P for the relative sensitivities of following functions: $S_x^{1/P}$; $S_{1/x}^P$; S_x^{kP}; S_x^{k+P}; S_{kx}^P, k is a constant not depending on x.

Exercise 2. Using the definition of relative sensitivity, give an expression as a function of S_x^P and S_x^Q for the relative sensitivity of following functions: $S_x^{P^2}$; $S_x^{P^n}$; S_x^{PQ}; $S_x^{P/Q}$; $S_x^{P[y(x)]}$; n is an integer not depending on x; Q is a function of x.

Exercise 3. The value of the transfer function $H(s)$, for $s = j\omega$, is written as follows: $H(\omega) = |H(\omega)|\exp(j\phi(\omega))$; $H(\omega)$ is a complex number. Show, using the sensitivity definition, the validity of the following formula: $S_x^H = S_x^{|H|} + jS_{sx}^{\phi}$. Note that the real part is a function of the relative sensitivity and the imaginary part is a function of the semirelative sensitivity.

Exercise 4. In the complex frequency plane, the first-order low pass transfer function is given by $H(s) = K/(1 + \tau s)$; for $s = j\omega$, determine the relative sensitivity of $|H(j\omega)|$ with respect to the time constant τ as a function of ω. Plot the variations of this sensitivity as a function of ω.

Exercise 5. For the first-order circuit ($\tau = RC$) shown in Fig. E5.1, using the program 'SENSITLU', determine the sensitivity of the output voltage V with respect to C for $R = 1\ \Omega$, $C = 1$ F, $\omega = 1$ rad/s. Verify this result using exercise 4.

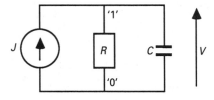

Figure E5.1

Exercise 6. Determine analytically the transfer function V_2/V_1 of the low pass cell shown in Fig. E5.2 using the following form:

$$H(s) = A_{\text{LP}} \frac{\omega_0^2}{s^2 + \dfrac{\omega_0}{Q}s + \omega_0^2}$$

Calculate analytically the following sensitivities:

$$S_{R_i}^{\omega_0}, S_{C_i}^{\omega_0}, S_K^{\omega_0}, S_{R_i}^Q, S_{C_i}^Q, S_K^Q \text{ and } S_{R_i}^{A_{\text{LP}}}, S_{C_i}^{A_{\text{LP}}}, S_K^{A_{\text{LP}}}, \text{i} = 1,2$$

Calculate analytically the sensitivity of H with respect to A_{LP}, ω_0 and Q. The values of the elements are $C_2 = C_1 = 1$ nF; $R_2 = R_1 = 10$ kΩ; $K = 0.5$.

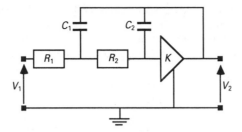

Figure E5.2

Exercise 7.   For the circuit shown in Fig. E5.3, determine analytically the expression giving V_o as a function of J. By differentiation, obtain the expression for the absolute sensitivity of V_o with respect to each resistance. Determine the numerical value of these sensitivities in the case where all resistances have a value of 1 Ω and the current source J is equal to 1 A.

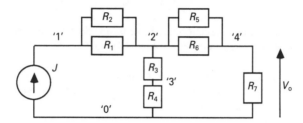

Figure E5.3

Exercise 8.   Using program 'SENSITLU', determine the absolute sensitivities of V_o with respect to all resistances for the circuit shown in Fig. E5.3; all resistances are equal to 1 Ω and the current source J is equal to 1 A; verify these results with exercise 7.

Exercise 9.   For the circuit shown in Fig. E5.4, use the program 'GAUSSJOR' in order to obtain the value of the voltage V_3 for the two following cases:

1. $R_1 = R_2 = R_3 = R_4 = R_5 = 1 \,\Omega$; $g_m = 0.1$ S.
2.   The values of all R_i increase by 10% and the g_m value decreases by 10%.

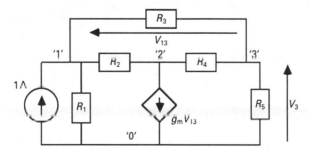

Figure E5.4

Exercise 10.   For the circuit shown in Fig. E5.4 and with numerical values the same as in exercise 9, use program 'SENSITLU' in order to determine the absolute sensitivities of V_3 with respect to R_1, R_2, R_3, R_4, R_5, g_m. Is it possible to compare these results with exercise 9?

Exercise 11.   Using program 'GAUSSJOR' (Chapter 4) verify Tellegen' S theorem for each network shown in Fig. E5.5(a, b). Calculate $\Sigma_k V_k I_k$ and $\Sigma_K V'_k I_k$. Determine whether the two circuits shown in Fig. E5.5(a, b) are adjoint networks, considering the definition given in the text. We assume that the independent voltage sources are in series with a resistor equal to $10^{-6}\,\Omega$.

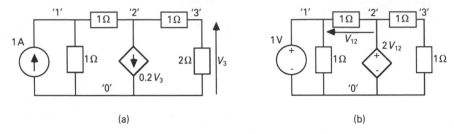

(a)   (b)

Figure E5.5

Exercise 12.   The discussions of Section 5.3 have allowed us to determine the adjoint network with a VCCS; repeat the same developments for a CCCS, a CCVS and a VCVS.

Exercise 13.   Consider Fig. 4.3(a) of Chapter 4. Using the program 'SENSITLU', determine the relative sensitivity of the transfer function V_5/V_1 with respect to the 40 kΩ resistor between the base of T_1 and the emitter of T_2.

Exercise 14.   Using the program 'SENSITLU' for the circuit shown in Fig. E5.6, determine the relative sensitivities of the transfer function V_4/V_1 with respect to the five parameters of the circuit: C_2, C_1, R_2, R_1, K. The numerical values are the same as those used in exercise 6 and $R_g = R_s = 1\,\Omega$; choose $k = 10^{+4}$ and $n = 10^{+5}$.
(Note: the VCVS has to be transformed in the same manner as is described in Section 2.2.3.)

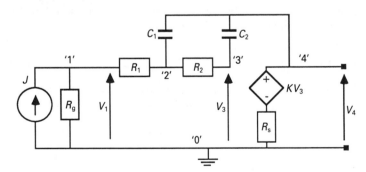

Figure E5.6

Exercise 15.   Fig. E5.7 represents the cascade connection of two-ports each having a given voltage transfer function $H_i(s)$; assuming that no interaction exists between these two-ports, determine the global transfer function $H(s)$.

An element x_{ki} influences the transfer function $H_i(s)$ only. Determine the relative sensitivity $S_{x_{ki}}^H$ as a function of the relative sensitivity $S_{x_{ki}}^{H_i}$. Give some conclusions regarding the order of the cascade connection.

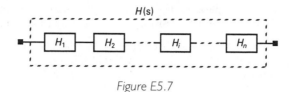

Figure E5.7

Exercise 16. Use the program 'SENSITLU' in order to obtain the absolute sensitivity of power dissipated in the resistance R_3 with respect to the parameters of the circuit shown in Fig. E5.8: $R_1 = R_2 = R_3 = 1\ \Omega$; $C_1 = C_2 = 1\ F$; $J = 1\ A$ and $\omega = 1\ rad/s$.

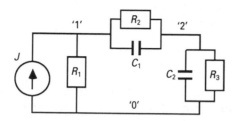

Figure E5.8

Exercise 17. For the second-order band pass network shown in Fig. E5.9, of general form:

$$H(s) = -A_{BP} \frac{\dfrac{\omega_0}{Q}s}{s^2 + \dfrac{\omega_0}{Q}s + \omega_0^2}$$

determine ω_0, A_{BP} and Q in terms of the four parameters R_1, R_2, C_1 and C_2. We assume that the temperature coefficients of the different elements are $\alpha_R = 200\ ppm/°C$ and $\alpha_C = -100\ ppm/°C$. The fabrication processes of these four elements are independent. When the temperature is increased by 50°C, determine the new values of ω_0, A_{BP} and Q using the variations calculation ($R_1 = R_2 = 10\ k\Omega$, $C_1 = 100\ nF$, $C_2 = 1\ nF$).

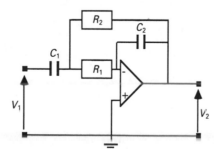

Figure E5.9

Exercise 18. Consider the circuit shown in Fig. E5.10, with $\omega = 1\ rad/s$; $R_g = 0.1\ \Omega$; $R = 1\ \Omega$; $L = 1\ H$; $C = 1\ F$. We want to take into account the parasitic elements modifying each element:

1. resistance R modified by a capacitor C' in parallel,

2.   capacitor C modified by a parasitic resistor in parallel defined by its conductance G',
3.   inductor L modified by a parasitic resistor in series defined by its resistance R'.

Determine analytically the absolute sensitivity of V_0 with respect to the three parasitic elements and compare the results using program 'SENSITLU'.

Consider the variations of these sensitivities as a function of frequency in the frequency range $\omega_1 = 0.1$ rad/s, $\omega_2 = 10$ rad/s.

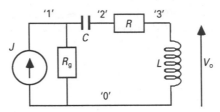

Figure E5.10

Exercise 19. Apply the program 'SENSITLU' in order to determine the Thevenin impedance for the circuits shown in Section 2.2.2 of Chapter 2.

Exercise 20.   Apply the program 'SENSITLU' in order to determine the Thevenin equivalent schema for the example of Section 5.3 shown in Fig. 5.5(a) with the same numerical values. Verify the result with the classical method described in Chapter 2.

Exercise 21.   Modify the program 'SENSITLU' in order to determine the Thevenin equivalent schema for a circuit containing many sources; this program will be named 'THEVENIN'. Apply the program 'THEVENIN' to the examples described in Section 2.2.2 of Chapter 2.

Exercise 22.   Apply the program 'THEVENIN' to the exercises 1, 2 and 3 of Chapter 2 concerning the Thevenin equivalent schema.

6 DC Simulation of Non-linear Circuits

6.1 Introduction

Elements having non-linear characteristics are encountered in many circuits. Diodes and transistors are the best-known examples. The most general current voltage relation of these components is of the form

$$i = g(v_0, v_1, \ldots, v_n) \tag{6.1}$$

where v_0 represents the terminal voltage of the non-linear component considered and i the current flowing through it; $v_1, \ldots, v_n$ are the voltages relative to other branches $1, 2, \ldots, n$ of the circuit.

Based on equation (6.1), many types of component can be identified:

1. $i = g(v_0)$: i is dependent only on the terminal voltage; the element considered is therefore a voltage controlled resistor (controlled by v_0). A typical example is the Boltzmann diode represented by the following relation (see Section 1.2):

    $$i = I_S \left[\exp\left(\frac{v_0}{V_T}\right) - 1 \right] \tag{6.2}$$

    where i is the current flowing through the diode and v_0 is the voltage across the diode.

2. $i = g(v_j)$, $j \neq 0$: the component is a non-linear voltage controlled current source and v_j is the controlling voltage. This is, for example, the case for the collector current of the bipolar transistor (see Section 1.3):

    $$i = i_C = I_S \left[\exp\left(\frac{v_{BE}}{V_T}\right) - \exp\left(\frac{v_{BC}}{V_T}\right) \right] - \frac{I_S}{\beta_R} \exp\left(\frac{v_{BC}}{V_T}\right) \tag{6.3}$$

    The current i_C flows from the collector to the emitter; the controlling voltages are the base–emitter v_{BE} and the base-collector v_{BC}.

3. $i = g(v_0, v_j)$: this is a more general case. For example, the drain current of a MOS transistor in the ohmic region is of the form (see Section 1.4)

    $$i_D = \beta \left(v_{GS} - V_{Th} - \frac{v_{DS}}{2} \right) v_{DS} \tag{6.4}$$

i_D represents the current flowing from the drain to the source. This current is controlled by the drain–source voltage, v_{DS}, and also by the gate–source voltage, v_{GS}.

These non-linear elements lead to a non-linear system for which an analytic solution is generally impossible.

In order to simplify this complex problem, we will consider two basic examples corresponding to non-linear one-dimensional equations:

1. for a diode biasing circuit (Fig. 6.1(a)),
2. for a MOS transistor biasing circuit (Fig. 6.1(b)).

In Section 6.2, we present the solution of non-linear equations and in the following sections, we present the 'iterative companion model' of non-linear elements forming part of a complex network.

For more complex circuits, as shown in Fig. 6.2(a), application of the Thevenin theorem between points A and B permits us to obtain the equation to be solved easily:

$$v = E_{Th} - R_{Th}i \quad \text{or} \quad v = F(v) = E_{Th} - R_{Th}i$$

with

$$E_{Th} = \frac{R_1 R_3}{R_1 + R_2 + R_3} J \quad \text{and} \quad R_{Th} = \frac{(R_1 + R_2)R_3}{R_1 + R_2 + R_3}$$

To find the bias point (i, v) of the diode shown in Fig. 6.1(a), it is necessary to

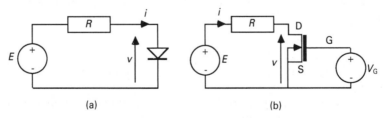

(a)   (b)

Figure 6.1   (a) Diode biasing circuit. (b) MOS transistor biasing circuit

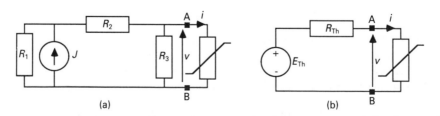

(a)   (b)

Figure 6.2   (a) Non-linear network. (b) Thevenin equivalent circuit

solve the following equations:

$$\begin{cases} i = I_S\left[\exp\left(\dfrac{v}{V_T}\right) - 1\right] \\[2ex] i = \dfrac{E - v}{R} \end{cases} \tag{6.5}$$

Exact analytical solution of this system is impossible. It is therefore necessary to consider an approximate numerical solution. If we consider the voltage as the main unknown, we obtain the following equations:

$$I_S\left[\exp\left(\frac{v}{V_T}\right) - 1\right] - \frac{E}{R} + \frac{v}{R} = 0 \tag{6.6a}$$

or

$$v = E - RI_S\left[\exp\left(\frac{v}{V_T}\right) - 1\right] \tag{6.6b}$$

In the same way as for the MOS transistor biasing circuit shown in Fig. 6.1(b), we obtain the following equations (equivalent to the previous form, with $v = v_{DS}$):

$$i - \frac{E}{R} + \frac{v}{R} = 0 \tag{6.7a}$$

$$v = E - Ri \tag{6.7b}$$

where i is given by the following relations (Section 1.4):

$$i = 0 \qquad\qquad\qquad \text{if } V_G < V_{Th} \tag{6.8a}$$

$$i = \beta\left(V_G - V_{Th} - \frac{v}{2}\right)v \quad \text{if } V_G \geqslant V_{Th} \text{ and } V_G - v \geqslant V_{Th} \tag{6.8b}$$

$$i = \frac{\beta}{2}(V_G - V_{Th})^2 \qquad \text{if } V_G \geqslant V_{Th} \text{ and } V_G - v < V_{Th} \tag{6.8c}$$

In the same way, for the circuit shown in Fig. 6.1(a), considering the current i as a main unknown, we obtain the following equations:

$$i = I_S\left[\exp\left(\frac{E - Ri}{V_T}\right) - 1\right] \quad \text{or} \quad i - I_S\left[\exp\left(\frac{E - Ri}{V_T}\right) - 1\right] = 0 \tag{6.9}$$

This chapter is dedicated to the methods used for the solution of non-linear systems: the fixed point method, the Newton–Raphson method and the third-order algorithms.

6.2 Iterative solution of a non-linear algebraic system

Consider the following non-linear equation of variable x:

$$x = F(x) \quad \text{or} \quad x - F(x) = f(x) = 0 \tag{6.10}$$

$F(x)$ and $f(x)$ represent mathematical functions of the variable x having derivative functions.

Let x_S be the solution of this equation. The problem is, therefore, to find an iterative method to calculate the solution, x_S, of equation (6.10).

An iterative method starts by choosing an initial value x_0. This value is then recalculated a finite number of times, N, according to an iterative rule, until an approximate solution $x_{N+1} \approx x_S$, having a small relative error ε, is reached. There are many algorithms for determining x_S.

The method's order depends on the computation complexity of each iteration:

1. Order 1: calculation of $f(x)$ only (fixed point algorithm).
2. Order 2: calculation of $f(x)$ and first derivative $f'(x)$ (Newton–Raphson algorithm and associated methods).
3. Order 3: calculation of $f(x)$, $f'(x)$ and second derivative $f''(x)$.

We will pay special attention, with some examples, to problems related to the initialization and convergence of iterative methods.

6.2.1 The fixed point algorithm

This algorithm is very interesting for its simplicity; moreover, it permits us to familiarize ourselves with problems concerning iterative algorithms: calculation time and convergence. It is defined, for the function $x = F(x)$, by the following iteration sequence:

```
Enter x₀ (initial value)
x₁ = F(x₀)
x₂ = F(x₁)
  .
xⱼ₊₁ = F(xⱼ)                                          (6.11)
  .
xₙ₊₁ = F(xₙ)
Stop if |xₙ₊₁ - xₙ| ≤ ε|xₙ|
Then xₛ = xₙ₊₁
```

Fig. 6.3(a–d) shows the graphical interpretation of this algorithm.

The program 'POINFIX1', shown in Program 6.1, permits us to examine the different steps of the calculation of the fixed point algorithm for the $F(x)$ functions corresponding to the circuits of Fig. 6.1, for different initial values. In the program

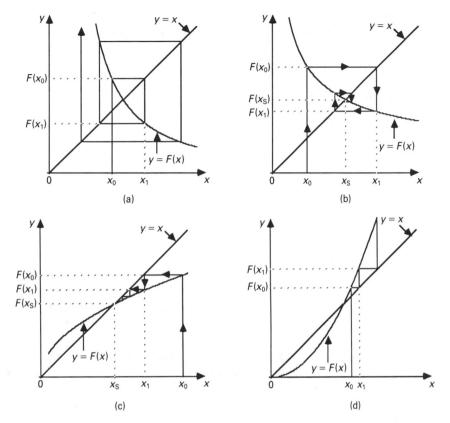

Figure 6.3 Graphical interpretations of the fixed point algorithm. (a) Divergence in spiral. (b) Convergence in spiral. (c) Convergence in steps. (d) Divergence in steps

the following two kinds of function are used:

1. the function describing the diode circuit, represented by equation (6.6),
2. the function describing the MOS transistor circuit, represented by equations (6.7) and (6.8).

See the listing of the program POINFIX1 Directory of Chapter 6 on attached diskette.

The results obtained show important convergence problems. For practical use of this algorithm, convergence must be improved. Indeed, the diode equation is impossible to solve (even when the initial condition is equal to the solution!). The MOS transistor equation, depending on the parameter values, sometimes converges but presents some problems. Note that this algorithm is rarely used.

Fixed point algorithm convergence depends on the initial value and the form of the function $F(x)$. Many methods exist for improving algorithm convergence but only a few are used in practice. The Wegstein method is, for example, a good one. It modifies the successive substitution method (equation (6.11)).

Instead of

$$x_{j+1} = F(x_j) \quad j = 0, 1, \ldots, N$$

the following sequence is used:

$$x_j = F(x_{j-1}) \tag{6.12a}$$

$$\Delta x = F(x_j) - x_j \tag{6.12b}$$

$$\Delta = \frac{F(x_j) - x_j}{x_j - X_{j-1}} \tag{6.12c}$$

$$\alpha = \frac{1}{1 - \Delta} \tag{6.12d}$$

$$x_{j+1} = x_j + \alpha \, \Delta x \tag{6.12e}$$

To improve the convergence, a relaxation parameter, α, is introduced. Considering relations (6.12), this parameter value has to be calculated for every iteration. To reduce the calculation time, α can be taken as constant for all iterations. The program 'POINFIX2' shown in Program 6.2 uses the Wegstein method.

> **See the listing of the program POINFIX2 Directory of Chapter 6 on attached diskette.**

Consider again the previous examples which present convergence problems in program 'POINFIX1'.

For the diode function analysis (the solution is $v_S = 0.6370$ V), we obtain the correct solution only for an initial value exactly equal to the solution, that is, $v_0 = v_S = 0.6370$ V. The convergence problems are still encountered since the 'Fdiode' function used in the program has an exponential form which varies quickly.

For the MOS function analysis (the solution is $v_S = 0.705$ V for $\beta = 8 \times 10^{-3}$; ohmic region) we obtain correct solutions after seventeen iterations for $v_0 = 0$ V, after six iterations for $v_0 = 1$ V, and after twenty-six iterations for $v_0 = 10$ V. In the case of the saturated region ($\beta = 10^{-3}$, $v_S = 8$ V) the solution is found with a maximum of three iterations for initial values between 0 and 10 V.

It can be seen that the Wegstein method permits us to improve initialization and convergence for the MOS transistor.

In conclusion, the fixed point algorithm can be used for functions with small variations such as quadratic functions.

6.2.2 The Newton–Raphson algorithm

The Newton–Raphson algorithm based on Taylor series expansion around an estimated x_j value is the most widely used. Let us consider the general equation

$$f(x) = x - F(x) = 0 \tag{6.13}$$

and the following Taylor series expansion:

$$f(x_S) = f(x_j) + \frac{df(x)}{dx}\bigg|_{x_j} (x_S - x_j) + \frac{1}{2!} \frac{d^2 f(x)}{dx^2}\bigg|_{x_j} (x_S - x_j)^2$$

$$+ \frac{1}{3!} \frac{d^3 f(x)}{dx^3}\bigg|_{x_j} (x_S - x_j)^3 + \dots \tag{6.14}$$

If x_j is near the solution x_S of the function $f(x) = 0$, we can make the following approximation:

$$f(x_S) = f(x_j) + f'(x_j)(x_S - x_j) \tag{6.15}$$

where

$$f'(x_j) = \frac{df(x)}{dx}\bigg|_{x_j}$$

Assuming that $f(x_S) = 0$, we can write

$$x_S = x_j - \frac{f(x_j)}{f'(x_j)} \tag{6.16}$$

Thus, we can assume that a closer value of x_S is

$$x_{j+1} = x_j - \frac{f(x_j)}{f'(x_j)} \tag{6.17}$$

Thus, the Newton–Raphson algorithm is defined from an initial value x_0 by the following iterative equation:

$$x_{j+1} = x_j - \frac{f(x_j)}{f'(x_j)} \qquad j = 0, 1, \dots, N \tag{6.18}$$

where $f'(x)$ represents the derivative function of $f(x)$. The iterative sequence will be as follows:

```
Enter x0
```

$$x_1 = x_0 - \frac{f(x_0)}{f'(x_0)}$$

$$x_2 = x_1 - \frac{f(x_1)}{f'(x_1)}$$

$$\cdot$$

$$x_{j+1} = x_j - \frac{f(x_j)}{f'(x_j)} \tag{6.19}$$

$$\cdot$$

$$x_{N+1} = x_N - \frac{f(x_N)}{f'(x_N)}$$

Stop if $|x_{N+1} - x_N| \leqslant \varepsilon |x_N|$
Thus $x_S = x_{N+1}$

From equation (6.18), we can see that if the value of the derivative (denoted by $f'(x_j)$) at the point $[x_j, f(x_j)]$ is calculated, then this straight line corresponding to $f'(x_j)$ crosses the horizontal axis, $y = f(x) = 0$ at the value x_{j+1}; hence, the graphical representation of the algorithm is shown in Fig. 6.4(a) and (b). The program 'NEWTRAP', shown in Program 6.3, permits us to consider the different steps of the Newton–Raphson algorithm.

> **See the listing of the program NEWTRAP Directory of Chapter 6 on attached diskette.**

As in the case of the fixed point algorithm, two kinds of function are used, describing the diode circuit and the MOS transistor circuit of Fig. 6.1.

Diode circuit analysis. The diode circuit equation takes the following form:

$$I_S \left[\exp\left(\frac{v}{V_T}\right) - 1 \right] - \frac{E}{R} + \frac{v}{R} = 0$$

(represented in the program by the function 'Fdiode(IS,E,R,v)'). With the following parameter values: $V_T = 0.026$ V, $I_S = 10^{-13}$ A, $E = 5$ V and $R = 1000\ \Omega$, the result to be obtained is $v_S = 0.637$ V.

Running the program we obtain the correct solution ($v_S = 0.637$ V) after 172 iterations for the initial value $v_0 = 0$ V, after six iterations for $v_0 = 0.6$ V and after seventeen iterations for $v_0 = 1$ V. The results obtained show that, in the case

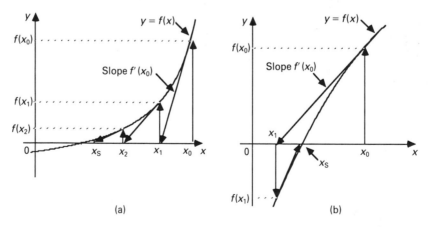

(a)   (b)

Figure 6.4 Graphical representation of the Newton–Raphson algorithm for (a) convex function, (b) concave function

of an exponential function, the Newton–Raphson algorithm gives an accurate solution.

MOS circuit analysis. The MOS circuit function takes the form $i(v) - E/R + v/R = 0$ (represented in the program by the function 'Fmos(BETA,VTh,VG,E,R,v)'). With the following parameter values: $V_{Th} = 1$ V, $V_G = 3$ V, $E = 10$ V and $R = 1000$ Ω, the results obtained are listed in Table 6.1.

In conclusion, we note that the diode function is solved even if the initial conditions are very different from the solution ('$v_0 = 0$ V' requires 172 iterations but convergence is obtained). The MOS function converges much more quickly. In the case where $\beta = 8 \times 10^{-3}$, only three iterations are necessary instead of six for the fixed point algorithm for '$v_0 = 1$ V'. If '$v_0 = 10$ V', the Newton–Raphson algorithm requires eight iterations, while twenty-six iterations are necessary for the fixed point algorithm.

These results are in agreement with the theory. Indeed, it can be demonstrated, considering equations (6.11) and (6.19), that the convergence of the fixed point algorithm is of order 1, and the convergence of the Newton–Raphson algorithm is of order 2.

An advantage of the Newton–Raphson algorithm is its quadratic convergence. However, many problems can exist. These problems are dependent on the form of $f(x)$. Some examples are shown in Fig. 6.5(a, b):

1. Fig. 6.5(a) shows a polynomial function (degree 4) with four 'zeros': x_1, x_2, x_3 and x_4. An initial value x_{i1} of the interval $[x_2, x_3]$ gives the solution x_1; an initial value x_{i2} of the interval $[x_3, x_4]$ gives the solution x_2; while an initial value x_{i3} near x_{i2} gives the solution x_4.
2. Fig. 6.5(b) shows the same function with other initial values chosen giving $f'(x_{02}) = 0$ and $f'(x_{02}') = 0$. Taking equation (6.17) into account, the algorithm cannot converge in these cases.

These examples show that it is very important to choose a 'good' initial value.

6.2.3 The associated Newton–Raphson algorithms

The Newton–Raphson algorithm is defined by the equation sequence (6.18). When the solution x_S is reached, $f(x_S) = 0$ and $x_{j+1} = x_j$ (with a relative error ε) are the two relations to be satisfied. These two conditions are independent of the accuracy of the computation of $f'(x_j)$. The derivative function often cannot be easily obtained analytically and the determination of this function requires a lot of calculation time. Thus, sometimes, approximations or numerical methods are necessary to calculate $f'(x_j)$. Some algorithms based on the Newton–Raphson algorithm are presented in this section.

The Chord method. In this method, the non-linear $f(x)$ function is modelled by a straight line with a slope M ('Chord') passing the previous iterative value. The

Table 6.1 Results of the MOS circuit of Fig. 6.1 analysis

| $\beta = 8 \times 10^{-3}$; $v_S = 0.705$ V (ohmic region) | | | | | | $\beta = 10^{-3}$; $v_S = 8$ V (saturated region) | | | | | |
| $v_0 = 0$ V | | $v_0 = 1$ V | | $v_0 = 10$ V | | $v_0 = 0$ V | | $v_0 = 1$ V | | $v_0 = 10$ V | |
iter.n^o	v_i value	iter.n^o	v_i value	iter.n^o	v_i value	iter.n^o	v_i value	iter.n^o	v_i value	iter.n^o	v_i value
0	0	0	0	0	10	0	0	0	1	0	10
1	0.5882	1	0.6667	1	−6	1	3.3333	1	4.7500	1	8
2	0.7008	2	0.7048	2	−2.0615	2	8	2	8	2	8
3	0.7053	3	0.7053	3	−0.2090	3	8	3	8		
4	0.7053	4	0.7053	4	0.5262						
				5	0.6352						
				6	0.7052						
				7	0.7053						

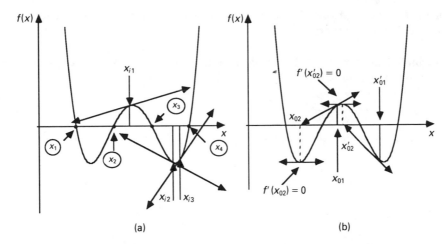

(a)   (b)

Figure 6.5 Examples showing a few convergence problems depending on the Newton–Raphson algorithm. (a) Different solutions for different initial values. (b) The derivative values $f'(x_{02}) = 0$ and $f'(x_{02}') = 0$ are obtained for initial values x_{01} and x_{01}', respectively

M value (chosen for the first iteration) does not vary for the other iterations. The iterative equation is as follows:

$$x_{j+1} = x_j - \frac{f(x_j)}{M}$$ (6.20)

where M is a constant value (correct values depend on the derivative of the function considered). The Chord method is easy to use because the derivative function is not calculated. However, its convergence is linear.

The von-Mises method. This algorithm is similar to the Chord algorithm, except that the M value is equal to $f'(x_0)$. The iterative equation is as follows:

$$x_{j+1} = x_j - \frac{f(x_j)}{f'(x_0)}$$ (6.21)

The secant method. In the 'secant method', the derivative function, $f'(x_j)$, of the Newton–Raphson algorithm is replaced by an approximation:

$$f'(x_j) = \frac{f(x_j) - f(x_{j-1})}{x_j - x_{j-1}}$$ (6.22a)

and the following iterative equation is obtained:

$$x_{j+1} = x_j - \frac{x_j - x_{j-1}}{f(x_j) - f(x_{j-1})} f(x_j)$$ (6.22b)

The convergence order of the secant method is fractional: 1.666. In the case of problems for which the derivative function is difficult to obtain, the secant method can be used to advantage.

The variable Chord method. In the variable Chord method, $f'(x_j)$ is approximated by the following expression:

$$f'(x_j) = \frac{f(x_j) - f(0)}{x_j - 0} \tag{6.23a}$$

and the iterative equation becomes

$$x_{j+1} = x_j - \frac{x_j}{f(x_j) - f(0)} f(x_j) \tag{6.23b}$$

6.2.4 Third-order algorithms

In this section, we present two 'order 3' algorithms: the Richmond and the Traub algorithms. However, these are difficult to use for complex non-linear circuit analysis. The program 'ORDER3' shown in Program 6.4 using the 'Fdiode' and 'Fmos' functions permits us to demonstrate the two algorithms presented here and to compare them with the Newton–Raphson method.

See the listing of program ORDER3 Directory of Chapter 6 on attached diskette.

The Richmond method. Assuming that $f(x)$ is continuous and $f'(x)$ and $f''(x)$ exist, the Richmond algorithm can be used for the solution of a non-linear equation. The iterative algorithm is as follows:

$$x_{j+1} = x_j - \frac{2f(x_j)f'(x_j)}{2f'^2(x_j) - f(x_j)f''(x_j)} \tag{6.24}$$

It is clear that it is an order 3 method because it uses first and second derivatives. Its convergence rate is bigger than the Newton–Raphson rate but it is necessary to determine $f''(x)$ analytically and to calculate its value for every iteration.

The procedure 'Richmond' in program 'ORDER3', applied to the diode circuit function, permits us to illustrate this method. Running the program, we obtain the correct solution after three iterations for $v_0 = 0.6$ V, after four iterations for $v_0 = 0.7$ V (six iterations are necessary using the Newton–Raphson algorithms for the two cases), and after eighty-eight iterations for $v_0 = 0$ V and -10 V (the Newton–Raphson algorithm requires 172 iterations). So this method converges more quickly than the Newton–Raphson method.

The Traub method. The iterative equations of the Traub algorithm are as follows:

$$z_j = x_j - \frac{f(x_j)}{f'(x_j)} \tag{6.25}$$

and

$$x_{j+1} = z_j - \frac{z_j - x_j}{2f(z_j) - f(x_j)} f(z_j) \tag{6.26}$$

The procedure 'Traub' in program 'ORDER3', applied to the diode circuit function, permits us to illustrate this method. Running the program, we obtain the correct solution after three iterations for $v_0 = 0.6$ V and $v_0 = 0.7$ V (six iterations are necessary using the Newton–Raphson algorithms for the two cases), after eighty-eight iterations for $v_0 = 0$ V and after fourteen iterations for $v_0 = -10$ V (the Newton–Raphson algorithm requires 172 iterations in the two cases).

This method seems to be especially well adapted to diode functions because it requires few iterations (compared with the Newton–Raphson or the Richmond algorithms). Moreover, it is easy to use.

6.2.5 Conclusion

The methods presented in Section 6.2 lead to the same conclusions: the convergence conditions strongly depend on the continuity and monotony of $f(x)$ and on the initial estimated value, x_0.

The Newton–Raphson method having quadratic convergence is used mostly. It is well suited to exponential functions. The modified methods are interesting only if $f'(x)$ cannot be calculated analytically. The order 3 methods require fewer iterations but more calculations for each iteration. They can only be used for simple circuit analysis.

6.3 Circuit interpretation of iterative algorithms

The numerical methods discussed so far for non-linear circuit analysis are based on linearization of the non-linear equation describing all the circuit. This approach is simple when we have only one non-linear element in the circuit. In such a case we can extract this single non-linear element and replace the rest of the linear part of the circuit by its Thevenin or Norton equivalent, as shown in Fig. 6.6. Then, for Thevenin and Norton circuits, we obtain the relations

$$\text{(KVL)} \quad E_{\text{Th}} - R_{\text{Th}} i_k = v_k \quad \text{and} \quad i_k = g_k(v_k) \tag{6.27a}$$

$$\text{(KCL)} \quad J_{\text{Nor}} - Y_{\text{Nor}} v_k = i_k \quad \text{and} \quad v_k = g_k^{-1}(v_k) \tag{6.27b}$$

However, the situation becomes very complex when we have many non-linear one-ports and two-ports. In such situations, it is much easier to linearize only non-linear circuit branches and to create 'iterative companion models' to obtain a fully linear circuit (in one iteration); then any linear equation solution method can be used to find the nodal voltages of the analyzed circuit, and, afterwards, any branch voltages and/or branch currents. After finding branch voltages and currents in one iteration

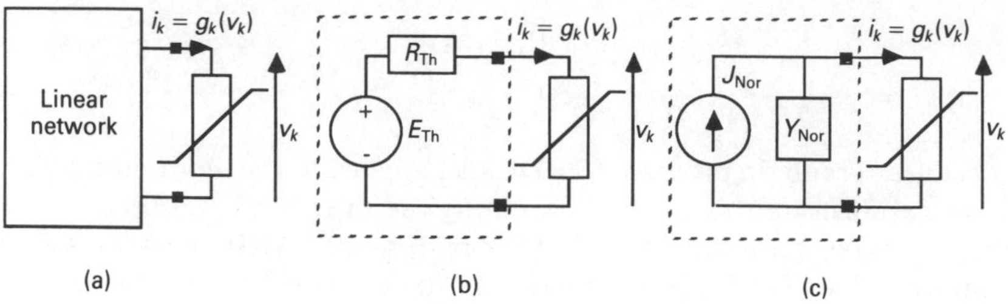

Figure 6.6   (a) Network containing non-linear resistor. (b) Thevenin Equivalent (from resistor nodes). (c) Norton Equivalent

step, the elements of the companion models are updated and the iterative process is repeated until the next iteration values are almost equal to the previous ones.

Since the Newton–Raphson algorithm is mostly used in non-linear circuit analysis and simulation, we restrict our discussion to this method.

Let us take into account the non-linear branch relationship (for the kth branch):

$$i_k = g_k(v_k) \tag{6.28}$$

If we denote by $v_k^{(j)}$ and $v_k^{(j+1)}$ the estimated branch voltages, on this non-linear element, in the jth and $(j + 1)$st iteration steps, and by $i_k^{(j)}$ and $i_k^{(j+1)}$ the currents corresponding to voltages $v_k^{(j)}$ and $v_k^{(j+1)}$, then relation (6.28) can be rewritten as

$$i_k^{(j)} = g_k(v_k^{(j)}) \tag{6.29a}$$

$$i_k^{(j+1)} = g_k(v_k^{(j+1)}) \tag{6.29b}$$

Taylor series expansion of $i_k^{(j+1)} = g_k(v_k^{(j+1)})$ around point $v_k^{(j)}$ (as in the Newton–Raphson method) gives

$$g_k(v_k^{(j+1)}) = i_k^{(j+1)} = g_k(v_k^{(j)}) + \frac{dg_k}{dv_k}\bigg|_{v_k^{(j)}} (v_k^{(j+1)} - v_k^{(j)})$$

$$+ \frac{1}{2!} \frac{d^2 g_k}{dv_k^2}\bigg|_{v_k^{(j)}} (v_k^{(j+1)} - v_k^{(j)})^2 + \dots \tag{6.30}$$

Neglecting the non-linear terms on the right-hand side of (6.30) and reordering the linear terms gives

$$i_k^{(j+1)} = g_k(v_k^{(j)}) - \frac{dg_k}{dv_k}\bigg|_{v_k^{(j)}} v_k^{(j)} + \frac{dg_k}{dv_k}\bigg|_{v_k^{(j)}} v_k^{(j+1)} \tag{6.31a}$$

or

$$i_k^{(j+1)} = J_k^{(j)} + G_k^{(j)} v_k^{(j+1)} \tag{6.31b}$$

where

$$J_k^{(j)} = g_k(v_k^{(j)}) - \frac{dg_k}{dv_k}\bigg|_{v_k^{(j)}} v_k^{(j)} = i_k^{(j)} - G_k^{(j)} v_k^{(j)} \tag{6.32a}$$

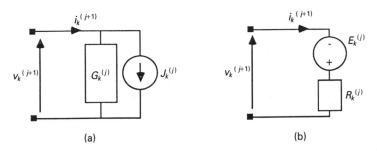

Figure 6.7   Iterative companion models. (a) Norton type or 'i type'. (b) Thevenin type or 'v type'

$$G_k^{(j)} = \left.\frac{\mathrm{d}g_k}{\mathrm{d}v_k}\right|_{v_k^{(j)}} - \text{ incremental conductance of the non-linear}$$
$$\text{one-port}$$
(6.32b)

Based on relation (6.31b) we can draw the 'iterative companion model' directly for the $(j + 1)$st iteration steps, as is shown in Fig. 6.7(a). This is the circuit interpretation of the Newton–Raphson method.

The Norton-type iterative companion model can be converted to the Thevenin type (if necessary – however, it is rarely used), as is shown in Fig. 6.7(b), with

$$R_k^{(j)} = \frac{1}{G_k^{(j)}}$$
(6.33a)

$$E_k^{(j)} = \frac{J_k^{(j)}}{G_k^{(j)}} = -v_k^{(j)} + R_k^{(j)}i_k^{(j)}$$
(6.33b)

To linearize any non-linear circuit with many non-linear one-ports, it is necessary to substitute each non-linear element by its iterative companion model with the corresponding $i = g(v)$ relation.

6.4 Application of the 'iterative companion model'

Let us consider the circuit shown in Fig. 6.8(a), in which branch 4 contains a non-linear component described by different constitutive equations $g(v_4)$:

Convex type: $i_4 = v_4^2$ for $v_4 > 0$; $i_4 = 0$ for $v_4 \leqslant 0$ (6.34a)

Concave type: $i_4 = \sqrt{v_4}$ for $v_4 > 0$; $i_4 = 0$ for $v_4 \leqslant 0$ (6.34b)

Hyperbolic type: $i_4 = \dfrac{0.1}{1.1 - v_4}$ for $v_4 < 1.1$; $i_4 = 0$ for $v_4 > 1.1$ (6.34c)

According to the definition of the 'iterative companion model' associated with the Newton–Raphson algorithm (see equation (6.32)), we can find the values for $G_4^{(j)}$ and $J_4^{(j)}$ at the jth iteration.

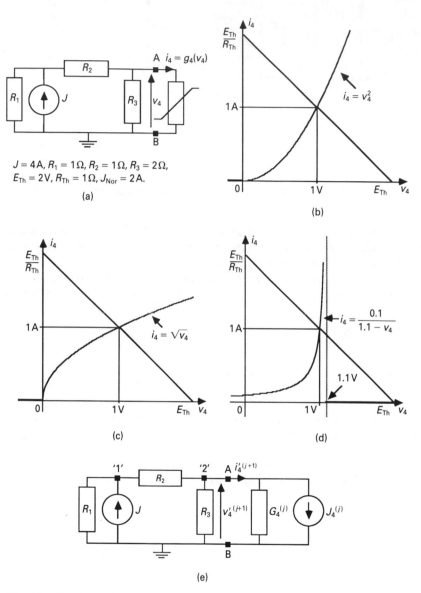

Figure 6.8 (a) Circuit with non-linear element described by functions $i_4 = g(v_4)$. (b) Convex type. (c) Concave type. (d) Hyperbolic type. (e) Equivalent circuit with 'iterative companion model'

For relation (6.34a)

$$G_4^{(j)} = 2v_4^{(j)} \quad \text{for } v_4 > 0; \quad G_4^{(j)} = 0 \quad \text{for } v_4 \leqslant 0 \qquad (6.35a)$$

$$J_4^{(j)} = -\{v_4^{(j)}\}^2 \quad \text{for } v_4 > 0; \quad J_4^{(j)} = 0 \quad \text{for } v_4 \leqslant 0$$

For relation (6.34b):

$$G_4^{(j)} = \frac{1}{2\sqrt{v_4^{(j)}}} \qquad \text{for } v_4 > 0; \qquad G_4^{(j)} = 0 \quad \text{for } v_4 \leqslant 0 \qquad (6.35b)$$

$$J_4^{(j)} = \sqrt{v_4^{(j)}} - \frac{v_4^{(j)}}{2\sqrt{v_4^{(j)}}} \qquad \text{for } v_4 > 0; \qquad J_4^{(j)} = 0 \quad \text{for } v_4 \leqslant 0$$

For relation (6.34c):

$$G_4^{(j)} = \frac{0.1}{\{1.1 - v_4^{(j)}\}^2} \qquad \text{for } v_4 > 0; \qquad G_4^{(j)} = 0 \qquad \text{for } v_4 \leqslant 0$$

$$(6.35c)$$

$$J_4^{(j)} = \frac{0.1}{1.1 - v_4^{(j)}} - \frac{0.1v_4^{(j)}}{\{1.1 - v_4^{(j)}\}^2} \qquad \text{for } v_4 > 0; \qquad J_4^{(j)} = 0 \qquad \text{for } v_4 \leqslant 0$$

These models are described in the 'Comp_Square', 'Comp_Root' and 'Comp_Hyperbo' procedures in the program 'COMP_NR' shown in Program 6.5. The computational method is based on the following algorithm:

```
Data Input : J = 4A , R₁ = 1Ω , R₂ = 1Ω , R₃ = 2Ω , ε = 10⁻⁴
Non-linear element initialization :{v₄⁽⁰⁾ and i₄⁽⁰⁾ =
g₄(v₄⁽⁰⁾)}or{i₄⁽⁰⁾ and v₄⁽⁰⁾ = g₄⁻¹(i₄⁽⁰⁾)}
j=0
Iterative calculation (from previous iteration or
                                  initialization) :
    Calculation of values of "iterative Companion Model"
                                  for {i₄⁽ʲ⁾, v₄⁽ʲ⁾}
    Calculation of operating point of linear circuit by
    solving a linear system (the resistor being replaced by
    its "iterative Companion Model" : {iℓ⁽ʲ⁺¹⁾, vℓ⁽ʲ⁺¹⁾}
    Determination of {i₄⁽ʲ⁺¹⁾, v₄⁽ʲ⁺¹⁾} : Choice of
    "Horizontal" or "Vertical" method
    Convergence test: |v₄⁽ʲ⁺¹⁾ - v₄⁽ʲ⁾|≤ε|v₄⁽ʲ⁾|or (and)
    |i₄⁽ʲ⁺¹⁾ - i₄⁽ʲ⁾|≤ε|i₄⁽ʲ⁾|
    if test = False j=j+1 : CONTINUE
    else : STOP
For N=j : Result {i₄⁽ᴺ⁾, v₄⁽ᴺ⁾}
```

The calculation of the operating point is performed using the nodal method

$$
\begin{bmatrix}
\dfrac{1}{R_1} + \dfrac{1}{R_2} & -\dfrac{1}{R_2} \\[2mm]
-\dfrac{1}{R_2} & \dfrac{1}{R_2} + \dfrac{1}{R_3} + G_4^{(j)}
\end{bmatrix}
\begin{bmatrix} v_{(1)}^{(j+1)} \\[1mm] v_{(2)}^{(j+1)} \end{bmatrix}
=
\begin{bmatrix} J \\[1mm] -J_4^{(j)} \end{bmatrix}
\tag{6.36}
$$

and is computed in the 'Calcul' procedure of the program 'COMP_NR'. In expression (6.36), $v_{(1)}^{(j+1)}$ and $v_{(2)}^{(j+1)}$ represent, respectively, the voltages at nodes 1 and 2 (referenced to the ground node) in Fig. 6.8(a). From this expression, the branch voltage $v_4'^{(j+1)}$ and branch current $i_4'^{(j+1)}$ can easily be calculated:

$$
v_4'^{(j+1)} = v_{(2)}^{(j+1)}
\tag{6.37a}
$$

$$
i_4'^{(j+1)} = J_4^{(j)} + G_4^{(j)} v_4'^{(j+1)}
\tag{6.37b}
$$

Depending on the choice of the method, we obtain

$$
\text{Vertical method:} \qquad v_4^{(j+1)} = v_4'^{(j+1)} \quad i_4^{(j+1)} = g(v_4^{(j+1)})
\tag{6.38a}
$$

$$
\text{Horizontal method:} \quad i_4^{(j+1)} = i_4'^{(j+1)} \quad v_4^{(j+1)} = g_4^{-1}(i_4^{(j+1)})
\tag{6.38b}
$$

The program 'COMP_NR' permits us to study the three functions (6.34) ('Square', 'Root' and 'Hyperbo' functions) choosing the horizontal or vertical method.

See the listing of the program COMP_NR Directory of Chapter 6 on attached diskette.

With the numerical values shown in Fig. 6.8(a), the operational point of the three circuits is $v_4 = 1$ V and $i_4 = 1$ A. Note that relation (6.34c) leads to additional operating points $v_4 = 2$ V and $i_4 = 0$ A. The two other relations imply a current equal to zero for negative or zero voltages. Table 6.2 summarizes the results obtained as a function of the method used and the initial point. The results show some problems that may occur in practice.

For the vertical method and for the three kinds of function discussed, different numerical problems may be encountered. Here, the solution is obtained quickly except for the case of the Hyperbolic equation, where, for chosen initial values, the solution obtained ($v_4 = 2$ V, $i_4 = 0$ A). If we want to obtain the correct solution ($v_4 = 1$ V, $i_4 = 1$ A), the initial point has to be near to 1 V; for example $v_4^{(0)} = 0.9$ V leads to the correct solution.

Two different problems appear for the horizontal method:

1. a false result is obtained (zero value),
2. numerical problems are encountered.

The incorrect result is dependent on the constitutive Square or Root equations because the current i_4 is equal to zero for every negative or zero value of the voltage v_4 for both functions. With $v_4^{(0)} = -2$ V as an initial value, we obtain $G_4^{(0)} = 0$;

Table 6.2 Results of program 'COMP_NR'

	'Vertical' method			'Horizontal' method		
	Initial.	*Iter.n°*	*Result*	*Initial.*	*Iter.n°*	*Result*
Constitutive	-10	5	1	-10	2	0
equation of	0	5	1	0	1	0
SQUARE type	0.5	4	1	0.5	4	1
eqn (6.34a)	2	4	1	2	4	1
	10	7	1	10	Num.Pb.*	STOP
Constitutive	-10	4	1	-10	2	0
equation of	0	4	1	0	1	0
ROOT type	0.5	3	1	0.5	3	1
eqn (6.34b)	2	3	1	2	3	1
	10	4	1	10	4	1
Constitutive	-10	2	2	-10	7	1
equation of	0	3	2	0	4	1
HYPERBOLIC	0.5	3	2	0.5	4	1
type eqn	2	1	2	2	Num.Pb.*	STOP
(6.34c)	10	2	2	10	Num.Pb.*	STOP

* Numerical problem

$J_4^{(0)} = 0$. Thus, the first 'iterative companion model' is an open circuit. Solving the linear system, we obtain $v_4'^{(0)} = 2$ V; $i_4'^{(0)} = 0$ A; $v_4'^{(0)}$ is equal to E_{Th}. Using definitions (6.38b), which characterize the horizontal method, the end of the first iteration leads to the following results: $i_4^{(1)} = 0$; $v_4^{(1)} = g_4^{-1}(i_4^{(1)}) = 0$. The second 'iterative companion model' is again characterized by $G_4^{(1)} = 0$; $J_4^{(1)} = 0$ (an open circuit, too). The solution of the linear system leads to: $v_4'^{(1)} = 2$ V; $i_4'^{(1)} = 0$ A, identical to $v_4'^{(0)} = 2$ V; $i_4'^{(0)} = 0$ A. The use of the horizontal method always leads to the point 0 V, 0 A. The points corresponding to two successive iterations are the same, so the convergence test is satisfied.

In order to solve this kind of problem, it is possible to modify the function considered and to replace '$i_4 = 0$ for $v_4 \leqslant 0$' by '$i_4 = (10^{-12}v_4)$ for $v_4 \leqslant 0$ V'. Indeed, a resistance value of 1000 Gigaohms (10^{12}) does not modify circuit behaviour.

Numerical problems can also occur in the case shown in Fig. 6.9. From $v_4^{(0)}$ shown in Fig. 6.9, the solution of the linear system leads to the point $v_4'^{(0)}$, $i_4'^{(0)}$. Applying the horizontal method, it is necessary to determine $i_4^{(1)} = i_4'^{(0)}$ and $v_4^{(1)} = g_4^{-1}(i_4^{(1)})$, but this is impossible when $i_4^{(1)}$ is negative, since $i_4 \geqslant 0$. This problem can be solved by replacing the condition $i_4 = 0$ for $v_4 \leqslant 0$ by the condition ($i_4 = 10^{-12}v_4$) for $v_4 \leqslant 0$ V. This modification is very important because it permits us to solve any numerical problems encountered.

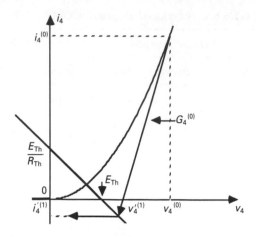

Figure 6.9 An illustration of problems depending on a bad choice of an initial value

6.5 Analysis of a diode bias circuit

The previous examples show that the use of the Newton–Raphson algorithm can generate some problems depending on the initial value. This kind of problem often exists when exponential functions describing diode or transistor models are used.

6.5.1 Description of method

When we want to compute the bias voltage of a diode, we can study the circuit shown in Fig. 6.6(b), with the following current–voltage diode equation:

$$i = I_S\left[\exp\left(\frac{v}{V_T}\right) - 1\right] = g(v) \tag{6.39}$$

We will consider the calculation of the operating point of the forward- and reverse-biased diodes in order to analyze precisely all the problems of initialization and convergence. Indeed, the Boltzmann equation (6.39) presents two regions with different characteristics: rapid variation of current for a voltage close to $V_d = 0.6$ V (for silicon diodes) for forward bias, and constant and very small current ($I \approx -I_S$) for the reverse bias.

The program 'COMP__DIODE', shown in Program 6.6, permits us to determine the bias point choosing the value of the initial point and the convergence method (that is vertical and horizontal).

See the listing of the program COMP__DIODE Directory of Chapter 6 on attached diskette.

In this program, the user can choose a manual or automatic initialization; in the case of manual initialization, only the horizontal and vertical methods are used. In the case of automatic initialization, methods more adapted to the Boltzmann diode equation are used; these methods are called 'intermediate methods'. A detailed analysis of these methods is described in Section 6.5.2.

The computation of the bias point is performed for the following values: $R = 1000 \ \Omega$ and $I_S = 10^{-13}$ A, and for two values of E:

1. $E = 5$ V: case of forward bias. The solution is $i_{ds} = 4.37$ mA; $v_s = 0.637$ V.
2. $E = -5$ V: case of reverse bias. The solution is $i_{ds} = -I_S$; $v_s = E = -5$ V.

According to the definition of the 'iterative companion model' associated with the Newton–Raphson algorithm (equations (6.31) and (6.32)), at the jth iteration we obtain the following values for $G^{(j)}$ and $J^{(j)}$:

$$G^{(j)} = \frac{I_S}{V_T}\left[\exp\left(\frac{v^{(j)}}{V_T}\right)\right] = \frac{i^{(j)} + I_S}{V_T} \quad \text{with} \quad i^{(j)} = I_S\left[\exp\left(\frac{v^{(j)}}{V_T}\right) - 1\right] \quad (6.40a)$$

$$J^{(j)} = i^{(j)} - G^{(j)}v^{(j)} = I_S\left[\exp\left(\frac{v^{(j)}}{V_T}\right) - 1\right] - \frac{I_S}{V_T}\left[\exp\left(\frac{v^{(j)}}{V_T}\right)\right]v^{(j)} \quad (6.40b)$$

The calculation method is based on the following algorithm:

```
Data input
Automatic initialization ? (Rep = yes or no)
    if Rep = no
        Choice of a current or voltage initial values
        Choice of the method : Horizontal or Vertical
        Enter {v (0) and i (0)=g(v (0))} or {i (0) and v (0) =
                                                g-1(i (0))}

    if Rep = yes
        1-Initialization as a function of bias
        2-Automatic initialization {v (0) = 0}
            choice of a method (4 possibilities) to improve
    convergence : see Section 6.5.2
j =0
Iterative calculation (from initialization or from
                            previous iteration) :
    Calculation of "iterative Companion Model" parameters
                                    for {i (j), v (j)}

    Calculation of bias point of linear circuit (the diode
    is replaced by its "iterative Companion Model") :
                                    {i ' (j+1), v ' (j+1)}

    Calculation of {i (j+1), v (j+1)}: Depends on the choice of
    the method ("Horizontal", "Vertical" or "Intermediate")
```

```
. Convergence test : |v⁽ʲ⁺¹⁾ - v⁽ʲ⁾| ≤ε|v⁽ʲ⁾| or(and)
                                      |i⁽ʲ⁺¹⁾ - i⁽ʲ⁾| ≤ε|i⁽ʲ⁾|
. if test = False j = j+1 : CONTINUE (new iteration)
. else : STOP
. for N=j : Result {i⁽ᴺ⁾, v⁽ᴺ⁾}
```

The computation of the bias point of the linearized circuit at iteration j is performed based on the circuit shown in Fig. 6.10 using the following equations:

$$v'^{(j+1)} = \frac{\left(\dfrac{E}{R} - J^{(j)}\right)}{\left(\dfrac{1}{R} + G^{(j)}\right)} = \frac{E - R J^{(j)}}{1 + R G^{(j)}} \tag{6.41a}$$

$$i'^{(j+1)} = \frac{E - v'^{(j)}}{R} \tag{6.41b}$$

where

$$G^{(j)} = \frac{I_S}{V_T}\left[\exp\left(\frac{v^{(j)}}{V_T}\right)\right] = \frac{i^{(j)} + I_S}{V_T} \tag{6.42a}$$

$$J^{(j)} = i^{(j)} - G^{(j)} v^{(j)} = I_S\left[\exp\left(\frac{v^{(j)}}{V_T}\right) - 1\right] - \frac{I_S}{V_T}\left[\exp\left(\frac{v^{(j)}}{V_T}\right)\right] v^{(j)} \tag{6.42b}$$

Using the program 'COMP_DIODE' and choosing a manual initialization, we obtain the results summarized in Table 6.3.

For the reverse bias, the horizontal method cannot be used (for any initial point) because the function $v = V_T \ln\{(i + I_S)/I_S\}$, the reverse function of the Boltzmann equation, is not defined for values smaller than $-I_S$. As is shown in Fig. 6.11(a), if the initialization value $v^{(0)}$ is point A, then using the 'iterative companion model' and solving the corresponding linear circuit, we obtain point B$\{v'^{(1)}, i'^{(1)}\}$. Applying the horizontal method, it is impossible to find $v^{(1)}$ so that $v^{(1)} = V_T \ln\{(i^{(1)} + I_S)/I_S\}$ because $i^{(1)} = i'^{(1)}$ is smaller than $-I_S$. The vertical method always permits us to obtain the correct results. In conclusion, for a reverse-biased diode, the vertical method is the only applicable method and it always converges.

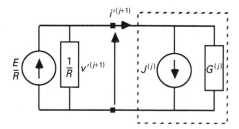

Figure 6.10 Linearized diode circuit permitting calculation of the values $(v'^{(j+1)}, i'^{(j+1)})$ using the 'iterative companion model'

Table 6.3   Results of program 'COMP__DIODE'

	Vertical method			Horizontal method		
	Initial.	*Iter.n°*	*Result*	*Initial.*	*Iter.n°*	*Result*
	−10	172	0.6370	−10	Num.Pb.*	STOP
	−5	172	0.6370	−5	Num.Pb.*	STOP
	0	172	0.6370	0	7	0.6370
$E = 5$ V	0.4	168	0.6370	0.4	3	0.6370
	0.5	91	0.6370	0.5	2	0.6370
	1	18	0.6370	1	3	0.6370
	2	55	0.6370	2	3	0.6370
	5	170	0.6370	5	4	0.6370
	−10	2	−5	0	Num.Pb.*	STOP
	−5	2	−5	0.6	Num.Pb.*	STOP
$E = -5$ V	0	2	−5	1	Num.Pb.*	STOP
	2	56	−5	2	Num.Pb.*	STOP
	5	172	−5	−2	Num.Pb.*	STOP

* Numerical problem

For the forward bias, both methods can be used. However, if the initial value is very different from the solution, a lot of iterations are required to converge to the result using the vertical method: 172 iterations are necessary if $v^{(0)} = 0$ V and only eighteen iterations if $v^{(0)} = 1$ V.

On the other hand, the horizontal method converges more quickly. Moreover, if $v^{(0)} = 0$ V is chosen, the value of the voltage $v^{(1)}$ after the first iteration is close

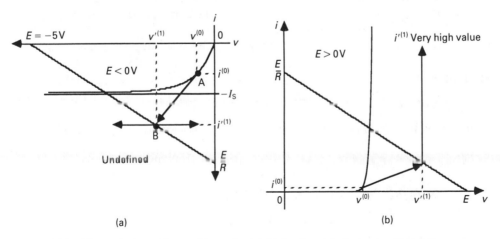

(a)   (b)

Figure 6.11   Numerical problems due to bad initialization of diode circuit. (a) For reverse bias. (b) For forward bias

to 5 V; then the use of the vertical method leads to very high currents, sometimes bigger than the capabilities of some computers, as explained in Fig. 6.11(b).

In this very simple case, we can choose an 'automatic' initialization with the following values:

1. If $E > 0$ then $v^{(0)} = V_T \ln\left(\dfrac{i^{(0)}}{I_S} + 1\right)$ with $i^{(0)} = \dfrac{E}{R}$ .... (6.43)

2. If $E < 0$ then $v^{(0)} = E$ and so $i^{(0)} = I_S\left(\exp\left(\dfrac{v^{(0)}}{V_T}\right) - 1\right)$ .... (6.44)

These values are used in the program 'COMP__DIODE' when the automatic initialization is chosen.

6.5.2 Methods permitting improvement of convergence speed

The exponential functions typically associated with diodes and bipolar transistors are monotonic and continuous. However, their variations are very fast. For the reverse bias, the function slope tends to zero, but for forward bias it tends to be infinite. So, in the 'iterative companion model', a conductance or a current source can be equal to zero, respectively. The convergence speed of the Newton–Raphson algorithm can vary a lot, whereas some numerical problems can stop the iteration process. The results presented in the previous section have shown that the vertical method is much better. We present some procedures that can be considered as a modified vertical method and that permit us to limit the variation of the voltage values during the iterative process in order to limit the value for the voltages and currents in the correct range.

Let us suppose that the exponential characteristic of the diode has been linearized at point $(i_k^{(j)}, v_k^{(j)})$ and that the solution of the corresponding linearized circuit is $\{v_k'^{(j+1)}, i_k'^{(j+1)}\}$. Now we want to find a new point $v_k^{(j+1)}$ and $i_k^{(j+1)} = g_k(v_k^{(j+1)})$ for the next iteration. This can be found using the following four methods.

Method 1. For small voltage variations, when $v_k'^{(j+1)}$ is close to $v_k^{(j)}$, that is, $|v_k'^{(j+1)} - v_k^{(j)}| \leqslant 2\,V_T$, and when voltages $v_k'^{(j+1)}$ and $v_k^{(j)}$ are smaller than $10\,V_T$, the vertical method is used, that is, $v_k^{(j+1)} = v_k'^{(j+1)}$. For voltages $v_k'^{(j+1)}$ and $v_k^{(j)}$ higher than $10\,V_T$, the next voltage value, $v_k^{(j+1)}$, is limited to $v_k^{(j)} \pm 2\,V_T$. These situations are summarized as follows:

1. for $|v_k'^{(j+1)} - v_k^{(j)}| \leqslant 2\,V_T$ choose $v_k^{(j+1)} = v_k'^{(j+1)}$

2. for $v_k'^{(j+1)} \leqslant 10\,V_T$ and $v_k^{(j)} \leqslant 10V_T$ choose $v_k^{(j+1)} = v_k'^{(j+1)}$

3. for $v_k'^{(j+1)} < v_k^{(j)}$ and $10\,V_T < v_k^{(j)}$ choose $v_k^{(j+1)} = v_k^{(j)} - 2\,V_T$

4. for $v_k'^{(j+1)} < v_k^{(j)}$ and $10\,V_T < v_k'^{(j+1)}$ choose
   $v_k^{(j+1)} = \max(10\,V_T, v_k^{(j)} + 2\,V_T)$

Method 2. In this method, we choose:

$$v_k^{(j+1)} = v_k^{(j)} + 10\, V_T \tanh\left(\frac{v_k'^{(j+1)} - v_k^{(j)}}{10\, V_T}\right)$$

Here, the properties of the hyperbolic tangent function are used (for large variations of its argument the value of the function is limited to $[-1,1]$). Thus, the values of $v_k^{(j+1)}$ are limited to $v_k^{(j)} \pm 10\, V_T$. For small voltage variations, that is, when $v_k'^{(j+1)}$ is close to $v_k^{(j)}$, we obtain $v_k^{(j+1)} = v_k'^{(j+1)}$, as in the vertical method.

Method 3. In this method, the current iteration (horizontal method) is used for increasing voltages and the voltage iteration (vertical method) for decreasing voltages. The use of the 'iterative companion model' associated with the Newton–Raphson algorithm and the solution of the linearized system lead to the determination of $v_k'^{(j+1)}$ and $i_k'^{(j+1)}$, as is shown in Fig. 6.12. The use of the vertical method leads one to choose $v_k^{(j+1)} = v_k'^{(j+1)}$. If a voltage iteration is required, using the horizontal method, some calculations have to be made. Point B in Fig. 6.12 is obtained from the previous point A, plotting a straight line of slope $G_k^{(j)}$.

The graphical construction of Fig. 6.12 involves the following relations:

$$i_k'^{(j+1)} = G_k^{(j)}\{v_k'^{(j+1)} - v_k^{(j)}\} + i_k^{(j)} \quad \text{with} \quad G_k^{(j)} = \frac{I_S}{V_T} \exp\left(\frac{v_k^{(j)}}{V_T}\right)$$

So for point B of Fig. 6.12:

$$i_k'^{(j+1)} = \frac{I_S}{V_T} \exp\left(\frac{v_k^{(j)}}{V_T}\right)\{v_k'^{(j+1)} - v_k^{(j)}\} + I_S\left[\exp\left(\frac{v_k^{(j)}}{V_T}\right) - 1\right] \tag{6.45a}$$

or

$$\frac{i_k'^{(j+1)} + I_S}{I_S} = \exp\left(\frac{v_k^{(j)}}{V_T}\right)\left\{\frac{v_k'^{(j+1)} - v_k^{(j)}}{V_T} + 1\right\} \tag{6.45b}$$

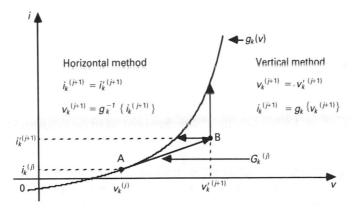

Figure 6.12 Explanation of mixed method with voltage and current iteration

Using the horizontal method we obtain

$$v_k^{(j+1)} = g_k^{-1}(i_k^{(j+1)}) = g_k^{-1}(i_k'^{(j+1)}) = V_T \ln\left[\frac{i_k'^{(j+1)} + I_S}{I_S}\right] \qquad (6.46\text{a})$$

and using relation (6.45b)

$$v_k^{(j+1)} = v_k^{(j)} + V_T \ln\left[\frac{v_k'^{(j+1)} - v_k^{(j)}}{V_T} + 1\right] \qquad (6.46\text{b})$$

When the expected voltage $v_k'^{(j+1)}$ is negative, $v_k'^{(j+1)} \leqslant 0$, or $v_k'^{(j+1)} \leqslant v_k^{(j)}$, then the next iteration voltage $v_k^{(j+1)}$ is chosen:

$$v_k^{(j+1)} = v_k'^{(j+1)}$$

On the other hand, when the expected voltage $v_k'^{(j+1)}$ is positive, $v_k'^{(j+1)} > 0$ and, additionally, $v_k'^{(j+1)} > v_k^{(j)}$, then $v_k^{(j+1)}$ is chosen according to formula (6.46b).

From equation (6.46a), we see that a voltage limitation is obtained.

Method 4. A modified version of method 3 is used in the SPICE simulator. The current iteration is used for high values of diode conductance $G_k^{(j)}$; the voltage iteration is used in the opposite case. The limit condition is defined as a voltage called v_{CRIT}. In the case of the diode equation, empirical choice gives the following value:

$$v_{CRIT} = V_T \ln\left(\frac{V_T}{I_S\sqrt{2}}\right) \qquad (6.47)$$

where $\sqrt{2}$ is in ohms. When the expected voltage $v_k'^{(j+1)}$ is equal to or smaller than v_{CRIT}, then the next iteration voltage is

$$v_k^{(j+1)} = v_k'^{(j+1)}$$

Otherwise, if $v_k'^{(j+1)} > v_{CRIT}$:

$$v_k^{(j+1)} = v_k^{(j)} + V_T \ln\left(\frac{v_k'^{(j+1)} - v_k^{(j)}}{V_T} + 1\right)$$

The value of the current i_{CRIT} flowing through the diode for $v = v_{CRIT}$ is easily obtained:

$$\frac{v_{CRIT}}{V_T} = \ln\left(\frac{V_T}{I_S\sqrt{2}}\right) \quad \text{so} \quad i_{CRIT} = I_S\left[\frac{V_T}{I_S\sqrt{2}} - 1\right]$$

and this practically equals $V_T/\sqrt{2} = 18.4$ mA.

For a standard diode $\{I_S = 10^{-13}$ A$\}$, the value of v_{CRIT} is 25.94 V_T. This value is of the order of 10 V_T, which corresponds to the condition required in methods 1 and 2.

The program 'COMP__DIODE' permits us to compare the different methods described. With a starting voltage $v^{(0)} = 0$ V, and $E = 5$ V (forward bias), we obtain a correct solution of the diode voltage $v_S = 0.637$ V, after twelve iterations for methods 1 and 2, after seven iterations for method 3, and after eight iterations for method 4. For the reverse bias, that is, $E = -5$ V, we obtain a correct solution for the diode voltage, $v_S = -5$ V, after two iterations for methods 1, 3 and 4, and after twenty-two iterations for method 2.

We note that all the methods give better results than does a basic method (horizontal or vertical). However, methods 3 and 4 are the best.

6.6 Generalization of the 'iterative companion model'

6.6.1 Model presentation

In previous sections we have studied the 'iterative companion model' associated with a non-linear element of the constitutive equation $i_k = g_k(v_k)$. Now, we can generalize the concept of the companion model for a controlled element, that is, the element for which the current is controlled by several voltages. Suppose that the branch k is composed of an element g_k where the current i_k is a non-linear function of different branch voltages, $\mathbf{v}^t = [v_1, v_2, ..., v_B]$, of the circuit:

$$i_k = g_k(\mathbf{v}) \tag{6.48}$$

where $\mathbf{v}$ is the vector of the circuit branch voltage. Using small variations, equation (6.48) becomes

$$i_k^{(j+1)} = i_k^{(j)} + G_{kk}^{(j)}(v_k^{(j+1)} - v_k^{(j)}) + \sum_{i=1,\, i \neq k}^{B} G_{ki}^{(j)}(v_i^{(j+1)} - v_i^{(j)}) \tag{6.49}$$

where B is the number of controlled branches, with

$$G_{kk}^{(j)} = \left. \frac{\partial i_k}{\partial v_k} \right|_{v_k = v_k^{(j)}}$$

and

$$G_{ki}^{(j)} = \left. \frac{\partial i_k}{\partial v_i} \right|_{v_i = v_i^{(j)}}$$

This equation can be modelled by the linear circuit shown in Fig. 6.13. This circuit represents the generalized 'iterative companion model' of an element in branch k,

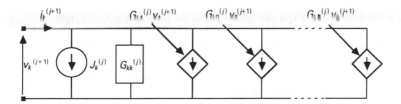

Figure 6.13 Generalized 'iterative companion model'

it is composed of a constant current source:

$$J_k^{(j)} = i_k^{(j)} - G_{kk}^{(j)} v_k^{(j)} - \sum_{i=1,\,i \neq k}^{B} G_{ki}^{(j)} v_i^{(j)}$$

representing the constant terms at the jth iteration, of a linear conductance $G_{kk}^{(j)}$ representing the derivative of the non-linear element function at point $v_k = v_k^{(j)}$, and $(B - 1)$ voltage controlled current sources representing the coupling of branch k with other controlling branches of the circuit.

6.6.2 'Iterative companion model' for bipolar transistor

To determine the bias points of a bipolar transistor circuit, we use the circuit shown in Fig. 6.14(a). The constitutive equations of the transistor depend on two voltages, v_{BE} and v_{BC} (base–emitter and base–collector voltages). We use the Ebers–Moll equivalent schema shown in Fig. 6.14(b); this model is described in Chapter 1 (Section 1.3.1).

The following equations are used in this model:

$$I_F = \frac{I_S}{\beta_F}\left[\exp\left(\frac{v_{BE}}{V_T}\right) - 1\right] = i_{DBE}; \quad I_R = \frac{I_S}{\beta_R}\left[\exp\left(\frac{v_{BC}}{V_T}\right) - 1\right] = i_{DBC} \quad (6.50)$$

$$i_{CT} = \beta_F I_F - \beta_R I_R = \beta_F i_{DBE} - \beta_R i_{DBC} = I_S\left[\exp\left(\frac{v_{BE}}{V_T}\right) - \exp\left(\frac{v_{BC}}{V_T}\right)\right] \quad (6.51)$$

The equivalent schema contains three non-linear elements: two diodes and a non-linear current source controlled by v_{BE} and v_{BC} voltages.

Each diode can be replaced by its 'iterative companion model' as described in Section 6.5.1. In the case of the base–emitter and base–collector diodes, the following linear equations are obtained (at iteration j):

$$G_{BE}^{(j)} = \frac{\partial i_{DBE}}{\partial v_{BE}}\bigg|_{v_{BE} = v_{BE}^{(j)}} = \frac{I_S}{\beta_F V_T}\exp\left(\frac{v_{BE}^{(j)}}{V_T}\right) = \frac{\left(i_{DBE}^{(j)} + \dfrac{I_S}{\beta_F}\right)}{V_T} \quad (6.52a)$$

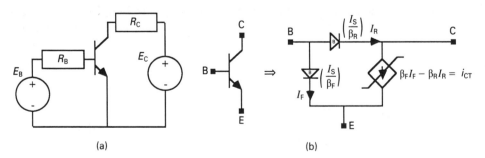

(a)   (b)

Figure 6.14   (a) Bipolar transistor biasing circuit. (b) Ebers–Moll equivalent schema of an NPN bipolar transistor

$$G_{BC}^{(j)} = \frac{\partial i_{DBC}}{\partial v_{BC}}\bigg|_{v_{BC}=v_{BC}^{(j)}} = \frac{I_S}{\beta_R V_T}\exp\left(\frac{v_{BC}^{(j)}}{V_T}\right) = \frac{\left(i_{DBC}^{(j)}+\dfrac{I_S}{\beta_R}\right)}{V_T} \tag{6.52b}$$

$$J_{BE}^{(j)} = i_{DBE}^{(j)} - G_{BE}^{(j)}\,v_{BE}^{(j)} \tag{6.53a}$$

$$J_{BC}^{(j)} = i_{DBC}^{(j)} - G_{BC}^{(j)}\,v_{BC}^{(j)} \tag{6.53b}$$

The non-linear controlled current source i_{CT} can also be modelled by the generalized 'iterative companion model' using the relation (6.51)

According to the general definition of the 'iterative companion model' given by equation (6.49), we obtain

$$G_{CBE}^{(j)} = \frac{\partial i_{CT}}{\partial v_{BE}}\bigg|_{v_{BE}=v_{BE}^{(j)}} = \frac{I_S}{V_T}\left[\exp\left(\frac{v_{BE}}{V_T}\right)\right] = \beta_F G_{BE}^{(j)} \tag{6.54a}$$

$$G_{CBC}^{(j)} = \frac{\partial i_{CT}}{\partial v_{BC}}\bigg|_{v_{BC}=v_{BC}^{(j)}} = -\frac{I_S}{V_T}\left[\exp\left(\frac{v_{BC}}{V_T}\right)\right] = -\beta_R G_{BC}^{(j)} \tag{6.54b}$$

$$J_{CT}^{(j)} = -G_{CBE}^{(j)}v_{BE}^{(j)} - G_{CBC}^{(j)}v_{BC}^{(j)} + i_{CT}^{(j)} \tag{6.54c}$$

Equations (6.52)–(6.54), describing the bipolar transistor at the iteration j, can be used in the bias circuit of Fig. 6.14(a) to form the iterative equivalent circuit shown in Fig. 6.15. Based on this equivalent circuit, we can calculate the values of the voltages and currents considered for the next iteration.

The calculation method of the transistor bias point is similar to the case of the diode. However, it is impossible to represent it graphically (two voltages and two currents: dimension 4). So only the calculation algorithm will be developed in this section. This algorithm contains an initialization, an iteration process, including the

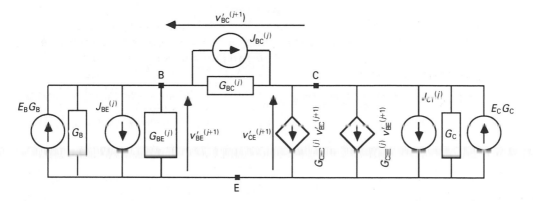

Figure 6.15 Equivalent iterative circuit of bipolar transistor biasing circuit of Figure 6.14(a),

$$\left(G_B = \frac{1}{R_B}\quad G_C = \frac{1}{R_C}\right)$$

definition of the 'iterative companion model' of each non-linear element, the calculation of the extrapolated values and a convergence test. The only difference in the calculations of the diode bias point, is the number of non-linear elements that lead to the following changes.

For initialization, two voltages (v_{BE} and v_{BC}) have to be initialized. In order to choose these values properly, we consider an active normal operation of the bipolar transistor for which the base–emitter diode is forward biased and the base–collector diode is reverse biased. Thus, we can use the following 'automatic' initialization:

$$1. \quad v_{BE}^{(0)} = 0.6 \text{ V if } E_B > 0.6 \text{ V}, \quad v_{BE}^{(0)} = E_B \text{ if } E_B < 0.6 \text{ V}$$

$$\text{(6.55)}$$

$$2. \quad v_{BC}^{(0)} = 0.6 \text{ V if } E_B > 0.6 \text{ V}, \quad v_{BC}^{(0)} = 0.6 - E_C \text{ if } E_B < 0.6 \text{ V}$$

The calculations are based on the following algorithm:

```
Input Data (Default values : R_B = 10 kΩ ; R_C = 2 kΩ ; E_C = 5 V) E_B
Automatic initialization ? (Rep = yes or no)
    If Rep = no
        Input v_BC^(0) and v_BE^(0)
    If Rep = yes
        initialization as a function of bias condition
                                    (see relations (6.55))
        automatic initialization : v_BC^(0) = v_BE^(0) = 0 V :
        choice of the convergence improving method : 1-2-3-4
j = 0
Iterative calculation performed based on previous
                                                iteration
    Calculation of "iterative Companion Model" of each
            non-linear element for {v_CE^(0), v_BE^(0), v_BC^(0)}
    Calculation of bias point of linearized circuit , with
    non-linear elements being replaced by their "iterative
                            Companion Model" respectively.
    Determination of : {v'_BE^(j+1), v'_BC^(j+1), v'_CE^(j+1)} and
        {i'_DBE^(j+1), i'_DBC^(j+1), i'_CT^(j+1)} using matrix method.
    Choice of the method for the determination of
                            {v_CE^(j+1), v_BE^(j+1), v_BC^(j+1)}
    Convergence test: |v_BE^(j+1) - v_BE^(j)| < ε |v_BE^(j)| and
                            |v_BC^(j+1) - v_BC^(j)| < ε |v_BC^(j)|
    If test = False ; j = j + 1 : CONTINUE
    Else : STOP
For N = j ; Result.
```

The calculation of the bias point of the linearized circuit at iteration j is based on the circuit shown in Fig. 6.15 using nodal analysis. For basic unknowns $v_{BE}^{'(j+1)}$ and

$v_{BC}^{\prime\,(j+1)}$, we obtain

$$\begin{bmatrix} G_B + G_{BE}^{(j)} + G_{BC}^{(j)} & -G_{BC}^{(j)} \\ G_{CBC}^{(j)} + G_{CBE}^{(j)} - G_{BC}^{(j)} & G_C + G_{BC}^{(j)} & -G_{CBC}^{(j)} \end{bmatrix} \begin{bmatrix} v_{BE}^{\prime\,(j+1)} \\ v_{CE}^{\prime\,(j+1)} \end{bmatrix}$$

$$= \begin{bmatrix} E_B G_B - J_{BE}^{(j)} - J_{BC}^{(j)} \\ -J_{CT} + E_C G_C + J_{BC}^{(j)} \end{bmatrix} \tag{6.56}$$

with $v_{BC}^{\prime\,(j+1)} = v_{BE}^{\prime\,(j+1)} - v_{CE}^{\prime\,(j+1)}$.

This linear system is solved using the procedure 'Calcul' in the program 'COMP_TRAN'.

We also obtain

$$i_{BE}^{\prime\,(j+1)} = G_{BE}^{(j)} v_{BE}^{\prime\,(j+1)} + J_{BE}^{(j)} \tag{6.57a}$$

$$i_{BC}^{\prime\,(j+1)} = G_{BC}^{(j)} v_{BC}^{\prime\,(j+1)} + J_{BC}^{(j)} \tag{6.57b}$$

For the calculation of $v_{BE}^{(j+1)}$ and $v_{BC}^{(j+1)}$, the same methods that are applied in the program 'COMP_DIODE' can be used.

See the listing of the program COMP_TRANS Directory of Chapter 6 on attached diskette.

Based on the circuit shown in Fig. 6.14(a), and choosing different values for E_B, E_C, R_B, R_C, we can consider all the operation regions of an NPN bipolar transistor, that is, direct region (saturated and linear), reverse region (saturated and linear) and off-region.

The computation results obtained from the program 'COMP_TRAN' are as follows:

1. For the direct saturated region with $E_B = E_C = 5$ V, $R_B = 10$ kΩ and $R_C = 2$ kΩ, we obtain the results $v_{BE} = 0.6296$ V and $v_{BC} = 0.5751$ V, for manual and automatic initialization, $v_{BE}^{(0)} = v_{BC}^{(0)} = 0.6$ V, after three iterations. Manual initialization with $v_{BE}^{(0)} = v_{BC}^{(0)} = 0$ V leads to divergence of the algorithm. However, with $v_{BE}^{(0)} = v_{BC}^{(0)} = 0$ V, and application of the convergence improvement methods, we obtain the correct results after twelve iterations for method 1, ten iterations for methods 2 and 3 and eight iterations for method 4.

2. For the direct linear region with $E_B = E_C = 5$ V, $R_B = 100$ kΩ and $R_C = 100$ Ω, we obtain the results $v_{BE} = 0.6372$ V and $v_{BC} = -3.927$ V for manual iteration with $v_{BE}^{(0)} - v_{BC}^{(0)} - 1$ V after twenty one iterations, for automatic initialization with $v_{BE}^{(0)} = v_{BC}^{(0)} = 0.6$ V after thirty-seven iterations. Manual initializations with $v_{BE}^{(0)} - v_{BC}^{(0)} - 0$ V, and $v_{BE}^{(0)} - 0$, $v_{BC}^{(0)} - 1$ V lead to divergence of the algorithm. However, with $v_{BE}^{(0)} = v_{BC}^{(0)} = 0$ V, and application of the convergence improvement methods, we obtain the correct results after nine iterations for method 1, eighteen iterations for method 2 and six iterations for methods 3 and 4.

3. For the off-region with $E_B = 0$ V, $E_C = 5$ V, $R_B = 10$ kΩ and $R_C = 2$ kΩ, we obtain the results $v_{BE} = 9.9 \times 10^{-10}$ V and $v_{BC} = -4.999$ V, for manual initialization with $v_{BE}^{(0)} = v_{BC}^{(0)} = 0$ V after two iterations, for manual initialization

with $v_{\mathrm{BE}}^{(0)} = v_{\mathrm{BC}}^{(0)} = 1$ V after twenty-four iterations, for automatic initialization with $v_{\mathrm{BE}}^{(0)} = 0$ V, $v_{\mathrm{BC}}^{(0)} = -5$ V after one iteration. With initialization $v_{\mathrm{BE}}^{(0)} = v_{\mathrm{BC}}^{(0)} = 0$ V, and application of the convergence improvement methods, we obtain the correct results after two iterations for methods 1, 3 and 4, and after twenty iterations for method 2.

The results for the reverse region are the same as for the direct region (in this case, it is necessary to assume that $E_{\mathrm{C}} = -5$ V).

6.6.3 'Iterative companion model' for MOS transistor

The constitutive equations of the MOS transistor depend on two voltages v_{DS} and v_{GS} (drain–source and gate–source voltage). In this section, the following equations are used:

$$i_{\mathrm{D}} = 0 \quad \text{if } v_{\mathrm{GS}} < V_{\mathrm{Th}} \tag{6.58a}$$

$$i_{\mathrm{D}} = \beta \left(v_{\mathrm{GS}} - V_{\mathrm{Th}} - \frac{v_{\mathrm{DS}}}{2} \right) v_{\mathrm{DS}} \quad \text{if } v_{\mathrm{GS}} \geqslant V_{\mathrm{Th}} \text{ and } v_{\mathrm{GS}} - v_{\mathrm{DS}} \geqslant V_{\mathrm{Th}} \tag{6.58b}$$

$$i_{\mathrm{D}} = \frac{\beta}{2} \left(v_{\mathrm{GS}} - V_{\mathrm{Th}} \right)^2 \quad \text{if } v_{\mathrm{GS}} \geqslant V_{\mathrm{Th}} \text{ and } v_{\mathrm{GS}} - v_{\mathrm{DS}} < V_{\mathrm{Th}} \tag{6.58c}$$

Figure 6.16 represents the simplified equivalent schema used for the MOS transistor. This schema contains only one non-linear source i_{D} dependent on v_{DS} and v_{GS}. This non-linear source can be modelled by an iterative companion model using

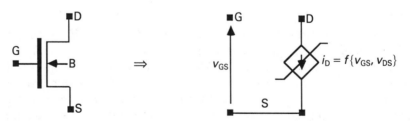

Figure 6.16   Equivalent schema of a MOS transistor

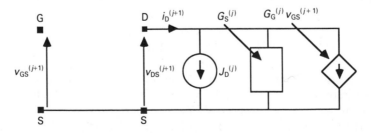

Figure 6.17   Iterative companion model of a MOS transistor

relations (6.58a)–(6.58c) and according to the definition given by equation (6.49). The iterative companion model is shown in Fig. 6.17. For different elements of the model, we obtain directly

$$G_S^{(j)} = \frac{\partial i_D}{\partial v_{DS}}\bigg|_{v_{DS} = v_{DS}^{(j)}} = 0 \quad \text{if } v_{GS} < V_{Th} \tag{6.59a}$$

$$G_S^{(j)} = \frac{\partial i_D}{\partial v_{DS}}\bigg|_{v_{DS} = v_{DS}^{(j)}} = \beta(v_{GS} - V_{Th} - v_{DS}) \quad \text{if } v_{GS} \geqslant V_{Th} \text{ and } v_{GD} \geqslant V_{Th} \tag{6.59b}$$

$$G_S^{(j)} = \frac{\partial i_D}{\partial v_{DS}}\bigg|_{v_{DS} = v_{DS}^{(j)}} = 0 \quad \text{if } v_{GS} \geqslant V_{Th} \text{ and } v_{GD} < V_{Th} \tag{6.59c}$$

$$G_G^{(j)} = \frac{\partial i_D}{\partial v_{GS}}\bigg|_{v_{GS} = v_{GS}^{(j)}} = 0 \quad \text{if } v_{GS} < V_{Th} \tag{6.60a}$$

$$G_G^{(j)} = \frac{\partial i_D}{\partial v_{GS}}\bigg|_{v_{GS} = v_{GS}^{(j)}} = \beta v_{DS} \quad \text{if } v_{GS} \geqslant V_{Th} \text{ and } v_{GD} \geqslant V_{Th} \tag{6.60b}$$

$$G_G^{(j)} = \frac{\partial i_D}{\partial v_{GS}}\bigg|_{v_{GS} = v_{GS}^{(j)}} = \beta(v_{GS} - V_{Th}) \quad \text{if } v_{GS} \geqslant V_{Th} \text{ and } v_{GD} < V_{Th} \tag{6.60c}$$

$$J_D^{(j)} = -G_S^{(j)}v_{DS}^{(j)} - G_G^{(j)}v_{GS}^{(j)} + i_D^{(j)} \tag{6.61}$$

The calculation method is similar to the method used in Section 6.6.2 in the case of the bipolar transistor. Only the network topology is different.

Exercises

The purpose of the following exercises is DC network analysis; computation of the operating point of these semiconductor circuits will be performed. However, some exercises contain time dependent sources; we assume that the time variations are slow with respect to the characteristics of the circuit. For a given time, a DC analysis can be performed. The capacitors are considered as open circuits and all the diodes are modelled by Boltzmann diodes (except when specifically stated otherwise).

Exercise 1. Run the program 'POINTFIX1'; note that, concerning the diode function, the program never converges. For the MOS transistor, run the program using the numerical values of Table 6.1 in Section 6.2.2. Note the convergence problems for the MOS ohmic region.

Exercise 2. Run the program 'POINTFIX2' using all the possibilities proposed in the program menu. Give a physical interpretation for the different functions and note the convergence problems.

Exercise 3. Verify that, using the program 'NEWTRAP' and the corresponding numerical values of Section 6.2.2, the diode function does not present convergence problems. Modify the program 'NEWTRAP' in order to obtain the operating point for the previous numerical values and also in the case of E varying from $-10 V_T$ to $10 V_T$.

Exercise 4.   Running the program 'ORDER3', compose a table to compare the convergence velocity of the three algorithms (Newton–Raphson, Richmond and Traub algorithms) for the diode function $E = 5$ V, $I_S = 10^{-13}$ A, $R = 1000$ Ω.

Exercise 5.   Construct the 'iterative companion model' of the non-linear device defined by the following function:

$$i_4 = Kv_4 \exp(-3v_4)$$

This non-linear device is placed in the circuit shown in Fig. 6.8(a) (same numerical values). Modify the program 'COMP_NR' in order to find the operating points for $K = 8$ and $K = 16$.

Exercise 6.   For the circuit shown in Fig. E6.1, plot the characteristic $(V_1 - V_2)$ versus V_1, V_1 varying linearly from -10 V to $+10$ V (consider fifty points of calculation). $I_S = 10^{-13}$ A; $R = 1$ kΩ.

   An iterative loop has to be added to the program 'COMP_DIODE'. Plot the characteristic $(V_1 - V_2)$ versus V_1, V_1 varying more precisely from -500 mV to $+500$ mV (consider fifty points of calculation) and compare the two sets of results.

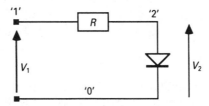

Figure E6.1

Exercise 7.   For the circuit shown in Fig. E6.2, plot the characteristic V_2 versus V_1, V_1 varying linearly from -10 V to $+10$ V (consider fifty points of calculation). Perform the same computations with V_1 varying from -500 mV to $+500$ mV (consider fifty points of calculation). $I_S = 10^{-13}$ A; $R = 1$ kΩ.

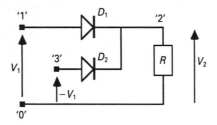

Figure E6.2

Exercise 8.   To demonstrate the DC characteristics of a Zener diode, the network shown in Fig. E6.3 is used.

   It is assumed that the Zener diode is a Boltzmann diode for which the I–V characteristic has been modified to take account of the reverse breakdown phenomenon (see the equations in Chapter 1). Develop a program 'COMP_ZENER' with the same approach as that of the program 'COMP_DIODE'. Numerical conditions: $e(t) = E + V \sin(\omega t)$ with $E = -3$ V and $V = +5$ V; $I_S = 10^{-15}$ A; $I_{bd} = 16 \times 10^{-3}$ A; $V_{bd} = 4.3$ V; $n_{bd} = 1.8$; $R = 1$ kΩ.

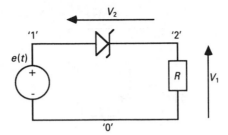

Figure E6.3

Exercise 9.   The circuit shown in Fig. E6.4 contains a PNP bipolar transistor with the following parameters: $\beta_F = 100$; $\beta_R = 1$; $I_S = 10^{-13}$ A. Determine its bias operating point.

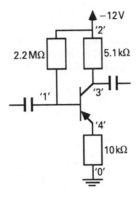

Figure E6.4

Exercise 10.   The circuit shown in Fig. E6.5 contains an NPN bipolar transistor with the following parameters: $\beta_F = 100$; $\beta_R = 1$; $I_S = 10^{-13}$ A. Determine its bias operating point.

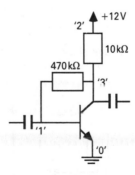

Figure E6.5

Exercise 11.   The circuit shown in Fig. E6.6 contains a Darlington bipolar transistor pair with the following parameters: $\beta_F = 100$; $\beta_R = 1$; $I_S = 10^{-13}$ A (the transistors are identical). Determine the bias point of each transistor.

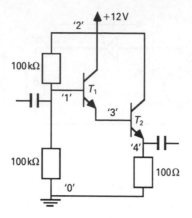

Figure E6.6

Exercise 12. The circuit shown in Fig. E6.7 contains two coupled NPN bipolar transistors. It is assumed that the transistors have the following same parameters: $\beta_F = 100$; $\beta_R = 1$; $I_S = 10^{-13}$ A. Determine the bias points of each transistor.

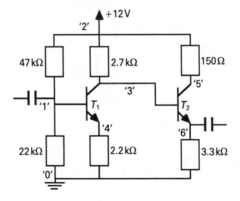

Figure E6.7

Exercise 13. The two-pole bipolar circuit shown in Fig. E6.8 is often used in integrated circuits; determine its $I-V$ characteristic for the bipolar transistor when V varies from -10 V to 10 V. $\beta_F = 100$; $\beta_R = 1$; $I_S = 10^{-13}$ A.

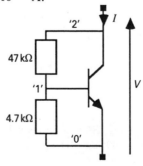

Figure E6.8

Exercise 14. The circuit shown in Fig. E6.9 contains a JFET transistor with the following parameters: $\beta = 5 \times 10^{-4}$ A/V^2; $V_P = -2.5$ V. Determine its bias point.

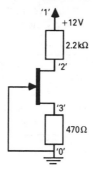

Figure E6.9

Exercise 15. The circuit shown in Fig. E6.10 contains two bipolar transistors: an NPN and a PNP type with the same parameters (gains and saturation current): $\beta_F = 100$; $\beta_R = 1$; $I_S = 10^{-13}$ A. Determine the bias point of each transistor.

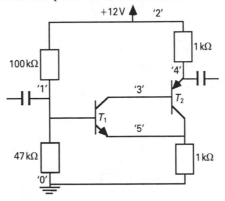

Figure E6.10

Exercise 16. The circuit shown in Fig. E6.11 contains two transistors: an NPN bipolar transistor with the following parameters: $\beta_F = 100$; $\beta_R = 1$; $I_S = 10^{-13}$ A and a JFET transistor with the following parameters: $\beta = 5 \times 10^{-4}$ A/V^2; $V_P = -2.5$ V. Determine the bias points of each transistor.

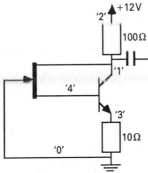

Figure E6.11

Exercise 17. The circuit shown in Fig. E6.12 represents an ECL logic gate. The two NPN bipolar transistors have the following parameters: $\beta_F = 100$; $\beta_R = 1$; $I_S = 10^{-13}$ A. Develop a program with the same approach as in the program 'COMP_TRAN' to plot the voltages V_{1o} and V_{2o} as a function of V_i for V_i varying from 0 to -5 V (consider fifty points of calculation).

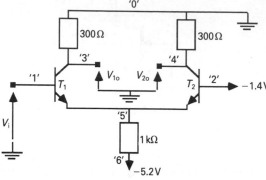

Figure E6.12

Exercise 18. The two bipolar NPN transistors of the circuit shown in Fig. E6.13 each have the following parameters: $\beta_F = 100$; $\beta_R = 1$; $I_S = 10^{-13}$ A. Develop a program with the same approach as in the program 'COMP_TRAN' to plot the output voltage V_o as a function of V_i, for V_i varying from 0 V to $+5$ V.

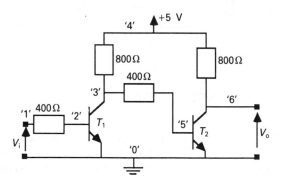

Figure E6.13

Exercise 19. The circuit shown in Fig. E6.14 represents a Schmitt trigger. The two NPN bipolar transistors each have the following parameters: $\beta_F = 100$; $\beta_R = 1$; $I_S = 10^{-13}$ A. Develop a program with the same approach as in the program 'COMP_TRAN' to plot the output voltage V_o as a function of V_i for V_i varying from -10 V to $+10$ V.

Exercise 20. The circuit shown in Fig. E6.15 represents a bistable network. The two NPN transistors each have the following parameters: $\beta_F = 100$; $\beta_R = 1$; $I_S = 10^{-13}$ A. Develop a program with the same approach as in the program 'COMP_TRAN' to plot the output voltage V_o as a function of V_i for V_i varying from -10 V to $+10$ V.

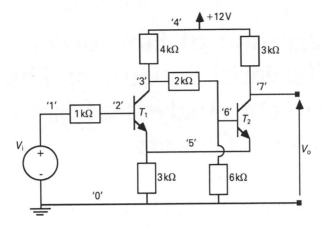

Figure E6.14

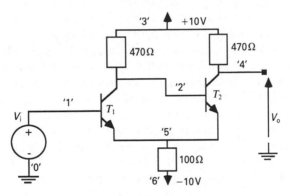

Figure E6.15

Exercise 21. The two-terminal circuit of Fig. E6.16 contains a diode and an NMOS transistor; determine the $I-V$ DC characteristic when E varies from -10 V to 10 V. The diode is modelled by a Boltzmann diode: $I_S = 10^{-13}$ A. The NMOS transistor has the following parameters: $K_P = 0.5 \times 10^{-4}$ A/V^2; $V_{Th} = 1$ V; $W = 10^{-5}$ m; $L = 5 \times 10^{-6}$ m.

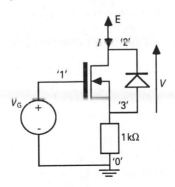

Figure E6.16

7 Integration of First-order Differential Equations; The Transient Analysis of Simple Dynamic Circuits

7.1 Introduction

The time or transient response of linear and non-linear dynamic circuits can be obtained using several methods. In this chapter we discuss the solution of first-order differential equations since higher-order differential equations can be transformed into a set of first-order equations. Numerical integration methods allow us to find the solutions of linear and non-linear differential equations simply, which is not so easy using analytical methods (especially for non-linear circuits).

Several algorithms for the integration of first-order differential equations are presented; these can generally be divided into one-step and multi-step algorithms, and additionally into explicit and implicit algorithms. The one-step algorithms are easier to understand and code into a program but multi-step algorithms are more accurate. The time domain solution of a dynamic circuit is simple if the differential equations describing the circuit are given in the normal state variable form.

7.2 Formulation and analytical solutions of first-order differential equations

To recall the formulation of first-order differential equations in the normal form let us consider two simple linear RC and RL circuits shown in Fig. 7.1(a, b).

The branch voltage–current relation in the time domain for the capacitor is

$$i_C(t) = C \frac{d v_C(t)}{dt} \tag{7.1}$$

and for the inductor

$$v_L(t) = L \frac{d i_L(t)}{dt} \tag{7.2}$$

For the RC circuit of Fig. 7.1(a) we can apply the KCL to obtain

$$i_S(t) = i_R(t) + i_C(t) \tag{7.3}$$

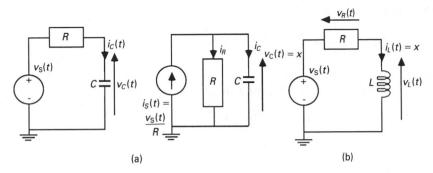

Figure 7.1   Simple linear circuits. (a) RC circuit. (b) RL circuit

Substitution of (7.1) into (7.3) and using Ohm's relation $i_R(t) = v_R(t)/R$ yields

$$\frac{v_S(t)}{R} = \frac{v_R(t)}{R} + C\frac{dv_C(t)}{dt}, \qquad v_C(0) = v_{C0} \tag{7.4}$$

and after dividing by C and rearranging we obtain

$$\frac{dv_C(t)}{dt} = \frac{-1}{RC}v_C(t) + \frac{1}{RC}v_S(t), \qquad v_C(0) = v_{C0} \tag{7.5}$$

where v_{C0} is the initial condition, that is, $v_C(t)$ for $t = 0$.

Similarly, for the RL circuit of Fig. 7.1(b) we can use the KVL to obtain

$$v_S(t) = v_R(t) + v_L(t) \tag{7.6}$$

Substitution of (7.2) and rearranging yields

$$\frac{di_L(t)}{dt} = \frac{-R}{L}i_L(t) + \frac{1}{L}v_S(t), \qquad i_L(0) = i_{L0} \tag{7.7}$$

where i_{L0} is the initial condition, that is, $i_L(t)$ for $t = 0$.

Introducing the state variable $x(t)$, which usually corresponds to the capacitor voltage $v_C(t)$, for capacitive circuits and to the inductor current $i_L(t)$, for inductive circuits and using the notation $u(t)$, for voltage or current excitations $v_S(t)$ or $i_S(t)$, we can rewrite relations (7.5) and (7.7) in the uniform format known as the normal state variable form:

$$\dot{x}(t) = \frac{dx(t)}{dt} = ax(t) + bu(t), \qquad x(0) = x_0 \tag{7.8}$$

where x_0 is called the initial state, with

$$x = v_C, \ a = -\frac{1}{RC} = \frac{-1}{\tau_{RC}}, \qquad b = \frac{1}{RC}, \qquad u = v_S \tag{7.9}$$

for the RC circuit, and

$$x = i_L, \qquad a = -\frac{R}{L} = \frac{-1}{\tau_{RL}}, \qquad b = \frac{1}{L}, \qquad u = v_S \tag{7.10}$$

for the RL circuit.

Relation (7.8) can be written in a more general form:

$$\dot{x}(t) = f[x(t), u(t)], \qquad x(0) = x_0 \tag{7.11}$$

with the equivalent integral equation

$$x(t) = x(0) + \int_0^t f[x(\tau), u(\tau)] \, d\tau \tag{7.12}$$

where τ is an integration variable.

It is known that the exact solution of equation (7.8) takes the form

$$x(t) = e^{at} \int_0^t bu(\tau) \, e^{-a\tau} \, d\tau + K \, e^{at} = x_{ss}(t) + x_{tr}(t) \tag{7.13}$$

where K is an integration constant, $x_{ss}(t)$ is the state variable in a steady state or steady state response, $x_{tr}(t)$ is the state variable in the transient state or the transient response of the circuit.

To investigate the transient properties of electronic circuits we very often use the voltage or current unit step function

$$1(t) = \begin{cases} 0, & \text{for } -\infty < t \leqslant 0^- \\ 1, & \text{for } 0^+ \leqslant t < \infty \end{cases} \tag{7.14}$$

as the excitation function $u(t)$. In such a case the voltage excitation $u(t) = U \, 1(t)$ is constant and equals U volts for the time $0^+ \leqslant t < \infty$. Using the voltage step function as $u(t)$ relation (7.13) is simplified considerably:

$$x(t) = \frac{-bU}{a} + \frac{bU}{a} \, e^{at} + K \, e^{at} = x_{ss}(t) + x_{tr}(t) \tag{7.15}$$

The first term of the right-hand side of (7.15) is the steady state response, that is, after the transient state has decayed at $t \to \infty$:

$$x_{ss}(t) = x(\infty) = \frac{-b}{a} \, U \tag{7.16}$$

The integration constant can be found from the initial condition, $x(0)$, at time $t = 0^+$, that is, at the beginning of observation of the circuit's properties, when $e^{at} = 1$:

$$x(0^+) = K \tag{7.17}$$

Substitution of (7.16) and (7.17) into relation (7.15) yields

$$x(t) = x(\infty) + [x(0^+) - x(\infty)] \, e^{at}, \qquad t > 0 \tag{7.18}$$

Thus, for the first-order RC or RL circuits (that is having a single time constant), we obtain for RC

$$v_C(t) = v_C(\infty) + [v_C(0^+) - v_C(\infty)]\ e^{-t/R_{eq}C}, \qquad t > 0 \tag{7.19}$$

and for RL

$$i_L(t) = i_L(\infty) + [i_L(0^+) - i_L(\infty)]\ e^{(-R_{eq}/L)t}, \qquad t > 0 \tag{7.20}$$

where the equivalent resistance, R_{eq}, is 'seen' from the terminals of the capacitor or inductor at $t > 0$.

As an example, consider the RC circuit shown in Fig. 7.2(a) in which the subcircuit with the voltage source (battery) is switched on at $t = 0$. From relation (7.19) we see that to find the transient response of the first-order RC circuit, excited by the voltage step function, it is necessary to compute three quantities: $v_C(0^+)$, $v_C(\infty)$ and the time constant $\tau = R_{eq}C$. For the circuit of Fig. 7.2(a) the voltage step function has the amplitude $U = 6R_2/(R_1 + R_2) = 3$ V.

The equivalent resistance, R_{eq}, seen from the capacitor terminals after switch S is closed is

$$R_{eq} = R_1 // R_2 = \frac{R_1 R_2}{R_1 + R_2} = 1\ k\Omega$$

Thus, the time constant τ equals (the internal resistance of the battery equal to zero)

$$\tau = R_{eq}C = 10^3 \times 2 \times 10^{-6} = 2 \times 10^{-3}\ s = 2\ ms$$

The values of $v_C(0^+)$ and $v_C(\infty)$ are easily found by inspection. Since the voltage across the capacitor cannot change instantly, then $v_C(0^+) = -1$ V and $v_C(\infty) = 3$ V (since, in the steady state, current does not flow through the capacitor). Then from relation (7.19) we) we have

$$v_C(t) = 3 - 4\ e^{(-t/2)\,[ms]}\ V, \qquad t > 0$$

where time, t, is expressed in milliseconds.

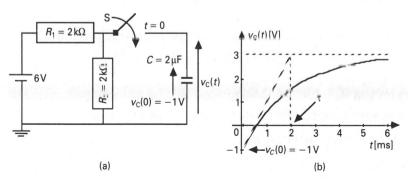

(a)   (b)

Figure 7.2   (a) The RC circuit with the switch closed at $t = 0$. (b) The transient response

The plot of the transient response $v(t)$ is shown in Fig. 7.2(b). From this example we see that the computation of the time response for a simple linear and time invariant circuit (that is, with constant parameters) excited by the constant function is simple and can easily be provided analytically. However, for non-linear or time varying circuits excited by more complex time functions, analytic solutions of equation (7.13) may be difficult. In such cases, approximate computer aided solutions based on numerical methods can be used.

7.3 Single-step algorithms for the solution of differential equations

In numerical approaches to the solution of differential equations the time t is quantized; the time increment, Δt, between the discrete time points is usually kept constant and is called the time step:

$$t_n - t_{n-1} = \Delta t = h \tag{7.21}$$

When the starting or initial time is denoted by t_0, then we can write

$$t_1 = t_0 + h, \ t_2 = t_1 + h = t_0 + 2h, \ ..., t_n = t_0 + nh$$

The solutions of the differential equation at the corresponding time points are denoted by

$$x(t_1), \qquad x(t_2), \qquad ..., \qquad x(t_n)$$

and their differences by

$$\Delta x_1 = x(t_1) - x(t_0), \ ..., \Delta x_n = x(t_n) - x(t_{n-1}) \tag{7.22}$$

Then the differential equation (7.11) can be approximated by the finite differences, starting from times t_0 and t_1:

$$\dot{x}(t_0) \cong \frac{\Delta x_1}{\Delta t} = \frac{x(t_1) - x(t_0)}{h} = f[x(t_0), u(t_0)] \tag{7.23}$$

or

$$x(t_1) = x(t_0) + hf[x(t_0), u(t_0)] \tag{7.24}$$

Thus, we have obtained the solution, $x(t_1)$, at time t_1, based on the initial state $x(t_0) = x(0)$ and the value of the right-hand side of the differential equation at t_0. Proceeding similarly, we obtain the solution at the next time t_2:

$$x(t_2) = x(t_1) + hf[x(t_1), u(t_1)]$$

or, generally,

$$x_{n+1} = x_n + hf[x_n, u_n] \tag{7.25}$$

where

$$x_{n+1} = x(t_{n+1})$$

$$x_n = x(t_n)$$

$$u_n = u(t_n)$$

Relation (7.25) is known as Euler's forward or Euler's extrapolation (explicit) algorithm for differential equation solutions. The accuracy of the solution obtained by this algorithm depends closely on the length of the time step h. For a smaller time step, we obtain better accuracy of the solution.

Euler's forward algorithm can be interpreted graphically, as is shown in Fig. 7.3(a) for a small time step and in Fig. 7.3(b) for a large time step, showing the differences in accuracy. As will be shown in the next section, the value of the time step, h, should be smaller than the time constant, τ, of the circuit, or $h < 1/|a|$.

The differential equation (7.11) can also be approximated in a slightly different way using the derivative at the time point considered (not at the previous time point as in the Euler forward algorithm):

$$\dot{x}(t_1) \cong \frac{x(t_1) - x(t_0)}{h} = f[x(t_1), u(t_1)] \tag{7.26a}$$

or

$$x(t_1) = x(t_0) + hf[x(t_1), u(t_1)] \tag{7.26b}$$

or, generally,

$$x_{n+1} = x_n + hf[x_{n+1}, u_{n+1}] \tag{7.27}$$

Relation (7.27) is known as the Euler backward or Euler interpolation (implicit) algorithm. Once again, the accuracy of the solution depends on the length of the

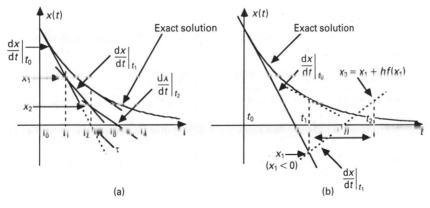

(a)   (b)

Figure 7.3 Interpretation of the Euler forward (extrapolation) algorithm. (a) For a small time step h. (b) For a large time step h

time step; however the choice of its value is not so critical as in the Euler forward algorithm, as will be shown in the next section. The graphical interpretations of this algorithm for small and large time steps are shown in Fig. 7.4(a, b). It should be noted, however, that this and other interpolation, or implicit, algorithms are generally not self-starting, that is, to find the solution at the next time step it is necessary to know the derivative at the next time step. Therefore, the interpolation algorithms are combined with extrapolation algorithms to create the predictor–corrector method. The extrapolation algorithms are used as predictors to find the approximate value of x_{n+1} and then the interpolation algorithms are used as correctors to correct the value of x_{n+1}.

However, for linear differential equations of the form, $dx/dt = ax(t) + bu(t) = f[x, u]$, application of Euler's interpolation formula gives

$$x_{n+1} = x_n + h(ax_{n+1} + bu_{n+1}) \tag{7.28}$$

This equation can be rearranged yielding

$$x_{n+1} = \frac{x_n + hbu_{n+1}}{1 - ha} \tag{7.29}$$

Now the value of x_{n+1} can be computed since the excitation (or forcing) function, $u(t)$, is usually known for all time; therefore, it is also known at the $n + 1$ time point. Such rearrangement is not possible for non-linear differential equations and thus other algorithms should be used.

The next simple approximation of the differential equation, giving better accuracy than Euler's algorithms, is the trapezoidal formula in which the average value of the derivatives at times t_n and t_{n+1} is used:

$$\frac{x_{n+1} - x_n}{h} = \frac{f[x_n, u_n] + f[x_{n+1}, u_{n+1}]}{2} \tag{7.30}$$

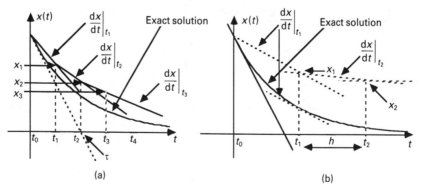

(a)   (b)

Figure 7.4   Graphical interpretation of the Euler backward (interpolation) algorithm. (a) For a small time step h. (b) For a large time step h

or

$$x_{n+1} = x_n + \frac{h}{2}\left(f[x_n, u_n] + f[x_{n+1}, u_{n+1}]\right) \tag{7.31}$$

The accuracy of the trapezoidal algorithm is much better than that of Euler's algorithms. It can be seen from Figs 7.3 and 7.4 that, for the same function, the use of forward approximation of a derivative gives values of x smaller than for the exact solution, and the use of backward approximation gives values of x greater than for the exact solution. Therefore, using the trapezoidal formula the errors are partially cancelled. A graphical interpretation of the trapezoidal algorithm is shown in Fig. 7.5.

Generally, the trapezoidal algorithm is not self-starting since on both sides of formula (7.31) we have the unknowns from the $(n + 1)$ time point. However, as for Euler's backward, this algorithm can be used for linear differential equations of the type

$$\dot{x} = ax + bu$$

written as

$$x_{n+1} = x_n + \frac{h}{2}\left(ax_{n+1} + bu_{n+1} + ax_n + bu_n\right) \tag{7.32}$$

or, after rearranging,

$$x_{n+1} = \frac{x_n\left(1 + \frac{h}{2}a\right) + \frac{h}{2}b(u_{n+1} + u_n)}{1 - \frac{h}{2}a} \tag{7.33}$$

Both interpolation algorithms, the Euler backward and the trapezoidal, can be modified and rendered self-starting by combining with the Euler forward algorithm.

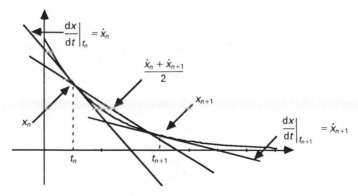

Figure 7.5 Graphical interpretation of the trapezoidal algorithm

1. Modified Euler:

$$x_{n+1} = x_n + hf[x_n + hf[x_n, u_n], u_{n+1}]$$
(7.34)

2. Modified trapezoidal:

$$x_{n+1} = x_n + \frac{h}{2} \left(f[x_n, u_n] + f[x_n + hf[x_n, u_n], u_{n+1}] \right)$$
(7.35)

The modified trapezoidal formula (7.35) is also known as the Heun algorithm. The modified Euler and Heun algorithms can be used for the solution of non-linear equations. The accuracy and stability of different integration algorithms is studied using the homogeneous differential equation

$$\frac{dx}{dt} = ax, \qquad x(0) = x_0$$
(7.36)

whose exact solution is known and equals

$$x(t) = x(0) \, e^{at}$$
(7.37)

To compare the algorithms described here, Euler's extrapolation (forward), modified Euler, interpolation Euler (backward) and Heun (modified trapezoidal), we can use the program 'DIFEQ1_0', shown in Program 7.1, in which the equation

$$f(x) = \frac{dx}{dt} = ax = -10x$$

is used for testing.

See the listing of the program DIFEQI_0 Directory of Chapter 7 on attached diskette.

The five algorithms can be tested repetitively with different time steps h. The exact values of solution (7.37) are also printed out for comparison. After running the program with $a = -10$, $h = 0.01$, $x_0 = 1$ we obtain the following results:

1. The Euler extrapolation:

```
t =       x =                exact solution
0.0       1.0                1.0
0.01      0.9                0.904837
0.02      0.81               0.817307
  .         .                  .
0.1       0.3486784          0.3678794
```

2. The Euler interpolation:

```
t =        x =                 exact solution
0.0        1.0                 1.0
0.01       0.90909             0.904837
0.02       0.826446            0.817307
.          .                   .
0.1        0.385543            0.3678794
```

3. The modified Euler:

```
t =        x =                 exact solution
0.0        1.0                 1.0
0.01       0.909999            0.904837
0.02       0.8280999           0.817307
.          .                   .
0.1        0.389416            0.3678794
```

4. The Heun (modified trapezoidal):

```
t =        x =                 exact solution
0.0        1.0                 1.0
0.01       0.9049999           0.904837
0.02       0.8190249           0.817307
.          .                   .
0.1        0.36854098          0.3678794
```

5. The trapezoidal:

```
t =        x =                 exact solution
0.0        1.0                 1.0
0.01       0.9047619           0.904837
0.02       0.8185941           0.817307
.          .                   .
0.1        0.3675725           0.3678794
```

However, after running the program with the time step $h = 0.1 = 1/|a|$ we obtain the results

1. The Euler extrapolation:

```
t =        x =                 exact solution
0.0        1.0                 1.0
0.1        0.0                 0.36787944
0.2        0.0                 0.13533528
.          .                   .
1.0        0.0                 4.5399929 E-5
```

2. The Euler interpolation:

```
t =        x =               exact solution
0.0        1.0               1.0
0.1        0.5               0.36787944
0.2        0.25              0.13533528
 .          .                 .
1.0        9.765625 E-4      4.5399929 E-5
```

3. The modified Euler:

```
t =        x =               exact solution
0.0        1.0               1.0
0.1        1.0               0.36787944
0.2        1.0               0.13533528
 .          .                 .
1.0        1.0               4.5399929 E-5
```

4. The Heun (modified trapezoidal):

```
t =        x =               exact solution
0.0        1.0               1.0
0.1        0.5               0.36787944
0.2        0.25              0.13533528
 .          .                 .
1.0        9.765625 E-4      4.5399929 E-5
```

5. The trapezoidal:

```
t =        x =               exact solution
0.0        1.0               1.0
0.1        0.33333           0.36787944
0.2        0.11111           0.13533528
 .          .                 .
1.0        1.6935 E-5        4.5399929 E-5
```

To see the influence of the value of the time step h on the accuracy of the calculations, let us run again the program with the time step $h = 0.2 = 2/|a|$. The results are

1. The Euler extrapolation:

```
t =        x =               exact solution
0.0         1.0              1.0
0.2        -1.0              1.3533528 E-1
0.4         1.0              1.83156388 E-2
0.6        -1.0              2.47875217 E-3
 .           .                .
2.0         1.0              2.06115362 E-9
```

2. The Euler interpolation:

t =	x =	exact solution
0.0	1.0	1.0
0.2	3.33333 E-1	1.3533528 E-1
0.4	1.11111 E-1	1.83156388 E-2
0.6	3.70370 E-2	2.47875217 E-3
.	.	.
2.0	1.69350 E-5	2.06115362 E-9

3. The modified Euler:

t =	x =	exact solution
0.0	1.0	1.0
0.2	3.0	1.3533528 E-1
0.4	9.0	1.83156388 E-2
0.6	27.0	2.47875217 E-3
.	.	.
2.0	5.9049 E+4	2.06115362 E-9

4. The Heun (modified trapezoidal):

t =	x =	exact solution
0.0	1.0	1.0
0.2	1.0	1.3533528 E-1
0.4	1.0	1.83156388 E-2
0.6	1.0	2.47875217 E-3
.	.	.
2.0	1.0	2.06115362 E-9

5. The trapezoidal:

t =	x =	exact solution
0.0	1.0	1.0
0.2	0.0	1.3533528 E-1
0.4	0.0	1.83156388 E-2
0.6		2.47875217 E-3
.	.	.
2.0	0.0	2.06115362 E-9

The results show that the Euler extrapolation and modified algorithms are useless for larger time steps; these will be discussed in the next section. Observe that for $h = 0.2 = -2/a$ the trapezoidal algorithm is also useless. This results from relation (7.33) with $u_{n+1} = u_n = 0$; inserting $h = -2/a$, always gives

$$x_{n+1} = x_n \left(\frac{1 - \frac{2}{a}\frac{a}{2}}{1 + \frac{2}{a}\frac{a}{2}} \right) = x_n \frac{0}{2} = 0$$

Note also that for $h = -2/a = 0.2$ the modified Euler algorithm is numerically unstable since the values of the solution are increasing instead of decreasing. For time steps larger than $2/|a|$ all extrapolation algorithms discussed here are numerically unstable.

For solutions of non-linear differential equations, only extrapolation algorithms can be used. Let us consider as an example a typical diode non-linear circuit as shown in Fig. 7.6 and apply the Heun algorithm to find the time response $v_C(t)$ for a given excitation voltage, $v_S(t)$.

The KCL for this circuit gives

$$i_D = i_C + i_R \tag{7.38}$$

with

$$i_D = I_S\left(\exp\left(\frac{v_d}{V_T}\right) - 1\right)$$

$$i_C = C\,\frac{dv_C}{dt}$$

$$i_R = \frac{v_C}{R}$$

$$v_D = v_S - v_C$$

$$V_T = \frac{kT}{q} \cong 0.026 \text{ V for } T = 300 \text{ K}$$

After rearrangement we obtain

$$\dot{v}_C = \frac{dv_C}{dt} = \frac{-1}{RC}\,v_C + \frac{I_S}{C}\left(\exp\left(\frac{v_S - v_C}{V_T}\right) - 1\right) \tag{7.39}$$

The program named 'HEUN' for the solution of the non-linear equation (7.39) is given in Program 7.2.

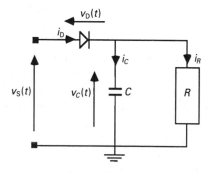

Figure 7.6   Typical diode dynamic circuit

> **See the listing of the program HEUN Directory of Chapter 7 on attached diskette.**

In this program the excitation function $v_S(t)$ is assumed to be periodic with time period $T_0 = 20$ ms:

$$v_S(t) = 10 \sin(2\pi \times 50t) \cong 10 \sin(314t)$$

Other constants in (7.39) can be entered into the program in the run-time. Assuming

$$I_S = 10^{-12} \text{ A}, \qquad R = 50 \text{ k}\Omega, \qquad C = 0.1 \text{ }\mu\text{F}$$

we obtain the time constant

$$\tau = 5 \times 10^{-3} \text{ s} = 5 \text{ ms, or } \tau(\text{ms}) = 5, \text{ and } v_S(t) = 10 \sin(0.314t)$$

Expressing the time, t, and time constant, τ, in milliseconds we can enter the resistance value R in kiloohms and the capacitance value C in microfarads:

$$\tau(\text{ms}) = 50 \times 0.1 = 5$$

To obtain accurate results we have to assume that the time step $h \ll \tau$ and $h \ll T_0$.

After running the program 'HEUN' let us enter $R = 50$ (kiloohms), $C = 0.1$ (microfarads), $I_S = 10^{-12}$ A, $h = 0.02$ (ms), $t_k = 30$ (ms) (t_k stands for the upper limit of the time range of calculation; $t_k = 30$ ms corresponds to 1.5 of the time period of the $v_S(t)$). The results printed out every one hundred time steps are

```
t =   0.0    v c = 0.0
t =   2.0    v c = 5.1544
t =   4.0    v c = 8.8077
t =   6.0    v c = 8.8762
t =   8.0    v c = 6.2035
t = 10.0    v c = 4.1583
t = 12.0    v c = 2.7874
t = 14.0    v c = 1.8684
t = 16.0    v c = 1.2524
t = 18.0    v c = 0.8395
t = 20.0    v c = 0.5627
t = 22.0    v c = 5.1670
t = 24.0    v c = 8.7977
t = 26.0    v c = 8.8855
t = 28.0    v c = 6.2161
t = 30.0    v c = 4.1668
```

Here we can observe the tracking of the sinusoidal input voltage in the first quarter of the sinusoidal time period and the exponential decay of $v_C(t)$ up to the beginning of the next sinusoidal time period (after 20 ms). It is necessary to note that keeping all the input data as before but increasing the amplitude of the input voltage from 10 V to 50 V leads to an incorrect time response.

After changing the amplitude in the 'Function g( )' of the program 'HEUN' and running the program we obtain the results every 100 time steps ($h = 0.02$)

```
t =  0.0   vc=    0.0
t =  2.0   vc= 297.4
t =  4.0   vc =  199.3
    .    .
t = 20.0   vc =    8.12
t = 22.0   vcC = 1088.2
```

However, a decrease in the time step, h, to the value $h = 0.01$ gives stable results: (the results are printed out every 200 time steps)

```
t =  0.0   vc =  0.0
t =  2.0   vc = 29.37
t =  4.0   vc = 46.85
t =  6.0   vc = 46.94
t =  8.0   vc = 32.80
    .    .
t = 20.0   vc =  2.97
t = 22.0   vc = 30.56
t = 24.0   vc = 46.85
```

One of the most accurate single-step extrapolation algorithms is the Runge–Kutta fourth-order algorithm. Besides high accuracy it allows us to use a relatively large time step, h (larger than in the algorithms discussed so far). The Runge–Kutta algorithm is used as a predictor for interpolation and multi-step algorithms.

In this algorithm the time step is divided into two parts.

Thus, for a known value of x_n and known excitation function u_n, $u_{n+1/2}$ and u_{n+1} at the time points t_n, $t_n + h/2$ and $t_n + h$, respectively, the x_{n+1} value is computed according to the formula

$$x_{n+1} = x_n + \frac{(K_1 + 2K_2 + 2K_3 + K_4)}{6} \tag{7.40a}$$

where the coefficients K_i are

$$K_1 = hf[x_n, u_n]$$
$$K_2 = hf[x_n + 0.5K_1, u_{n+0.5}] \tag{7.40b}$$
$$K_3 = hf[x_n + 0.5K_2, u_{n+0.5}]$$
$$K_4 = hf[x_n + K_3, u_{n+1}]$$

The program 'RUNGEKUT' with the Runge–Kutta fourth-order extrapolation algorithm for the test differential equation solution $dx/dt = ax$, $x(0) = x_0$, is given in Program 7.3.

> **See the listing of the program RUNGEKUT Directory of Chapter 7 on attached diskette.**

After running the program with $a = -10$, $h = 0.01$, $x_0 = 1$ we obtain the results

```
t =        x =                    exact solution
0.0        1.0                    1.0
0.01       9.0483750 E-1          9.04837418 E-1
0.02       8.1873090 E-1          8.18730753 E-1
.          .
0.1        3.6787977 E-1          3.6787944 E-1
```

Repeating the computation with $h = 0.1$ gives the results

```
t =        x =                    exact solution
0.0        1.0                    1.0
0.1        0.3750                 3.6787944 E-1
0.2        0.140625               1.3533528 E-1
.          .
1.0        5.49936667E-5          4.5399929 E-5
```

For $h = 0.2 = 2/|a|$ we also obtain stable results:

```
t =        x =                    exact solution
0.0        1.0                    1.0
0.2        3.33333 E-1            1.3533528 E-1
0.4        1.11111 E-1            1.8315638 E-2
.          .
2.0        1.62350878 E-5         2.061153 E-9
```

Comparison of these results with those obtained from the trapezoidal algorithm show much better accuracy for the Runge–Kutta algorithm.

7.4 Accuracy and stability of single-step algorithms

The accuracy of the solutions of differential equations using different numerical algorithms can be expressed in terms of error ε_{n+1} introduced at time $t = t_{n+1} = t_0 + (n+1)h$. The error ε_{n+1} is usually defined as the difference of the exact value x_{n+1}^e of the solution and the computed approximate value x_{n+1}^a. Based on the results of Section 7.3 we can say that the best accuracy is obtained using the Runge–Kutta algorithm, while the worst accuracy or the largest error is produced by the Euler extrapolation algorithm. The error ε_{n+1} is different for different algorithms and is inherent in the method. To show the dependence of the error on the time step h and the method let us express the exact value x_{n+1}^e of the solution at time

$t = t_{n+1}$ in the Taylor series:

$$x_{n+1}^e = x_n + hx_n' + \frac{h^2}{2!} x_n'' + \frac{h^3}{3!} x_n''' + O(h^4) \tag{7.41}$$

where $O(h^4)$ represents the fourth- and higher-order terms of the expression. Now we can use formulae for different algorithms to express the approximate values x_{n+1}^a.

For the Euler extrapolation algorithm we have

$$x_{n+1}^a = x_n + hx_n'$$

Subtracting this formula from (7.41) we obtain the error for Euler's forward algorithm:

$$\varepsilon_{n+1} = x_{n+1}^e - x_{n+1}^a = \frac{h^2}{2} x_n'' + O(h^3) \tag{7.42a}$$

or

$$\varepsilon_{n+1} = \frac{h^2}{2} x_\xi'' \tag{7.42b}$$

where

$$x_\xi'' = x''(\xi) \qquad \text{for } nh \leqslant \xi \leqslant (n+1)h$$

This error arises in the Euler forward method because in this method the Taylor series expansion (7.41) is truncated leaving only the first two terms. Therefore, the error ε_{n+1} is referred to as the truncation error. The approximate value x_{n+1}^a for the Euler backward algorithm is

$$x_{n+1}^a = x_n + hx_{n+1}'$$

but the Taylor series expansion of the derivative x_{n+1}' is

$$x_{n+1}' = x_n' + hx_n'' + \frac{h^2}{2} x_n''' + O(h^4) \tag{7.43}$$

Thus

$$x_{n+1}^a = x_n + hx_n' + h^2 x_n'' + O(h^3) \tag{7.44}$$

Subtracting this relation from (7.41) gives the error for the Euler backward method:

$$\varepsilon_{n+1} = x_{n+1}^e - x_{n+1}^a = \frac{h^2}{2} x_n'' - h^2 x_n'' + O(h^3) \tag{7.45a}$$

or

$$\varepsilon_{n+1} = -\frac{h^2}{2} x_\xi'' \tag{7.45b}$$

where

$$x''_\xi = x''(\xi) \qquad \text{for } nh \leqslant \xi \leqslant (n+1)h$$

Note that the magnitudes of the errors for Euler's methods are equal but the signs are different. Similarly, we proceed by inserting relation (7.43) into the trapezoidal formula

$$x^a_{n+1} = x_n + \frac{h}{2} x'_n + \frac{h}{2} x'_{n+1}$$

$$= x_n + \frac{h}{2} x'_n + \frac{h}{2} x'_n + \frac{h^2}{2} x''_n + \frac{h^3}{4} x'''_n + O(h^4) \tag{7.46}$$

Substitution of (7.46) from (7.41) gives the error ε_{n+1} for the trapezoidal method:

$$\varepsilon_{n+1} = x^e_{n+1} - x^a_{n+1} = \left[\frac{h^3}{6} - \frac{h^3}{4}\right] x'''_n + O(h^4) \tag{7.47a}$$

$$\varepsilon_{n+1} = -\frac{h^3}{12} x'''_\xi \tag{7.47b}$$

where

$$x'''_\xi = x'''(\xi) \qquad \text{for } nh \leqslant \xi \leqslant (n+1)\,h$$

The truncation error can be found similarly for other algorithms.

Another important property of the solution methods discussed is the stability of numerical algorithms. Stability is a global property of a given algorithm and is related to the decay or growth of errors generated at each time point which are transferred (or propagated) to successive time points during the computation of the solution.

Again, based on the results of Section 7.3 we can observe that the stability depends on the method and the time step h. Let us return back to the program 'DIFEQ1__0' of Program 7.1 and select the Euler extrapolation algorithm. To show the details of the instability let us modify the 'Exteuler' procedure to be able to print out the results at each time step (remove the line: if(n mod $10 = 0$)then) and run the program with $a = -10$, $h = 0.2$ and $h = 0.25$. The results will be as follows:

```
for : h = 0.2              for h = 0.25

t = 0.0    x =    +1.0     t = 0.0     x =    +1.0
t = 0.2    x =    -1.0     t = 0.25    x =    -1.5
t = 0.4    x =    +1.0     t = 0.5     x =    +2.25
t = 0.6    x =    -1.0     t = 0.75    x =    -3.375
t = 0.8    x =    +1.0     t = 1.0     x =    +5.0625
t = 1.0    x =    -1.0     t = 1.25    x =    -7.59375
  .          .      .        .           .      .
  .          .      .        .           .      .
```

We see that for $h = 2/|a|$ in the Euler forward algorithm, we obtain the relaxation of the solution between $x = +1.0$ and $x = -1.0$. For $h > 2/|a|$ we obtain the relaxation and increase of the absolute value $|x|$, instead of a decay to zero as $n \to \infty$. Similar situations exist for other algorithms.

Therefore, it is necessary to consider what are the limits for the time step h value for different algorithms. These considerations are usually based on the homogeneous test equation

$$\dot{x}(t) = \frac{dx(t)}{dt} = ax(t), \qquad x(0) = x_0$$

which is typical for the simple RC circuit excited by the initial condition $x(0) = x_0$, with $a = -1/RC = -1/\tau$. For the Euler forward formula we can write the values of x for the successive time points:

$$x_1 = x_0 + hax_0 = (1 + ha)x_0$$
$$x_2 = x_1 + hax_1 = (1 + ha)^2 x_0$$
$$\vdots$$
$$x_n = (1 + ha)^n x_0$$

But we know that the exact solution of the test equation is

$$x(t) = x_0 e^{at} = x_0 e^{-t/\tau}$$

and for $a < 0, x(t) \to 0$ for $t \to \infty$. Therefore we should have $x_n \to 0$ for $n \to \infty$. Thus, the stability condition requires

$$|1 + ha| < 1 \tag{7.48a}$$

or

$$\left| 1 - \frac{h}{\tau} \right| < 1 \tag{7.48b}$$

The time step h is always positive but the coefficient a can generally be complex with negative and positive both real and imaginary parts. Restricting for simplicity to real values of a we can state that fulfilment of stability condition (7.48a) requires a to be negative, such that

$$-2 < ha < 0 \tag{7.49a}$$

or

$$-2 < \sigma' < 0 \tag{7.49b}$$

where $\sigma' = ha$. This can be expressed graphically, as shown in Fig. 7.7(a).

Note, however, that for $ha = -1$ this algorithm becomes useless since we obtain

$$x_1 = (1 + ha)x_0 = (1 - 1)x_0 = 0; \quad x_2 = 0, \ldots$$

Proceeding, we can find in a similar way the stability requirements for other algorithms already discussed.

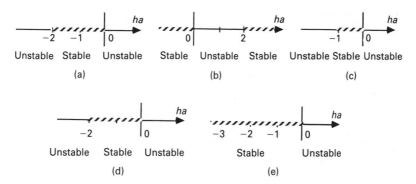

Figure 7.7   Graphical interpretations of stability regions for algorithms. (a) Euler forward.
(b) Euler backward. (c) Modified Euler. (d) Heun. (e) Trapezoidal

1. Euler backward:

$$x_n = \frac{x_0}{(1 - ha)^n}$$

or stability condition:

$$|1 - ah| > 1 \tag{7.50}$$

This requires that ah should be either negative, $ah < 0$, or positive and greater than 2, that is,

$$\sigma' = ah < 0 \quad\text{and}\quad 2 < ah = \sigma' \tag{7.51}$$

Thus, the instability region for this algorithm is bound to

$$0 < ah < 2 \tag{7.52}$$

which is shown in Fig. 7.7(b).

2. Modified Euler (combination of forward and backward):

$$x_n = [1 + ha + (ha)^2]^n x_0$$

or stability condition:

$$|1 + ha + (ha)^2| < 1 \tag{7.53}$$

This condition requires ha to be negative in the range (see Fig. 7.7(c))

$$-1 < ha < 0 \tag{7.54}$$

3. Heun (modified trapezoidal):

$$x_n = \left[1 + ha + \frac{(ha)^2}{2}\right]^n x_0$$

with stability condition

$$\left| 1 + ha + \frac{(ha)^2}{2} \right| < 1 \tag{7.55}$$

This condition requires ha to be negative in the range

$$-2 < ha < 0 \tag{7.56}$$

which is shown in Fig. 7.7(d).

    4. Trapezoidal:

$$x_n = \left[\frac{1 + \dfrac{ha}{2}}{1 - \dfrac{ha}{2}} \right]^n x_0$$

for which the stability condition is

$$\left| \frac{1 + \dfrac{ha}{2}}{1 - \dfrac{ha}{2}} \right| < 1 \tag{7.57}$$

This condition is satisfied for all negative values of ha (see Fig. 7.7(e)):

$$ha < 0 \tag{7.58}$$

Note, however, that for $ha = -2$ the trapezoidal method becomes useless since in the first and other time steps we obtain

$$x_1 = \frac{1 + \dfrac{ha}{2}}{1 - \dfrac{ha}{2}} x_0 = \frac{0}{2} x_0 = 0$$

$$x_2 = 0, \ldots, x_n = 0$$

    If ha is complex, with $ha = \sigma' + j\omega'$ then we obtain simple regular figures on the $\sigma' + j\omega'$ plane for stable or unstable regions for Euler forward, Euler backward and trapezoidal algorithms, as shown in Fig. 7.8.

    For the Euler forward algorithm relation (7.48a) can be rewritten as

$$|1 + \sigma' + j\omega'| < 1 \tag{7.59a}$$

or

$$(1 + \sigma')^2 + (\omega')^2 < 1 \tag{7.59b}$$

which corresponds to the region inside a circle with centre at $(-1, 0)$ and radius $r = 1$ at the $\sigma' + j\omega'$ plane.

    For the Euler backward algorithm relation (7.50) becomes

$$1 < |1 - \sigma' - j\omega| \tag{7.60a}$$

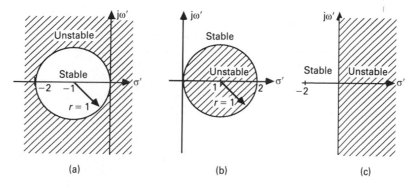

Figure 7.8 The stability regions for the algorithms. (a) Euler forward. (b) Euler backward. (c) Trapezoidal

or

$$1 < (1 - \sigma')^2 + (\omega')^2 \tag{7.60b}$$

which corresponds to the region outside a circle with centre at $(1, 0)$ and radius $r = 1$ at the $\sigma' + j\omega'$ plane.

For the trapezoidal algorithm, relation (7.57) becomes

$$\left| \frac{2 + \sigma' + j\omega'}{2 - \sigma' - j\omega'} \right| < 1 \tag{7.61a}$$

or

$$\frac{(2 + \sigma')^2 + (\omega')^2}{(2 - \sigma')^2 + (\omega')^2} < 1 \tag{7.61b}$$

After rearrangement we obtain

$$\sigma' < 0 \tag{7.61c}$$

which shows that the stability region corresponds to the open left half plane.

The stability condition for the modified Euler algorithm can be rewritten as

$$\{1 + \sigma' + (\sigma')^2 - (\omega')^2\}^2 + (\omega')^2 (1 + 2\sigma')^2 < 1 \tag{7.62}$$

and for the Heun algorithm as

$$\left\{ 1 + \sigma' + \frac{(\sigma')^2}{2} - \frac{(\omega')^2}{2} \right\}^2 + (\omega')^2 (1 + \sigma')^2 < 1 \tag{7.63}$$

The stability regions for these two algorithms are shown in Fig. 7.9(a, b).

It should be noted that fulfilment of the stability condition means that the errors in calculation do not grow in subsequent time steps; but this does not imply that the solution is correct. The accuracy of the solution is another problem, as was discussed earlier in this section.

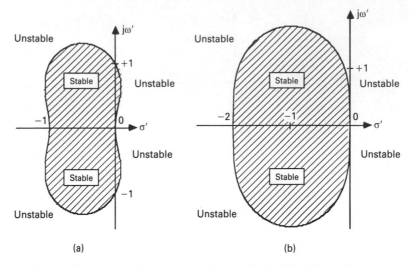

Figure 7.9   The stability regions for the algorithms. (a) Modified Euler. (b) Heun (modified trapezoidal)

7.5 Predictor–corrector method

As it was mentioned in Section 7.3 the interpolation algorithms are generally not 'self-starting' since on the right-hand side of the formulas we have derivatives at the next time point $n + 1$ but the value x_{n+1} is not known yet. Therefore, we use the extrapolation algorithm to compute or predict the approximate value x_{n+1} and use it in the interpolation algorithm to correct the value of x_{n+1}. These corrections may be repeated several times to increase the accuracy. For example, assume that the Euler forward algorithm is a predictor and the Euler backward algorithm is a corrector in order to solve the equation

$$\frac{\mathrm{d}x}{\mathrm{d}t} = -10x, \quad \text{with } x_0 = 1 \text{ and time step } h = 0.02$$

1. The predicted x_{1p} value is ($t = 0.02$)

    $$x_{1p} = x_0 + hax_0 = 1 - 0.2 \times 1 = 0.8$$

2. The corrected x_{1c} value is ($t = 0.02$)

    $$x_{1c} = x_0 + hax_{1p} = 1 - 0.2 \times 0.8 = 1 - 0.16 = 0.84$$

3. The second predicted x_{2p} value is ($t = 0.04$)

    $$x_{2p} = x_{1c} + hax_{1c} = 0.84 - 0.2 \times 0.84 = 0.672$$

4.   And the second corrected x_{2c} value is ($t = 0.04$)

$$x_{2c} = x_{1c} + hax_{2p} = 0.84 - 0.2 \times 0.672 = 0.7056$$

5.   And so on.

To increase the accuracy we may repeat the corrections in steps (2) and (4) to compute doubly corrected values of x_{1cc} and x_{2cc}:

2a. The doubly corrected value of x_{1cc} is ($t = 0.02$)

$$x_{1cc} = x_0 + hax_{1c} = 1 - 0.2 \times 0.84 = 0.832$$

3.   The second predicted x_{2p} value is ($t = 0.04$)

$$x_{2p} = x_{1cc} + hax_{1cc} = 0.832 - 0.2 \times 0.832 = 0.6656$$

4.   The second corrected x_{2c} value is ($t = 0.04$)

$$x_{2c} = x_{1cc} + hax_{2p} = 0.832 - 0.2 \times 0.6656 = 0.698\ 88$$

4a. The doubly corrected value of x_{2cc} is ($t = 0.04$)

$$x_{2cc} = x_{1cc} + hax_{2c} = 0.832 - 0.2 \times 0.698\ 88 = 0.692\ 22$$

etc.

An illustration of the predictor–corrector method is given in program 'PRED-COR' shown in Program 7.4.

See the listing of the program **PRED-COR** Directory of Chapter 7 on attached diskette.

This program contains two procedures. In the first one 'Predcoreu' the Euler extrapolation algorithm is used as a predictor and the Euler interpolation algorithm as a corrector. In the second procedure 'Predcortr', the Euler extrapolation algorithm is again used as a predictor and the trapezoidal algorithm as a corrector. The corrections can be iterated two or more times by proper entry of the 'ite' parameter. The exact solution values and the error $eps = x_e - x_a (x_e - exact, x_a$ *approximated values of the solution*) are printed out for comparison. After running this program with $a - -10$, $x_0 - 1$, $h = 0.02$ and $ite = 1$ we obtain the results:

1.   Euler backward correction

```
t =       x =          x₀*exp(at)        eps = xₑ -xₐ
0.0      1.0           1.0               0.0
0.02     0.84          0.81879           -2.1269 E-2
0.04     0.7056        0.67032           -3.52799 E-2
 .        .             .                 .
0.2      0.1749        0.13533528        -3.95659 E-2
```

2.   Trapezoidal correction

t =	x =	$x_0 * exp(at)$	$eps = x_e - x_a$
0.0	1.0	1.0	0.0
0.02	0.798	0.81879	2.073 E-2
0.04	0.6368	0.67032	3.3516 E-2
.	.	.	.
0.2	0.1047198	0.13533528	3.0615 E-2

Repeating the computation with $ite = 2$ yields:

3.   Euler backward correction

t =	x =	$x_0 * exp(at)$	$eps = x_e - x_a$
0.0	1.0	1.0	0.0
0.02	0.832	0.81879	-1.3269 E-2
0.04	0.692224	0.67032	-2.19 E-2
.	.	.	.
0.2	0.15894	0.13533528	-2.36047 E-2

4.   Trapezoidal correction

t =	x =	$x_0 * exp(at)$	$eps = x_e - x_a$
0.0	1.0	1.0	0.0
0.02	0.8202	0.81879	-1.4692 E-3
0.04	0.672728	0.67032	-2.40799 E-3
.	.	.	.
0.2	0.13778	0.13533528	-2.448355 E-3

Increasing the iterations to $ite = 3$ does not decrease the error in the Euler backward correction, but it does decrease the error in the trapezoidal correction a little.

7.6 Multi-step algorithms for the solution of differential equations

In previous sections of this chapter only single-step algorithms have been considered, in which only x_n, x_n' and x_{n+1}, x_{n+1}' for the time points t_n and t_{n+1} have been taken into account.

Generally, more values of x and derivatives x' with different weights, from previous time points, can be included in numerical integration algorithms, leading to the multi-step formula

$$x_{n+1} = \sum_{i=0}^{k} a_i x_{n-i} + h \sum_{j=-1}^{m} b_j x_{n-j}' \qquad (7.64)$$

Based on relation (7.64) we can make a classification of different methods. The formulas for which only $a_0 \neq 0$, $b_0 \neq 0$ and $b_{-1} \neq 0$ are the single-step algorithms discussed earlier. For example, for $a_0 = 1$, $b_0 = 1$ and $b_{-1} = 0$ we obtain the Euler

extrapolation algorithm and for $a_0 = 1$, $b_0 = 0$ and $b_{-1} = 1$ we obtain the Euler interpolation algorithm; for $a_0 = 1$, $b_0 = 1/2$ and $b_{-1} = 1/2$ we obtain the trapezoidal
algorithm. The formulas where only $b_{-1} = 0$ represent extrapolation or explicit
multi-step algorithms, whereas the formulas $b_{-1} \neq 0$ represent interpolation or
implicit multi-step algorithms. The special case of interpolation algorithms where
only $b_{-1} \neq 0$ represents integration methods and which are referred to as backward
differentiation formulas.

7.6.1 Multi-step extrapolation algorithms

The most popular multi-step extrapolation algorithms for the solution of differential equations are the Adams–Bashforth formulas where only $a_0 \neq 0 = 1$ and
$b_{-1} = 0$ (in relation (7.64)). Other weights b_i have different positive and negative
values. The algorithms for the first three orders have the forms:

1. The first order:

$$x_{n+1} = x_n + hf[x_n, u_n] \tag{7.65}$$

2. The second order:

$$x_{n+1} = x_n + h\left(\frac{3}{2} f[x_n, u_n] - \frac{1}{2} f[x_{n-1}, u_{n-1}]\right) \tag{7.66}$$

3. The third order:

$$x_{n+1} = x_n + h\left(\frac{23}{12} f[x_n, u_n] - \frac{4}{3} f[x_{n-1}, u_{n-1}] + \frac{5}{12} f[x_{n-2}, u_{n-2}]\right) \tag{7.67}$$

Note that the first-order Adams–Bashforth algorithm is a single-step Euler extrapolation algorithm. The higher-order algorithms are multi-step and are not self-
starting. To start the second-order algorithm the first value of x_1 should be computed using any of the single-step extrapolation algorithms.

For example, to use the Adams–Bashforth second-order algorithm to solve the
test equation $x' = f(x) = -10x$, with $x_0 = 1$ and $h = 0.02$, we can use the
Adams–Bashforth first-order (Euler) algorithm as the starting time point t_1; for
higher time points, the second-order algorithm can be used in a loop:

1. First order:

$$x_1 = x_0 + h(-10x_0) = 1 - 0.2 = 0.8$$

2. Second order:

$$x_2 = x_1 + h\left(\frac{3}{2} (-10x_1) - \frac{1}{2} (-10x_0)\right) = 0.66$$

3. Second order:

$$x_3 = x_2 + h\left(\frac{3}{2} (-10x_2) - \frac{1}{2} (-10x_1)\right) = 0.542$$

4. Second order:

$$x_4 = \text{etc.}$$

The example program 'AD-BASH' showing the application of the second-order Adams–Bashforth algorithm with forward Euler formula as a starting point is given in Program 7.5.

See the listing of the program AD-BASH Directory of Chapter 7 on attached diskette.

The error $eps = x_e - x_a$, which represents the difference of exact solution and computed solution, is also printed out for comparison.

After running the program for the test equation $x' = ax$, $x(0) = x_0$, with $a = -10$, $x_0 = 1$, $h = 0.02$ we obtain the results:

```
t=        x a=               x0*exp(at)        eps = x e -x a
0.0       1.0               1.0               0.0
0.02      0.8               0.81873           1.873 E-2
0.04      0.66              0.67032           1.032 E-2
 .         .                 .                 .
0.12      0.300726          0.301194          4.6821 E-4
 .         .                 .                 .
0.2       0.37095           0.135335          -1.7598 E-3
 .         .                 .                 .
0.3       5.135538 E-2      4.9787 E-2        -1.5683 E-3
 .         .                 .                 .
0.4       1.92375 E-2       1.83156 E-2       -9.21917 E-4
```

7.6.2 Multi-step interpolation algorithms

A family of popular multi-step interpolation algorithms, with only $a_0 \neq 0 = 1$ (in relation (7.64)), are known as the Adams–Moulton formulas.

1. The first order:

$$x_{n+1} = x_n + hf[x_{n+1}, u_{n+1}] \tag{7.68}$$

2. The second order:

$$x_{n+1} = x_n + \frac{h}{2}(f[x_{n+1}, u_{n+1}] + f[x_n, u_n]) \tag{7.69}$$

3. The third order:

$$x_{n+1} = x_n + \frac{h}{12} \left(5f[x_{n+1}, u_{n+1}] + 8f[x_n, u_n] - f[x_{n-1}, u_{n-1}] \right) \tag{7.70}$$

Note that the Adams–Moulton first- and second-order algorithms correspond to the Euler interpolation and trapezoidal algorithms, respectively. The Adams–Moulton algorithms are not self-starting since they are implicit, and starting from the third order they require additional starting points as in Adams–Bashforth formulas. However, compared to Adams–Bashforth algorithms, Adams–Moulton algorithms require one less starting point for a given order. For example, the second-order (trapezoidal) algorithm can start using only one single-step extrapolation algorithm (see the Heun algorithm in Section 7.3) and the third-order algorithm requires two extrapolation algorithms.

Slight modification of the program 'AD-BASH' of Program 7.5 leads to the program 'AD-MOULT' shown in Program 7.6, in which the third-order Adams–Moulton algorithm is used with the Euler extrapolation and the Adams–Bashforth second-order algorithms are used as starting points.

> **See the listing of the program AD-MOULT Directory of Chapter 7 on attached diskette.**

Running this program for the differential test equation with $a = -10$, $x_0 = 1$, $h = 0.02$ gives the results:

t=	x_a=	$x_0 * exp(at)$	eps = $x_e - x_a$
0.0	1.0	1.0	0.0
0.02	0.8	0.81873	1.873 E−2
0.04	0.655	0.67032	1.532 E−2
.	.	.	.
0.12	0.239984	0.301194	7.19577 E−3
.	.	.	.
0.2	0.13196	0.135335	3.37286 E−3
.	.	.	.
0.3	4.8482 E−2	4.9787 E−2	1.3049 E−3
.	.	.	.
0.4	1.7812 E−2	1.83156 F−2	5.03609 E−4

Comparison of these results with those of the Adams–Bashforth second-order algorithm shows that the errors are of similar order. The high errors at the beginning of the computations are due to the inaccuracy of the Euler forward algorithm which is used as the first starting point. Insertion of the Runge–Kutta algorithm instead of the Euler forward algorithm as the first starting point is used in the program 'AD-MOURK' shown in Program 7.7.

> **See the listing of the program AD-MOURK Directory of Chapter 7 on attached diskette.**

Running this program with $a = -10$, $x_0 = 1$, $h = 0.02$ gives the results:

t =	$x_a =$	$x_0*\exp(at)$	eps = $x_e - x_a$
0.02	0.8187333	0.81873	-2.58 E-6
0.04	0.67014	0.67032	1.77268 E-4
.	.	.	.
0.12	0.300796	0.301194	3.9785 E-4
.	.	.	.
0.2	0.1350137	0.135335	3.2157 E-4
.	.	.	.
0.3	4.960317 E-2	4.9787 E-2	1.83897 E-4
.	.	.	.
0.4	1.822388 E-2	1.83156 E-2	9.1752 E-5

Here we have seen the great importance of starting point accuracy on the overall computations.

A very important group of interpolation algorithms, belonging to the backward differentiation formulas (with only $b_{-1} \neq 0$ in relation (7.64)), are known as the Gear algorithms. These are especially useful in solutions of so-called stiff differential equations (in which very small and very large time constants may exist at the same time). The first three orders of the Gear algorithms have the forms:

1. The first order:

$$x_{n+1} = x_n + hf[x_{n+1}, u_{n+1}] \tag{7.71}$$

2. The second order:

$$x_{n+1} = \frac{4}{3} x_n - \frac{1}{3} x_{n-1} + \frac{2h}{3} f[x_{n+1}, u_{n+1}] \tag{7.72}$$

3. The third order:

$$x_{n+1} = \frac{18}{11} x_n - \frac{9}{11} x_{n-1} + \frac{2}{11} x_{n-2} + \frac{6h}{11} f[x_{n+1}, u_{n+1}] \tag{7.73}$$

The first-order algorithm corresponds to Euler's interpolation formula. The example program 'GEAR-2' uses the Gear second-order algorithm with the Runge–Kutta formula as a starting point and the Adams–Bashforth second-order algorithm as a predictor. This can be found in Program 7.8.

> **See the listing of the program GEAR-2 Directory of Chapter 7 on attached diskette.**

The stability of multi-step algorithms is widely discussed in the literature.

7.7 Circuit representation of interpolation algorithms; associated or companion models for capacitors and inductors

In the preceding sections of this chapter we have applied various algorithms to a differential equation describing given RC or RL circuits. Another approach is to apply any interpolation algorithm only to the capacitive branch or inductive branch in the RLC circuit. As a result we obtain the equivalent circuits for capacitors and/or inductors which are referred to as companion models or associated models with given interpolation algorithms. In such cases, associated models have constant parameters for one time step, converting the dynamic circuit into an equivalent direct-current circuit, allowing us to use the solution methods applicable to the DC circuit, such as writing nodal equations and solving them using the LU decomposition method or others.

After solving the set of equations for a given time step, the parameters of the associated models are updated, and the solution of the equations is performed again, etc. Thus, the analysis of dynamic circuits is converted into multiple analysis of equivalent (associated) DC circuits.

7.7.1 Companion or associated models for capacitor

The capacitor branch current–voltage relation (7.1) can be rewritten as

$$\frac{dv}{dt} = f(t) = \frac{1}{C} i(t) \tag{7.74}$$

Then, for the nth and $(n+1)$th time step we can write

$$f_n = \frac{1}{C} i_n, \qquad f_{n+1} = \frac{1}{C} i_{n+1} \tag{7.75}$$

Applying, for example, the Euler interpolation algorithm to equation (7.74) with the notation used in (7.75), we obtain

$$v_{n+1} = v_n + h f_{n+1} = v_n + h \frac{1}{C} i_{n+1} = v_n^* + R_C i_{n+1} \tag{7.76}$$

Relation (7.76) can be interpreted as a description of the electronic circuit at the $(n+1)$st time point, where $R_C = h/C$ is the equivalent resistance and v_n^* is the voltage source representing the voltage at the nth or previous time point, as is shown in Fig. 7.10(a).

The associated circuit with voltage source or Thevenin circuit can be transformed into a Norton equivalent circuit by rewriting relation (7.76):

$$i_{n+1} = \frac{C}{h} v_{n+1} - \frac{C}{h} v_n = G_C v_{n+1} - i_n^* \tag{7.77}$$

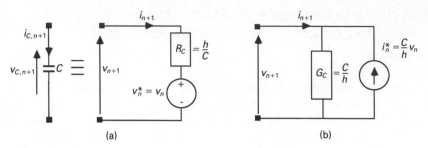

Figure 7.10   Model of capacitor associated with Euler's interpolation algorithm. (a) Thevenin's circuit. (b) Norton's circuit

where

$$G_C = \frac{C}{h} \text{ and } i_n^* = \frac{C}{h}\, v_n$$

as is shown in Fig. 7.10(b).

When we apply the trapezoidal (second-order Adams–Moulton) algorithm to relation (7.74) we obtain

$$v_{n+1} = v_n + \frac{h}{2}\,(f_{n+1} + f_n) = v_n + \frac{h}{2C}\, i_n + \frac{h}{2C}\, i_{n+1} \tag{7.78}$$

or

$$v_{n+1} = v_n^* + R_C i_{n+1} \tag{7.79a}$$

with

$$v_n^* = v_n + \frac{h}{2C}\, i_n, \qquad R_C = \frac{h}{2C} \tag{7.79b}$$

which corresponds to the circuit of Fig. 7.10(a). Relation (7.78) can be transformed into the form

$$i_{n+1} = \frac{2C}{h}\, v_{n+1} - \left(\frac{2C}{h}\, v_n + i_n\right) \tag{7.80}$$

or

$$i_{n+1} = G_C v_{n+1} - i_n^* \tag{7.81a}$$

with

$$G_C = \frac{2C}{h}, \qquad i_n^* = \frac{2C}{h}\, v_n + i_n \tag{7.81b}$$

which corresponds to the Norton equivalent circuit of Fig. 7.10(b).

Similarly, we can use the higher-order Adams–Moulton algorithms to find the parameters R_C, G_C, v_n^* and i_n^* of the associated models of the capacitor. However, these algorithms have some drawbacks, since, besides the knowledge of voltages in the past time points, it is also necessary to compute the currents in these time points. Better in this sense, due to improved stability, are the Gear algorithms.

Application of the second-order Gear algorithm to the capacitive branch relation gives

$$v_{n+1} = \frac{4}{3}\, v_n - \frac{1}{3}\, v_{n-1} + \frac{2h}{3}\, f_{n+1} \tag{7.82}$$

or

$$v_{n+1} = \frac{4}{3}\, v_n - \frac{1}{3}\, v_{n-1} + \frac{2h}{3C}\, i_{n+1} \tag{7.83a}$$

or

$$v_{n+1} = v_n^* + R_C i_{n+1} \tag{7.83b}$$

with

$$v_n^* = \frac{1}{3}\,(4v_n - v_{n-1}), \qquad R_C = \frac{2h}{3C} \tag{7.83c}$$

which corresponds to the Thevenin equivalent of Fig. 7.10(a). Transformation into the Norton equivalent gives

$$i_{n+1} - \frac{3C}{2h}\, v_{n+1} - \frac{C}{2h}\,(4v_n - v_{n-1}) \tag{7.84a}$$

or

$$i_{n+1} = G_C v_{n+1} - i_n^* \tag{7.84b}$$

with

$$G_C = \frac{3C}{2h}, \qquad i_n^* = \frac{C}{2h}\,(4v_n - v_{n-1}) \tag{7.84c}$$

As we see in this model, only the voltages on the capacitor are taken into account.

Applying the third-order Gear algorithm we obtain the relation

$$v_{n+1} = \frac{18}{11}\, v_n - \frac{9}{11}\, v_{n-1} + \frac{2}{11}\, v_{n-2} + \frac{6}{11}\frac{h}{C}\, i_{n+1} \tag{7.85a}$$

Thus the elements of the associated model of Thevenin and Norton equivalents are

$$R_C = \frac{6}{11}\frac{h}{C}, \qquad v_n^* = \frac{18}{11}\, v_n - \frac{9}{11}\, v_{n-1} + \frac{2}{11}\, v_{n-2} \tag{7.85b}$$

and

$$G_C = \frac{11}{6}\frac{C}{h}, \qquad i_n^* = \frac{3C}{h}\, v_n - \frac{3C}{2h}\, v_{n-1} + \frac{C}{3h}\, v_{n-2} \tag{7.85c}$$

Application of higher-order multi-step algorithms requires the use of lower-order formulae to initialize computations. Note, however, that computations based on the Euler interpolation and trapezoidal associated models are self-starting.

An example program 'ASSOCMOD' with two procedures 'AssoEuler' and 'AssGear3' is given in Program 7.9. In the procedure 'AssoEuler', the capacitor model associated with the Euler algorithm is used and in the procedure 'AssGear3', the capacitor model associated with the Gear third-order algorithm, is used.

> **See the listing of the program ASSOCMOD Directory of Chapter 7 on attached diskette.**

A simple RC circuit shown in Fig. 7.11(a) is considered where $e_S(t) = E_S u(t)$ ($u(t)$ is the unit step function). Its associated circuit is shown in Fig. 7.11(b). The exact solution for $v_{Ce}(t)$ for the circuit of Fig. 7.11(a) is known:

$$v_{Ce}(t) = E_S + [v_C(0) - E_S]\, \exp\left(-\frac{t}{RC}\right) \tag{7.86}$$

and is included in the program for printing out and for the computation of the error $eps = v_{Ce} - v_C$.

The approximate calculation of $v_C(t)$ based on the associated circuit of Fig. 7.11(b) is from KCL:

$$-e_{n+1}G + v_{Cn+1}G + v_{Cn+1}G_C - i_n^* = 0 \tag{7.87}$$

or

$$v_{Cn+1} = \frac{Ge_{n+1} + i_n^*}{G + G_C} \tag{7.88}$$

1. For Euler's algorithm we have

$$G_C = \frac{C}{h}, \qquad i_n^* = G_C v_{Cn}$$

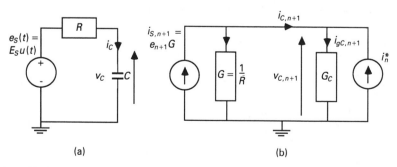

(a)   (b)

Figure 7.11   (a) Simple RC circuit excited by the step voltage function. (b) Its associated model

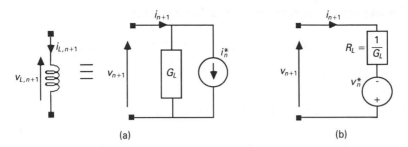

Figure 7.12 Model of inductor associated with interpolation algorithms. (a) Thevenin's equivalent. (b) Norton's equivalent

2.   And for Gear's third-order algorithm:

$$G_C = \frac{11C}{6h}, \quad i_n^* = \frac{C}{3h}\left(9v_{Cn} - \frac{9}{2}\,v_{Cn-1} + v_{Cn-2}\right)$$

After running the program with $E_S = 1$ V, $R = 1\ \Omega$, $C = 1$ F, $v_C(0) = 0$ V, $h = 0.2$ s and the end of the time interval $= 4$ s we obtain the results (if we enter R in kiloohms and C in nanofarads, then time will be in microseconds):

1.   For the Euler associated model:

$t=$	$v_C=$	v-accurate $=$	eps $= v_{ac} - v_C$
0.0	0.0	0.0	0.0
0.2	0.16666	0.181269	1.460258 E-2
.	.	.	.
2.0	0.838494	0.86466	2.6170 E-2
.	.	.	.
4.0	0.97391	0.98168	7.76841 E-3

2.   For the Gear third-order associated model:

$t=$	$v_C=$	v-accurate $=$	eps $= v_{ac} - v_C$
0.0	0.0	0.0	0.0
0.2	0.16666	0.181269	1.460258 E-2
.	.	.	.
2.0	0.861543	0.86466	3.12137 E-3
.	.	.	.
4.0	0.981167	0.98168	5.16938 E-4

Application of capacitor associated models for large linear and non-linear circuits will be discussed in the next chapter.

7.7.2 Companion or associated models for inductor

The inductor branch voltage–current relation (7.2) can be rearranged as

$$\frac{di}{dt} = f(t) = \frac{1}{L}\,v(t) \tag{7.89}$$

Then, for the nth and $(n+1)$st time point we can write

$$f_n = \frac{1}{L}\,v_n; \qquad f_{n+1} = \frac{1}{L}\,v_{n+1} \tag{7.90}$$

Proceeding as we did for the capacitor, we can use different interpolation algorithms to obtain associated models for the inductor.

Starting with Euler's interpolation algorithms we obtain

$$i_{n+1} = i_n + hf_{n+1} = i_n + \frac{h}{L}\,v_{n+1} \tag{7.91}$$

or

$$i_{n+1} = i_n^* + G_L v_{n+1} \tag{7.92a}$$

with

$$i_n^* = i_n, \qquad G_L = \frac{h}{L} \tag{7.92b}$$

Relation (7.92a) leads to the associated Norton equivalent shown in Fig. 7.12(a).

Transformation of equation (7.91) into

$$v_{n+1} = \frac{L}{h}\,i_{n+1} - \frac{L}{h}\,i_n = R_L i_{n+1} - v_n^* \tag{7.93a}$$

with:

$$R_L = \frac{L}{h}, \qquad v_n^* = \frac{L}{h}\,i_n \tag{7.93b}$$

gives the associated Norton equivalent circuit shown in Fig. 7.12(b). Similar associated models are valid for other interpolation algorithms; only model parameters are different for each algorithm.

1. For the trapezoidal algorithm we obtain

$$i_{n+1} = i_n + \frac{h}{2L}\,v_n + \frac{h}{2L}\,v_{n+1} = i_n^* + G_L v_{n+1} \tag{7.94a}$$

with

$$i_n^* = i_n + \frac{h}{2L}\,v_n, \qquad G_L = \frac{h}{2L} \tag{7.94b}$$

and

$$v_n^* = \frac{2L}{h} i_n + v_n, \qquad R_L = \frac{2L}{h} \tag{7.94c}$$

2. For the Gear second-order algorithm:

$$i_{n+1} = \frac{4}{3} i_n - \frac{1}{3} i_{n-1} + \frac{2h}{3L} v_{n+1} = i_n^* + G_L v_{n+1} \tag{7.95a}$$

with

$$i_n^* = \frac{4}{3} i_n - \frac{1}{3} i_{n-1}, \qquad G_L = \frac{2h}{3L} \tag{7.95b}$$

and

$$v_n^* = \frac{2L}{h} i_n - \frac{L}{2h} i_{n-1}, \qquad R_L = \frac{3L}{2h} \tag{7.95c}$$

3. For the Gear third-order algorithm:

$$i_{n+1} = \frac{18}{11} i_n - \frac{9}{11} i_{n-1} + \frac{2}{11} i_{n-2} + \frac{6h}{11L} v_{n+1} = i_n^* + G_L v_{n+1} \tag{7.96a}$$

with

$$i_n^* = \frac{18}{11} i_n - \frac{9}{11} i_{n-1} + \frac{2}{11} i_{n-2}, \qquad G_L = \frac{6h}{11L} \tag{7.96b}$$

and

$$v_n^* = \frac{3L}{h} i_n - \frac{3L}{2h} i_{n-1} + \frac{L}{3h} i_{n-2}, \qquad R_L = \frac{11L}{6h} \tag{7.96c}$$

The application of associated models for inductors is analogous to that for capacitors. For RLC circuits we first replace each capacitor and each inductor by its associated model, then we rewrite the nodal equilibrium equations for the DC equivalent circuit and then solve them repetitively. For this reason the most efficient method is the LU decomposition method since in the iterative solutions only i_n^* is changing while other parameters remain constant. Therefore, only one LU decomposition is required for all iterations.

Exercises

Exercise 1. Using the method described in Section 7.2 (equation (7.13)), solve the following differential equations with initial conditions indicated:

$$3 \frac{dx}{dt} + 4x(t) = 0; \qquad x(0) = 0 \text{ and } x(0) = 1$$

$$\frac{dx}{dt} + 2x(t) = 2; \qquad x(0) = 0 \text{ and } x(0) = 1$$

$$2\frac{dx}{dt} + 3x(t) = e^{-t}; \qquad x(0) = 0 \text{ and } x(0) = 1$$

$$3\frac{dx}{dt} + 4x(t) = \sin(2t); \quad x(0) = 0 \text{ and } x(0) = 1$$

Determine the transient and steady state responses.

Exercise 2. For the circuit shown in Fig. E7.1, the switch has been in position 1 for a very long time (all transients have died away). The switch is put in position 2 at time $t = 0$. Determine the voltage across resistor R_2 for $t \geqslant 0$. $R_1 = R_2 = 1\ \Omega$; $L_1 = 1$ H; $L_2 = 2$ H.

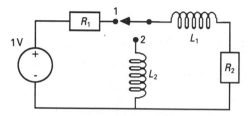

Figure E7.1

Exercise 3. For the circuit shown in Fig. E7.2, determine the differential equation for $v(t)$; solve this equation analytically. The capacitor is uncharged at $t = 0$. $R_1 = R_2 = R_3 = 1\ \Omega$; $C = 1$ F; $j(t) = tu(t)$; $e(t) = \sin(t)u(t)$. $u(t)$ is the unit step function.

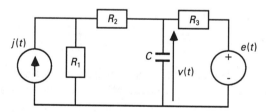

Figure E7.2

Exercise 4. Consider the following differential equation:

$$(t + 1)\frac{dx}{dt} + (-2\alpha t^2 - 2\alpha t + 1)x = 0; \ x(0) = 1$$

Write the iterative equation for the Euler extrapolation algorithm and trapezoidal algorithm. Develop a program with the same approach as in program 'DIFEQ1__0' and compare, for the two cases, the results with the exact solution $\{\exp(\alpha t^2)/(t + 1)\}$. Consider several negative and positive values of $\alpha : \alpha = -2, -1, 1, 2$.

Exercise 5. For the differential equation given in exercise 4, write the iterative equation corresponding to the modified Euler algorithm and Heun algorithm. Develop a program with the same approach as in program 'DIFEQ1__0' and compare, for the two cases, the results with the exact solution $\{\exp(\alpha t^2)/(t+1)\}$. Again, some values of $\alpha(>0$ and $<0)$ will be considered.

Exercise 6.   Write the iterative equation corresponding to the five algorithms of Section 7.2 permitting us to solve the differential equation $\alpha \, dx/dt + x^2 = 0$, with $x(0) = 1$. Develop a program with the same approach as in the program 'DIFEQ1__0' and compare, for all the cases, the results with the exact solution $\{\alpha/(t+\alpha)\}$.

Exercise 7.   For the circuit shown in Fig. E7.3, determine the differential equation and solve with respect to voltage $v_C(t)$ across capacitor C.

Modify the program 'DIFEQ1__0' in order to compare the results obtained from the forward Euler algorithm, the backward Euler algorithm, and the trapezoidal algorithm with the exact solution. Numerical conditions: $R_1 = 1 \, \Omega$; $C = 1 \, \text{F}$; $R_2 = 2 \, \Omega$; $v_S(t) = \{e^{-at} - 1\}u(t)$; $v_C(0) = 0$; $a = 0.5, 1, 2, 10$. $u(t)$ is the unit step function.

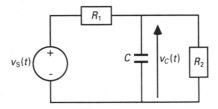

Figure E7.3

Exercise 8.   For the circuit shown in Fig. 7.6, modify the input data of the program 'HEUN' in order to obtain the response $v_C(t)$ when the input source is represented by a square periodic signal as follows:

$$v_S(t) = + V \qquad \text{for } 0 < t < T_0/2;$$

$$v_S(t) = - V \qquad \text{for } T_0/2 < t < T_0$$

$T_0 = 20$ ms; $V = 10$ V. Is it possible to observe an instability of the algorithm if the magnitude V varies?

Exercise 9.   In the circuit shown in Fig. 7.6, the diode is replaced by a quadratic resistor: $i = kv^2$ with $k = 2 \times 10^{-5}$. Modify the program 'HEUN' in order to compare the Heun and modified Euler algorithms. The input source is described by $v_S(t) = \{e^{-at} - e^{-bt}\}u(t)$; use several couples (a, b): $(0, 1)$, $(-1, -1)$, $(-10, 0)$. $u(t)$ is the unit step function.

Exercise 10.   The sine and cosine functions satisfy the following second-order differential equation: $d^2x/dt^2 + x = 0$. Assuming that $y = dx/dt$, determine the equivalent first-order system, apply the trapezoidal algorithm, and develop a program to obtain the solution in such a way that $x(0) = 0$ and $y(0) = 1$.

Exercise 11.   Repeat the problem of exercise 10 for the following second-order differential equation:

$$\frac{d^2x}{dt^2} = \frac{8x^2}{1+2t}; \quad x(0) = 1 \text{ and } \frac{dx}{dt} = -2 \text{ for } t = 0$$

Develop a program to obtain the solution and to compare the Heun and modified Euler algorithms.

Exercise 12.   Consider the circuit shown in Fig. E7.4. Apply Kirchhoff's current law to the two nodes 1 and 2; determine the second-order linear system having as unknowns $v_1(t)$ and $v_2(t)$. Apply the modified Euler algorithm and deduce the matrix equation to be solved in order to calculate the voltages $v_1(t)$ and $v_2(t)$ at time $(n+1)h$, assuming that the electrical state of the circuit is known at time (nh). Develop a solving procedure for a linear second-order system and determine the variations of $v_1(t)$ and $v_2(t)$ in the time range $0 \leqslant t \leqslant 10$ s. Numerical conditions: $R_1 = 1\ \Omega$; $R_2 = 2\ \Omega$; $C = 1$ F; and $e(t) = 1$ for $t > 0$.

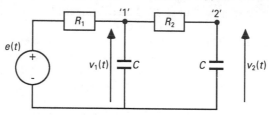

Figure E7.4

Exercise 13.   Give a graphical interpretation of the fourth-order Runge–Kutta equation (see equations (7.40a) and (7.40b)).

Exercise 14.   Modify the program 'AD-BASH' using the third-order algorithm; at the beginning use the first- and second-order Adam–Bashford algorithms. Compare the results obtained with results given by the second-order Adams–Bashford algorithm for the test function $x_0 \exp(at)$.

Exercise 15.   Modify the program 'AD-BASH' using the third-order algorithm; use the Runge–Kutta algorithm at the beginning. Compare the results obtained for the test function $x_0 \exp(at)$.

Exercise 16.   Modify the program 'AD-MOULT' using the fourth-order algorithm defined by the following iterative equation:

$$x_{n+1} = x_n + \frac{h}{24}\,(9f[x_{n+1}, u_{n+1}] + 19f[x_n, u_n] - 5f[x_{n-1}, u_{n-1}] + f[x_{n-2}, u_{n-2}])$$

At the beginning, use lower-order Adams–Bashforth algorithms. Compare the results with the second-order algorithm presented in the Section 7.6.2.

Exercise 17.   Run the algorithms of the program 'DIFEQ1__0' for which the stability is studied in Section 7.4 and determine the stability range as indicated in Fig. 7.7.

Exercise 18.   Use the program 'ASSOCMOD' using the following voltage source: $e_S(t) = E_S \cos(\omega t)u(t)$. Use several values of ω smaller or bigger than $\omega_0 = 1/RC$.

Exercise 19.   The program 'ASSOCMOD' analyzes the circuit shown in Fig. 7.11. In this circuit, the capacitor $(C = 1$ F$)$ is replaced by an inductor $(L = 1$ H$)$; the source has the same value. Modify the program 'ASSOCMOD' and compare the results obtained for the third-order Gear algorithm and for the Euler algorithm. Consider several values for the initial condition: $i_L(0) = -0.5$ A, 0 A, 0.5 A.

Exercise 20.   Consider the influence of the reference phase ϕ for the input voltage $e_S(t) = E_S \cos(\omega t + \phi)u(t)$ for the circuit shown in Fig. E7.5. Develop a program with the same approach as in the program 'ASSOCMOD' in order to calculate the output voltage $v(t)$ on the resistor. Note the maximum value of the transient state. The inductor has zero initial conditions. $L = 1$ H; $R = 1$ Ω; $E_S(t) = 1$ V; $\omega = 1$ rad/s.

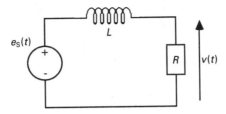

Figure E7.5

Exercise 21. Modify the program 'GEAR-2', including the third-order algorithm. Compare the results obtained with 'GEAR-2' results.

Exercise 22. Modify the procedures of the program 'ASSOCMOD' in order to compare the first-, second-, third- and fourth-order Gear algorithms. The fourth-order algorithm is defined by the following equation:

$$x_{n+1} = \frac{48}{25} x_n - \frac{36}{25} x_{n-1} + \frac{16}{25} x_{n-2} - \frac{3}{25} x_{n-3} + \frac{12h}{25} f[x_{n+1}, u_{n+1}]$$

Exercise 23. Consider the circuit shown in Fig. E7.6, often used as an oscilloscope probe. The capacitors have zero initial conditions. Replace the capacitors by their associated models and develop a program permitting the analysis of the voltage $v(t)$ across the resistor R_2. The voltage source $e_S(t)$ will be defined in a procedure. Verify that for $\tau_1 = \tau_2 = R_i C_i = C_i/G_i$, $i = 1, 2, v(t) = G_1/(G_1 + G_2)e(t)$. If $e_S(t)$ is a periodic signal, give some examples, choosing the time period with respect to τ_1 and τ_2 appropriately.

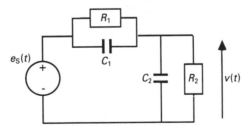

Figure E7.6

Exercise 24. Consider the resonant circuit shown in Fig. E7.7(a). The initial conditions are zero for the inductor and the capacitor; the input voltage source is a unit step function $u(t)$. Show that the voltage $v_C(t)$ on the capacitor satisfies the following differential equation:

$$\frac{d^2 v_C(t)}{dt^2} + \frac{\omega_0}{Q} \frac{d v_C(t)}{dt} + \omega_0^2 v_C(t) = \omega_0^2 e_S(t); \quad \text{with: } \omega_0 = \frac{1}{\sqrt{LC}} \text{ and } Q = \frac{\sqrt{\frac{L}{C}}}{R}$$

Solve this second-order differential equation analytically and apply the fourth-order Gear algorithm described in exercise 22 (the differential equation has to be transformed in a first-order system).

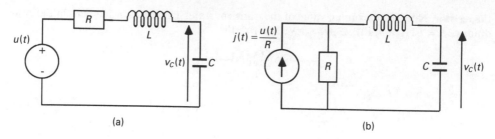

Figure E7.7

Exercise 25. In the circuit shown in Fig. E7.7(b), replace the inductor and the capacitor by their associated models using the third-order Gear algorithm. The initial conditions for the inductor and capacitor are equal to zero. Using the nodal description, determine the voltage on the capacitor when the input voltage source is a unit step function $u(t)$. Use the approach used in exercise 24. Verify the results obtained for several values of Q: $Q = 1$, 10, etc.

8 Transient Analysis of Larger Dynamic Circuits

8.1 Introduction

The application of the state variable method to an analysis of higher-order dynamic circuits allows the description of the properties of such circuits by a set of first-order differential equations of similar form to those described in Chapter 7:

$$\dot{x}_1 = a_{11}x_1 + a_{12}x_2 + \cdots + a_{1n}x_n + b_{11}u_1 + b_{12}u_2 + \cdots + b_{1m}u_m$$
$$\vdots$$
$$\dot{x}_n = a_{n1}x_1 + a_{n2}x_2 + \cdots + a_{nn}x_n + b_{n1}u_1 + b_{n2}u_2 + \ldots + b_{nm}u_m$$

(8.1)

or, in matrix form,

$$\dot{\mathbf{x}} = \mathbf{A}\mathbf{x} + \mathbf{B}\mathbf{u}, \mathbf{x}(0) = \mathbf{x}_0 \qquad (8.2)$$

where the square matrix [**A**]:

$$\mathbf{A} = \begin{bmatrix} a_{11} & a_{12} & . & . & . & a_{1n} \\ . & . & . & . & . & . \\ . & . & . & . & . & . \\ a_{n1} & a_{n2} & . & . & . & a_{nn} \end{bmatrix} \qquad (8.3)$$

with n equal to the number of state variables, and the rectangular matrix **B**:

$$\mathbf{B} = \begin{bmatrix} b_{11} & b_{12} & . & . & . & b_{1m} \\ . & . & . & . & . & . \\ . & . & . & . & . & . \\ b_{n1} & b_{n2} & . & . & . & b_{nm} \end{bmatrix} \qquad (8.4)$$

with m equal to the number of inputs, that is, excitation independent voltage and/or current sources.

Vector **x** represents the state variables (usually voltages on capacitors and currents flowing through inductors), and vector $\mathbf{x}_0$ represents the initial conditions (initial states).

Selecting the output variables, that is, voltages between selected nodes (voltage output ports) or currents flowing through selected branches (current output ports), and denoting them as

$$\mathbf{y} = [y_1 y_2 \ ... \ y_k]^t \tag{8.5}$$

where k is the number of output variables, we can write the relation between the output variables, state variables and input excitation sources:

$$\mathbf{y} = \mathbf{Cx} + \mathbf{Du} + (\mathbf{D}_1\dot{\mathbf{u}} + \mathbf{D}_2\ddot{\mathbf{u}} + \cdots) \tag{8.6}$$

where the rectangular matrix $\mathbf{C}$:

$$\mathbf{C} = \begin{bmatrix} c_{11} & c_{12} & . & . & . & c_{1n} \\ . & & . & . & . & . \\ . & & . & . & . & . \\ c_{k1} & c_{k2} & . & . & . & c_{kn} \end{bmatrix} \tag{8.7}$$

relates the output y variables with the state variables, and the rectangular matrix $\mathbf{D}$:

$$\mathbf{D} = \begin{bmatrix} d_{11} & d_{12} & . & . & . & d_{1m} \\ . & & . & . & . & . \\ . & & . & . & . & . \\ d_{k1} & d_{k2} & . & . & . & d_{km} \end{bmatrix} \tag{8.8}$$

relates the output variables with the input excitation voltages or currents; the matrices $\mathbf{D}_1$, $\mathbf{D}_2$, ..., if they exist, relate the derivatives of the input vector with the output variables.

The matrix equations (8.2) and (8.6) are called the normal state and output equations, respectively.

The matrix $\mathbf{A}$ is referred to as the system matrix.

Solution of the state equation in the normal form (8.1) or (8.2), for given $\mathbf{u}$ and $\mathbf{x}_0$, gives us the state variable vector $\mathbf{x}$ and then the output variables $\mathbf{y}$ can easily be computed from equation (8.6).

The solution of the state equation (8.2) has a form similar to the solution of a single first-order differential equation (see Chapter 7):

$$\mathbf{x} = e^{\mathbf{A}t} \int_0^t e^{-\mathbf{A}\tau}\mathbf{Bu} \ d\tau + e^{\mathbf{A}t}\mathbf{x}_0 \tag{8.9}$$

The square matrix $e^{\mathbf{A}t}$, defined by the infinite series:

$$e^{\mathbf{A}t} = 1 + \mathbf{A}t + \frac{1}{2!} [\mathbf{A}t]^2 + \cdots \frac{1}{n!} [\mathbf{A}t]^n + \cdots \tag{8.10}$$

is referred to as the transition matrix.

There exist many numerical methods for solutions of normal state equations with the use of the $e^{\mathbf{A}t}$ matrix computation. These methods are especially useful for analysis of large linear dynamic circuits and are described in Chapter 9.

The time relations for the state variable $\mathbf{x}$ vector can also be obtained directly from equations (8.1). This approach allows us to find the state variables for linear and non-linear circuits, since equations (8.1) can be written in the form

$$\dot{x}_1 = f_1(x_1, x_2, \ldots, x_n, u_1, u_2, \ldots, u_m)$$

$$\dot{x}_2 = f_2(x_1, x_2, \ldots, x_n, u_1, u_2, \ldots, u_m)$$

$$\vdots$$

$$\dot{x}_n = f_n(x_1, x_2, \ldots, x_n, u_1, u_2, \ldots, u_m)$$

(8.11)

with linear or non-linear function relations, or in matrix form

$$\dot{\mathbf{x}} = \mathbf{f}(\mathbf{x}, \mathbf{u}) \qquad \text{with } \mathbf{x}(t_0) = \mathbf{x}_0$$

(8.12)

Solutions of the set of equations (8.11) can be performed using any of the algorithms described in Chapter 7, as will be shown in subsequent sections.

A separate problem is the formulation of the state equation in normal form. For small circuits this can be done by inspection using KCL and KVL for nodal and mesh equations, as is outlined in the next section of this chapter.

For larger circuits, automatic computer formulation of the state equations can be performed using the methods based on topological circuit matrices (incidence, cut-set, loop, etc., matrices), and their manipulations, as will be described later.

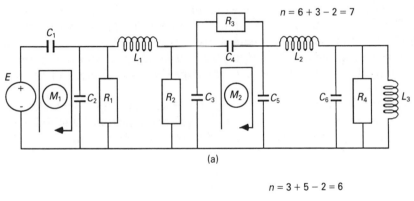

(a)

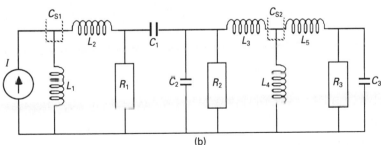

(b)

Figure 8.1 Dynamic circuits containing (a) two independent meshes composed of capacitors and voltage source, (b) two independent cut-sets composed of inductors and current source

As was mentioned earlier, we usually choose voltages on capacitors and currents flowing through inductors as state variables, and write n independent equations, with n referred to as the order of the circuit.

The order n of the circuit is given by the relation

$$n = \Sigma C + \Sigma L - \Sigma MCE - \Sigma C_S LI \tag{8.13}$$

where

1. ΣC denotes the number of capacitors in a given circuit,
2. ΣL denotes the number of inductors in the circuit,
3. ΣMCE denotes the number of independent meshes composed of capacitors and voltage sources (see Fig. 8.1(a)),
4. $\Sigma C_S LI$ denotes the number of independent cut-sets composed of inductors and current sources, as shown in Fig. 8.1(b).

8.2 Formulation of normal state equations for simple dynamic circuits

For simple dynamic circuits we can write mesh and nodal equations, by inspection, using KVL for meshes containing inductors and KCL for nodes with capacitors.

These equations can be written directly in the time domain, or in the frequency domain, using the Laplace transform (with zero initial conditions) and complex frequency variable s and, after proper rearrangement, substitution of the derivative symbol $\mathrm{d}/\mathrm{d}t$ in place of $s [s \equiv \mathrm{d}/\mathrm{d}t]$.

To illustrate these methods let us consider several simple dynamic circuits.

For the simple RLC circuit of Fig. 8.2, the two equations required, containing inductive and capacitive branch descriptions in the time domain, are

$$- v_L + v_C = 0 \quad \text{with } v_L = L\,\frac{\mathrm{d}i_L}{\mathrm{d}t} \tag{8.14a}$$

$$- i_S + i_L + i_C + i_R = 0, \quad \text{with } i_C = C\,\frac{\mathrm{d}v_C}{\mathrm{d}t}, \quad i_R = \frac{v_C}{R} \tag{8.14b}$$

After denoting

$$x_1 = i_L, \qquad x_2 = v_C, \qquad u = i_S$$

we obtain directly

$$\dot{x}_1 = \frac{\mathrm{d}i_L}{\mathrm{d}t} = \frac{v_C}{L} = \frac{1}{L}\,x_2 \tag{8.15a}$$

$$\dot{x}_2 = \frac{\mathrm{d}v_C}{\mathrm{d}t} = -\frac{i_L}{C} - \frac{v_C}{RC} + \frac{i_S}{C} = -\frac{1}{C}\,x_1 - \frac{1}{RC}\,x_2 + \frac{1}{C}\,u \tag{8.15b}$$

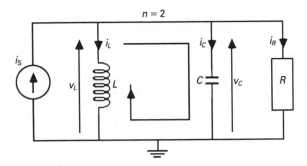

Figure 8.2 Simple second-order RLC circuit

or, in matrix form,

$$\dot{\mathbf{x}} = \begin{bmatrix} 0 & \dfrac{1}{L} \\[2mm] -\dfrac{1}{C} & -\dfrac{1}{RC} \end{bmatrix} \mathbf{x} + \begin{bmatrix} 0 \\[2mm] \dfrac{1}{C} \end{bmatrix} \mathbf{u} \tag{8.16}$$

Now, if we choose as three outputs

$$y_1 = v_C = x_2$$

$$y_2 = i_R = \frac{1}{R} x_2 \tag{8.17}$$

$$y_3 = i_L = x_1$$

then the matrix form of the output equations is

$$\mathbf{y} = \begin{bmatrix} 0 & 1 \\[1mm] 0 & \dfrac{1}{R} \\[1mm] 1 & 0 \end{bmatrix} \mathbf{x} + \mathbf{0u} \tag{8.18}$$

For the first order circuit of Fig. 8.3 containing two capacitors and having one mesh, composed of E, C_1 and C_2, the initial KCL equation is

$$i_{G1} + i_{C1} + i_{G2} + i_{C2} = 0 \tag{8.19}$$

with

$$i_{G1} = G_1 v_1, \qquad i_{C1} = C_1 \frac{dv_1}{dt}, \qquad i_{G2} = G_2(v_1 - E), \qquad i_{C2} = C_2 \frac{d(v_1 - E)}{dt} \tag{8.20}$$

After substitution of relations (8.20) into (8.19) and rearrangement we obtain

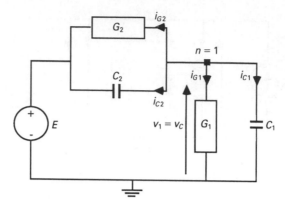

Figure 8.3 Dynamic first-order circuit with two capacitors and one mesh composed of E, C_1 and C_2

the initial state equation

$$\dot{v}_1 = -\frac{G_1 + G_2}{C_1 + C_2}\, v_1 + \frac{G_2}{C_1 + C_2}\, E + \frac{C_2}{C_1 + C_2}\, \dot{E} \tag{8.21a}$$

or (with $x = v_1$, $u = E$):

$$\dot{x} = ax + bu + b_1\dot{u} \tag{8.21b}$$

Selecting two outputs $y_1 = -i_{C2}$ and $y_2 = v_1$, we obtain

$$y_1 = -C_2\dot{v}_1 + C_2\dot{E} \tag{8.22}$$

Substituting (8.21a) into (8.22) yields

$$y_1 = \frac{C_2(G_1 + G_2)}{C_1 + C_2}\, v_1 - \frac{C_2 G_2}{C_1 + C_2}\, E + \frac{C_1 C_2}{C_1 + C_2}\, \dot{E} \tag{8.23a}$$

$$y_2 = v_1 \tag{8.23b}$$

or, in matrix form,

$$\mathbf{y} = \begin{bmatrix} y_1 \\ y_2 \end{bmatrix} = \begin{bmatrix} \dfrac{C_2(G_1 + G_2)}{C_1 + C_2} \\ 1 \end{bmatrix} v_1 + \begin{bmatrix} \dfrac{-C_2 G_2}{C_1 + C_2} \\ 0 \end{bmatrix} E + \begin{bmatrix} \dfrac{C_2 C_1}{C_1 + C_2} \\ 0 \end{bmatrix} \dot{E} \tag{8.24a}$$

or

$$\mathbf{y} = \mathbf{Cx} + \mathbf{Du} + \mathbf{D}_1\dot{\mathbf{u}} \tag{8.24b}$$

Equation (8.21) is not in normal form since it contains the derivative $dE/dt = \dot{E}$. To eliminate the derivative $\dot{E}$ we can substitute

$$\mathbf{z} = \mathbf{x} - \mathbf{B}_1\mathbf{u} \tag{8.25}$$

and obtain

$$\mathbf{x} = \mathbf{z} + \mathbf{B}_1\mathbf{u} \tag{8.26}$$

and

$$\dot{\mathbf{x}} = \dot{\mathbf{z}} + \mathbf{B}_1\dot{\mathbf{u}} \tag{8.27}$$

Substituting (8.26) and (8.27), in scalar form, into (8.21b) and (8.24b) we obtain

$$\dot{z} = az + (ab_1 + b)u \tag{8.28}$$

which has normal form, and

$$\mathbf{y} = \mathbf{Cz} + (\mathbf{CB}_1 + \mathbf{D})\mathbf{u} + \mathbf{D}_1\dot{\mathbf{u}} \tag{8.29}$$

Computation of z from (8.28) and substitution into (8.29) gives the required output variables y_1 and y_2.

It can be seen from this simple example that a lot of computations are required when the initial state equations are not in normal form.

Let us now start with the frequency domain description. For the second-order active circuit of Fig. 8.4, the nodal admittance equations are

$$(G_1 + sC_1 + sC_2)V_1 - sC_2V_2 = I_s \tag{8.30}$$
$$(-sC_2 + g_m)V_1 + (G_2 + sC_2)V_2 = 0$$

Substituting $V_1 = V_{C1}$ and $V_2 = V_{C1} - V_{C2}$ into (8.30) and rearranging gives

$$G_1V_{C1} + sC_1V_{C1} + sC_2V_{C2} = I_s \tag{8.31a}$$
$$g_mV_{C1} + G_2V_{C1} - sC_2V_{C2} - G_2V_{C2} = 0 \tag{8.31b}$$

Elimination of the term sC_2V_{C2} from (8.31a) and transition from the frequency

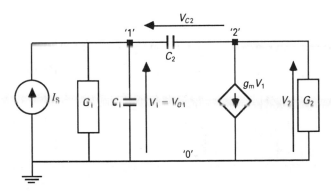

Figure 8.4 Linear active second-order circuit

domain to time domain ($s \equiv d/dt$) yields

$$\dot{v}_{C1} = -\frac{G_1 + g_m + G_2}{C_1} v_{C1} + \frac{G_2}{C_1} v_{C2} + \frac{1}{C_1} i_S \qquad (8.32a)$$

$$\dot{v}_{C2} = \frac{g_m + G_2}{C_2} v_{C1} - \frac{G_2}{C_2} v_{C2} \qquad (8.32b)$$

Selecting v_2 as the output variable y, we obtain an output equation

$$y = v_{C1} - v_{C2} \qquad (8.33)$$

For the third-order linear RLC circuit of Fig. 8.5, the KCL equations for the two nodes and the KVL equation for the mesh containing the inductor L are

$$-I_1 + G_1 V_1 + sC_1 V_1 + I_L = 0$$
$$-I_2 + G_2 V_2 + sC_2 V_2 - I_L = 0 \qquad (8.34)$$
$$-V_1 + V_L + V_2 = 0$$

After transition from the frequency domain to the time domain ($s \equiv d/dt$) and rearrangement we obtain

$$\dot{v}_1 = -\frac{G_1}{C_1} v_1 - \frac{1}{C_1} i_L + \frac{1}{C_1} i_1 \qquad (8.35a)$$

$$\dot{v}_2 = -\frac{G_2}{C_2} v_2 + \frac{1}{C_2} i_L + \frac{1}{C_2} i_2 \qquad (8.35b)$$

$$\dot{i}_L = \frac{1}{L} v_1 - \frac{1}{L} v_2 \qquad (8.35c)$$

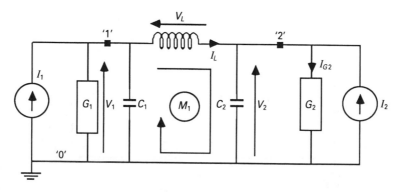

Figure 8.5   Linear third-order circuit

or, in matrix form,

$$
\begin{bmatrix} \dot{v}_1 \\ \dot{v}_2 \\ \cdot \\ i_L \end{bmatrix} = \begin{bmatrix} -\dfrac{G_1}{C_1} & 0 & -\dfrac{1}{C_1} \\ 0 & -\dfrac{G_2}{C_2} & \dfrac{1}{C_2} \\ \dfrac{1}{L} & -\dfrac{1}{L} & 0 \end{bmatrix} \begin{bmatrix} v_1 \\ v_2 \\ i_L \end{bmatrix} + \begin{bmatrix} \dfrac{1}{C_1} & 0 \\ 0 & \dfrac{1}{C_2} \\ 0 & 0 \end{bmatrix} \begin{bmatrix} i_1 \\ i_2 \end{bmatrix} \tag{8.36}
$$

Selecting, for example, the output variables

$$
y_1 = v_2 \quad \text{and} \quad y_2 = i_{G2} = G_2 v_2
$$

we obtain the output matrix equation

$$
\mathbf{y} = \begin{bmatrix} 0 & 1 & 0 \\ 0 & G_2 & 0 \end{bmatrix} \begin{bmatrix} v_1 \\ v_2 \\ i_L \end{bmatrix} \tag{8.37}
$$

For time-variant or non-linear capacitors and inductors, we choose the electrical charge q in the capacitors and magnetic flux ϕ in the inductors as state variables. In such cases the current through a capacitor is given by the relation

$$
i_C = \frac{dq}{dt} = \frac{dq}{dv}\frac{dv}{dt} = C(v)\frac{dv}{dt} \tag{8.38}
$$

and the voltage across an inductor is described by the formula

$$
v_L = \frac{d\phi}{dt} = \frac{d\phi}{di}\frac{di}{dt} = L(i)\frac{di}{dt} \tag{8.39}
$$

For linear time-variant capacitors and inductors the relations between variables and parameters are

$$
C(t) = \frac{q}{v_C} \tag{8.40a}
$$

and

$$
L(t) = \frac{\phi}{i_L} \tag{8.40b}
$$

For example, let us consider the linear time variant circuit shown in Fig. 8.6, for which the node and mesh equations are

$$
-i_S + i_C + i_L = 0 \tag{8.41}
$$

$$
-v_C + v_L + v_R = 0
$$

or

$$
i_C = -i_L + i_S \tag{8.42}
$$

$$
v_L = v_C - Ri_L
$$

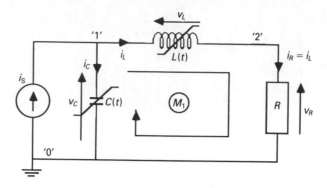

Figure 8.6   Second-order linear time variant circuit

Substituting relations (8.38), (8.39) and (8.40) into (8.42) we obtain

$$\dot{q} = \frac{dq}{dt} = -\frac{1}{L(t)} \phi + i_S \tag{8.43a}$$

$$\dot{\phi} = \frac{d\phi}{dt} = \frac{1}{C(t)} q - \frac{R}{L(t)} \phi \tag{8.43b}$$

or, in matrix form,

$$\begin{bmatrix} \dot{q} \\ \dot{\phi} \end{bmatrix} = \begin{bmatrix} 0 & -\dfrac{1}{L(t)} \\ \dfrac{1}{C(t)} & -\dfrac{R}{L(t)} \end{bmatrix} \begin{bmatrix} q \\ \phi \end{bmatrix} + \begin{bmatrix} 1 \\ 0 \end{bmatrix} i_S \tag{8.44}$$

Assuming, for example, the following time relations for capacitance $C(t)$ and inductance $L(t)$:

$$C(t) = C_0(1 + 0.25 \cos(\omega t)) \tag{8.45a}$$

and

$$L(t) = L_0(1 + 0.3 \sin(\omega t)) \tag{8.45b}$$

where C_0 and L_0 are quiescent-point values for capacitance and inductance, we can compute the values of q and ϕ. Then, using relations (8.40a) and (8.40b), we can find v_C and i_L for given time relations $C(t)$ and $L(t)$.

Selecting as an output variable, y, the voltage on resistor v_R:

$$y = v_R = R i_L = \frac{R}{L(t)} \phi$$

we obtain the output matrix equation

$$y = \begin{bmatrix} 0 & \dfrac{R}{L(t)} \end{bmatrix} \begin{bmatrix} q \\ \phi \end{bmatrix} \tag{8.46}$$

For non-linear circuits, with non-linear capacitors and inductors, the state variable equations are written in functional form, as in (8.12).

Also, voltage–charge and current–flux relations for capacitors and inductors, respectively, are given by the functions

$$v_C = g_C(q) \tag{8.47a}$$

and

$$i_L = g_L(\phi) \tag{8.47b}$$

For example, the voltage–charge relation for the non-linear capacitive branch C_i can be

$$v_{Ci} = g_{Ci}(q_i) = q_i + 2q_i^2 + q_i^3$$

The current–flux relation for the non-linear inductor can be expressed in a similar way.

Let us analyze the non-linear circuit with non-linear capacitors and inductors, shown in Fig. 8.7, for which the non-linear voltage–charge, and current–flux relations are described by the functions

$$v_{C1} = g_{C1}(q_1) \tag{8.48a}$$

$$v_{C2} = g_{C2}(q_2) \tag{8.48b}$$

$$i_L \ = g_L(\phi_1) \tag{8.48c}$$

The two nodal and one mesh equations for the circuit are

$$-i_S + i_{R1} + i_{C1} + i_L = 0$$

$$-i_L + i_{C2} + i_{R2} = 0 \tag{8.49}$$

$$-v_{C1} + v_L + v_{C2} = 0$$

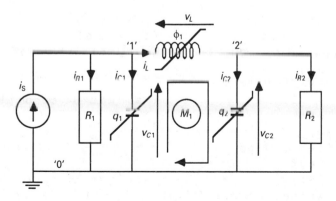

Figure 8.7 Non-linear circuit with non-linear capacitors and inductors

Substituting relations (8.38), (8.39) and (8.48) into (8.49) yields

$$\dot{q}_1 = \frac{dq_1}{dt} = -\frac{1}{R_1} g_{C1}(q_1) - g_L(\phi_1) + i_S$$

$$\dot{q}_2 = \frac{dq_2}{dt} = g_L(\phi_1) - \frac{1}{R_2} g_{C2}(q_2) \qquad\qquad (8.50)$$

$$\dot{\phi}_1 = \frac{d\phi_1}{dt} = g_{C1}(q_1) - g_{C2}(q_2)$$

These equations can be written in the general form

$$\dot{q}_1 = f_1[q_1, \phi_1, i_S]$$

$$\dot{q}_2 = f_2[q_2, \phi_1] \qquad\qquad (8.51)$$

$$\dot{\phi}_1 = f_3[q_1, q_2]$$

For larger and more complicated circuits automatic computer methods for state equation formulation are used.

8.3 Numerical solution of the set of state equations

The algorithms for the first-order differential equation solutions described in Chapter 7, can be applied directly to the set of differential state variable solutions. The difference is only that, in a single time step h not one but a set of equations is solved.

As an example, let us consider the second-order RLC circuit of Fig. 8.2 for which the state equations are given by (8.15); let us assume additionally that $i_S = u = t$. Then the procedure using Euler's extrapolation algorithm can be written as follows (assume here that global variables are declared at the beginning of the program):

```
Procedure Euler ;
  Begin
     readln (R,C,L) ;
     readln (h,tk) ;              {tk-end of the time interval}
     for i:=1 to n do             {n-number of equations}
        begin
        readln(x0) ;
        x[i]:=x0                  {x0-initial conditions }
        end;
     t:=0;
     while t<=tk do
        begin
        iS:= t;                   {excitation function}
        writeln(t:2:2,' ',iS:2:2, x[1],x[2]);
        f[1] := x[2]/L;           {equations (8.15)}
```

```
f[2] := -x[1]/C - x[2]/(R*C) + iS/C;
for i:=1 to n do
x[i]:= x[i] + h*f[i];{Euler's alg. applied to
                                both equations}
    t:= t + h ;
  end;                          {while}
End;                            {procedure Euler}
```

Similarly, we can write a procedure with Heun's algorithm in which we first have to write a formula for Euler's extrapolation predictor, applied to both state equations, and then a formula for the trapezoidal corrector and also for both equations.

The procedure 'Euler' and the procedure 'Heun' are included in the program, Program 8.1 'MULDIFSO', for the solution of multiple differential equations (n-number of equations) using Euler's forward and Heun's algorithms.

See the listing of the program MULDIFSO Directory of Chapter 8 on attached diskette.

When we run this program, we have to enter the values of R, C, L, the time step h, the end of time interval tk, and initial conditions $x\emptyset1$ and $x\emptyset2$ (recall that $x[1] = i_L$ and $x[2] = v_C$).

$$\text{Entering: } R = 4, \qquad C = 0.5, \qquad L = 2;$$

$$h = 0.25, \qquad tk = 5;$$

$$x\emptyset1 = 0, \qquad x\emptyset2 = 0;$$

then the resulting print-outs are (every four steps)

1. For Euler's algorithm:

t =	i_S =	x[1] =	x[2] =
0.0	0.0	0.0	0.0
1.0	1.0	0.11575	1.06616
2.0	2.0	0.099261	2.6729
3.0	3.0	2.50949	3.34974
4.0	4.0	4.11046	2.79064
5.0	5.0	5.27240	1.64899

2. For Heun's algorithm:

t =	i_S =	x[1] =	x[2] =
0.0	0.0	0.0	0.0
1.0	1.0	0.13642	0.80273
2.0	2.0	0.88552	2.16213
3.0	3.0	2.18589	2.85799
4.0	4.0	3.59634	2.64568
5.0	5.0	4.77168	2.04479

With given R, C, L values the quality factor Q for the circuit considered equals

$$Q = \frac{\omega_0 C}{G} = R\omega_0 C = R \sqrt{\frac{C}{L}} = 2$$

Therefore, we observe visible oscillations for $x[2] = v_C$. Also, we observe that for a relatively large time step, $h = 0.25$, the results from Euler's algorithm are significantly less accurate, giving larger oscillations in $v_C(t)$.

In the second example program 'MULDIFVA', presented in Program 8.2, Heun's algorithm is adapted to analyze the time variant circuit of Fig. 8.6, described by two state equations (8.43), in which it is assumed that the excitation current $i_S(t)$:

$$i_S(t) = 1(t) \text{ (unit time-step function)}$$

the resistance $R = 1$, and the time varying capacitance and inductance are given by relations

$$C(t) = 1 \times (1 + 0.4 \times \sin(t))$$

$$L(t) = 1 \times (1 + 0.4 \times \cos(t))$$

Here it is assumed that $x[1] = q$, $x[2] = \phi$, and the voltage on resistor v_R is computed as

$$v_R = R \times i_L = \frac{Rx[2]}{L(t)} = R\frac{\phi}{L(t)}$$

See the listing of the program MULDIFVA Directory of Chapter 8 on attached diskette.

The program 'MULDIFVA' computes the electric charge q and magnetic flux ϕ and voltage on R for the time variant circuit of Fig. 8.6.

Running the program with the time step $h = 0.25$, time interval $tk = 5$, zero initial states $x\emptyset 1 = 0$, $x\emptyset 2 = 0$, we obtain the resulting print-outs:

t =	x[1] =	x[2] =	$v_R = \dfrac{R\,x[2]}{L(t)}$
0.0	0.0	0.0	0.0
1.0	0.91387	0.30159	0.24795
2.0	1.40284	0.66725	0.80050
3.0	1.33514	0.75419	1.24866
4.0	1.08672	0.91130	1.23392
5.0	0.88968	1.27796	1.14773

Program 'MULDIFSO' or 'MULDIFVA' can easily be adapted to analyze non-linear circuits with non-linear capacitors, inductors and resistors; in the case of non-linear reactances the electrical charge and magnetic flux associated with the capacitors and inductors are chosen as the state variables.

Now let us consider a non-linear circuit with linear capacitors shown in Fig. 8.8, where the non-linear resistor is composed of an MOS n-channel transistor operating in a triode region.

The current–voltage relation for this non-linear resistance is given by the formula (triode operating region)

$$i_D = \frac{\beta}{2} \left((v_X - v_S)^2 - (v_X - v_D)^2 \right), \qquad \text{with } v_X > v_S, \qquad v_X \geqslant v_D \qquad (8.52)$$

where

1. β = transconductance parameter,
2. $v_X = v_G - V_{Th}$,
3. V_{Th} = threshold voltage,
4. v_G, v_S, v_D voltages between transistor electrodes (gate, source, drain) and reference ground node.

For simplicity, it is assumed here that $\beta = 20\ \mu A/V^2$, $V_{Th} = const = 1\ V$.

The KCL equations for this circuit are

$$-i_R + i_{C1} + i_D = 0 \qquad\qquad\qquad\qquad (8.53)$$

$$-i_D + i_{C2} = 0$$

with

$$i_R = \frac{E - v_1}{R}, \qquad i_{C1} = C_1 \frac{dv_1}{dt}, \qquad i_{C2} = C_2 \frac{dv_2}{dt} \qquad (8.54)$$

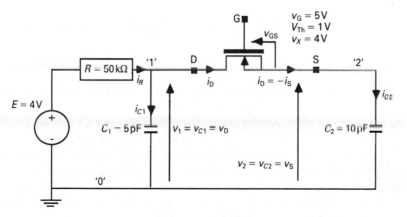

Figure 8.8 Non-linear circuit with n-channel MOS transistor operating in a triode region

After rearranging we obtain

$$\frac{dv_1}{dt} = f_1 = -\frac{v_1}{RC_1} - \frac{i_D}{C_1} + \frac{E}{RC_1}$$

$$\frac{dv_2}{dt} = f_2 = \frac{i_D}{C_2} \tag{8.55}$$

with i_D given by relation (8.52).

Modifications of Euler and Heun procedures in the program 'MULDIFSO' require the entry of the circuit parameters R, C_1, C_2, E, v_x, and writing of the circuit equations:

```
iD := 20*(vx-x[2])*(vx-x[2]) - (vx-x[1])*(vx-x[1])/2;
f[1] := -x[1]/R/C1 - iD/C1 + E/R/C1;
f[2] := iD/C2;
```

with

$$x[1] = v_{C1} = v_1 = v_D; \qquad x[2] = v_{C2} = v_2 = v_S$$

Expressing the current, i_D in micro-amperes (hence $\beta = 20$), resistance in megaohms ($R = 0.05$) and capacitances in picofarads ($C_1 = 5$, $C_2 = 10$), we obtain the voltages expressed in volts and time in microseconds.

After running the program with these modifications and entering

$$h = 0.05, \qquad tk = 8, \text{ and initial conditions } x\varnothing 1 = 0, \qquad x\varnothing 2 = 0,$$

we obtain the results for Euler's algorithm:

```
t =      x[1] =      x[2] =
0.5      2.1825      1.6114
1.0      3.0691      2.5354
2.0      3.7226      3.3041
 .          .           .
 .          .           .
 .          .           .
4.0      3.9483      3.6995
 .          .           .
 .          .           .
8.0      3.9900      3.8633
```

The results for Heun's algorithm are similar; the values of $x[1]$ and $x[2]$ are slightly lower, giving the end results for $t = 8 \mu s$: $x[1] = v_1 = 3.9896$ V, $x[2] = v_2 = 3.8606$ V. These results show that the voltage on C_1 almost reaches the value of the input voltage, E. It can also be observed from relation (8.52), that, for fixed value of v_x and increasing value of $v_2 = v_S$, the value of current i_D decreases to zero, thereby not allowing v_2 to reach the input voltage $E = 4$ V.

Using the same set of non-linear equations for the same circuit of Fig. 8.8 we can find the time relations for voltages $v_{C1} = x[1]$ and $v_{C2} = x[2]$ when the circuit is excited by the voltage pulse defined by the relation

$$E = 4 \times 1(t) - 4 \times 1(t - 1)$$

which is written in a program as

```
if t<=1 then E:=4
else E:=0;
```

Running the program with

$$R = 0.05 \text{ M}\Omega, \ C_1 = 5 \text{ pF}, \ C_2 = 10 \text{ pF}$$

(that is, enter $R = 0.05$, $C_1 = 5$, $C_2 = 10$), we obtain the print-outs for Euler's algorithm:

t =	E =	x[1] =	x[2] =
0.0	4.0	0.0	0.0
0.125	4.0	0.96878	0.33737
0.25	4.0	1.4424	0.84275
0.5	4.0	2.1623	1.5804
0.75	4.0	2.6744	2.1143
1.0	4.0	3.0387	2.5032
1.125	0.0	1.9958	2.4901
1.25	0.0	1.6382	2.2176
1.5	0.0	1.2827	1.6651
2.0	0.0	0.7524	0.9243
2.5	0.0	0.2459	0.5100

From these results we see that at the beginning C_1 charges up faster than C_2 due to the non-linear resistance between C_1 and C_2 and then, when the input voltage $E = 0$, C_1 discharges faster than C_2; however, C_1 discharges slower than it would be isolated from C_2, since the charge from C_2 flows to C_1 through the non-linear resistance (MOS transistor).

We can emphasize this effect by disconnecting C_2 from C_1 by making $i_D = 0$ at a time greater than 1. This requires slight modification of the program, that is, the addition of a conditional statement to the previous set of equations:

```
if t<= 1 then
iD:= 20*((vx-x[2])*(..........))/2
else iD := 0;
```

Running the program with the same data as before we obtain the same results as previously for $t \leqslant 1$, and the results for $t \geqslant 1$ are

t =	E =	x[1] =	x[2] =
1.0	4.0	3.0387	2.5032
1.125	0.0	1.9247	2.5197
1.25	0.0	1.1524	2.5197
1.5	0.0	0.4131	2.5197
2.0	0.0	0.05309	2.5197
2.5	0.0	0.00682	2.5197

Now we can see clearly that C_1 discharges much faster than before and the voltage on capacitor C_2 is constant, since $i_D = 0$ for $t > 1$.

Using the same set of non-linear equations again, we can observe charging of C_2 from C_1 (non-zero initial condition $v_{C1}(0) \neq 0$), when the input voltage $E = 0$, as is shown in Fig. 8.9.

Removing the 'if...else' conditions, which were used in the previous example, inserting $E = 0$, and entering the data according to the values shown in Fig. 8.9: $R = 10$, $C_1 = 5$, $C_2 = 10$, $x\emptyset 1 = 4$, $x\emptyset 2 = 0$, $h = 0.01$, $tk = 6$, we obtain the results (for Heun's algorithm):

```
t =      x[1] =     x[2] =
0.0      4.0        0.0
0.05     2.7204     0.6381
0.1      2.0047     0.9948
0.2      1.4731     1.2589
0.5      1.3277     1.3275
1.0      1.3221     1.3237
1.5      1.3177     1.3193
2.0      1.3133     1.3149
3.0      1.3046     1.3062
4.0      1.2959     1.2975
6.0      1.2788     1.2804
```

These results indicate that at $t = 0^+$ (that is, at the start of our observations), when $v_{C1} = 4$ V, the initial value of current i_R flowing through resistance 10 MΩ is $i_R = 4 \times 10^{-7} = 0.4$ µA, and the initial value of $i_D = 160$ µA, the capacitor C_2 is charged from C_1; thus the voltage across C_1 drops and the voltage across C_2 rises. After $t \cong 1$ µs the voltage v_{C2} exceeds the voltage v_{C1}, since C_1 is connected directly to the discharging resistance R.

After $t > 1$ µs both capacitors are discharged with $v_{C2} > v_{C1}$.

Similarly, we can simulate time responses for different situations and for

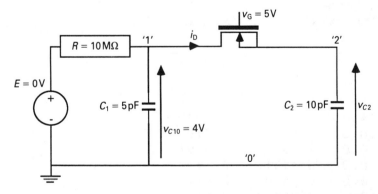

Figure 8.9 Non-linear circuit with zero input voltage $E = 0$ and non-zero initial condition $v_{C10} = 4$ V

different circuits: linear, time varying and non-linear. We can also write more complex and more accurate procedures using multi-step algorithms.

8.4 Application of companion models for capacitors associated with different interpolation algorithms for computation of the time domain response

Chapter 7 and Sections 1, 2 and 3 of this chapter present the numerical algorithms used for the computation of the transient response of circuits containing only linear resistors, capacitors and inductors (the values of resistance, capacitance and inductance are independent of bias conditions). In this section, we consider non-linear circuits containing the following elements:

1. Semiconductor elements (diodes, for example) associated with a static equivalent schema, with non-linear resistors and sources. So iterative companion models must be used to analyze the circuits with these elements.
2. Non-linear capacitors modelling the dynamic behaviour of semiconductors. The capacitances considered are non-linear and are functions of the voltage. The associated companion models have to be modified in order to be used for non-linear capacitors.

The examples presented in this section use the algorithms described in Chapter 7 for transient analysis, in particular,

1. the interpolation Euler algorithm for its simplicity,
2. the trapezoidal algorithm,
3. the order 2 and 3 Gear algorithms.

Trapezoidal and Gear algorithms are the most frequently used in simulators such as SPICE. For the computation of non-linear resistive circuits, the Newton–Raphson algorithm with improved convergence is used (see Section 6.5). Two kinds of circuit are considered in this section:

1. circuits containing Boltzmann diodes and linear capacitors,
2. circuits containing semiconductor diodes modelled by their dynamic equivalent schema with non-linear capacitors.

8.4.1 Circuits containing ideal diodes and linear capacitors

Let us consider the circuit shown in Fig. 8.10. In this circuit, besides the voltage source and the linear resistors, we have two elements:

1. a linear capacitor described by the differential equation $i_C(t) = C \dfrac{\mathrm{d}v_C(t)}{\mathrm{d}t}$,

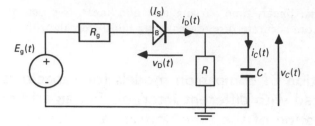

Figure 8.10   Non-linear dynamic circuit (ideal diode and linear capacitor)

2.   a Boltzmann diode described by its equation $i_D = I_S \left[\exp\left(\dfrac{v_D}{V_T}\right) - 1 \right]$.

Suppose that the capacitor initial voltage v_{C0} is known at $t = 0$. Its current i_{C0} is always equal to zero.

In order to analyze the non-linear dynamic circuit, we decompose the procedure into two parts.

First, let us consider the capacitor. Suppose that the current i_{Cn} and the voltage v_{Cn} of the capacitor are known at $t = nh$ (h is the computation time step). So the associated model of the capacitor can be defined for the calculation of the circuit voltages at $t = (n + 1)h$. One of the standard algorithms presented in Table 8.1 can be used.

The state of the circuit has to be computed for $t = h, 2h, ..., nh, (n + 1)h$. Suppose that the state of the circuit has been computed for $t = nh$. Next we have to calculate the values at $t = (n + 1)h$, and so on. Replacing the capacitor by its associated model, we obtain the circuit shown in Fig. 8.11. In order to simplify, we adopt the

Table 8.1   Capacitor models associated with different algorithms

Capacitor voltage and current at $\mathrm{t} = \mathrm{nh}$: v_{Cn} i_{Cn}

Interpolation Euler algorithm	$G_C = \dfrac{C}{h}$	$i_n^* = \dfrac{C}{h} v_{Cn}$
Trapezoidal algorithm	$G_C = \dfrac{2C}{h}$	$i_n^* = \dfrac{2C}{h} v_{Cn} + i_{Cn}$
Gear2 algorithm	$G_C = \dfrac{3C}{2h}$	$i_n^* = \dfrac{C}{2h} (4v_{Cn} - v_{Cn-1})$
Gear3 algorithm	$G_C = \dfrac{11C}{6h}$	$i_n^* = \dfrac{3C}{h} v_{Cn} - \dfrac{3C}{2h} v_{Cn-1} + \dfrac{C}{3h} v_{Cn-2}$

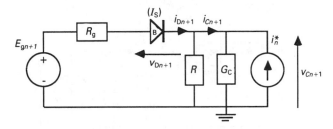

Figure 8.11   Non-linear circuit of Fig. 8.10 with capacitor replaced by its associated model

following notation:

$$E_g[(n + 1)h] = E_{gn+1};\ v_D[(n + 1)h] = v_{Dn+1};\ v_C[(n + 1)h] = v_{Cn+1}$$
$$i_D[(n + 1)h] = i_{Dn+1};\ i_C[(n + 1)h] = i_{Cn+1}...$$

(8.56)

The capacitor voltage v_{Cn+1} depends on the previous time values (v_{Cn} and i_{Cn}). So the circuit of Fig. 8.11 contains independent sources (E_{gn+1} and i_n^*), linear resistors (R_G, R and G_C) and a non-linear element (the Boltzmann diode). In order to calculate the currents and voltages of this non-linear circuit, the use of an iterative companion model for the diode is necessary. So, we obtain the circuit shown in Fig. 8.12, where the superscripts (k) correspond to the Newton–Raphson iterations.

Thus, the general algorithm used for the analysis of a circuit containing non-linear resistive elements and linear capacitors or inductors is as follows:

```
Data input
nmax=Tmax/h
For n=0 to nmax do
   t=(n+1)h
   Associated model calculation of linear capacitors
                                   and inductors.
   Iterative calculation of voltages of the circuit for
        t = (n+1)h using an iterative companion model for
                                   non-linear elements.
   Results at t = (n+1)h
   end.
```

The program 'FILT' shown in Program 8.3 permits us to simulate the circuit of Fig. 8.10.

See the listing of the program FILT Directory of Chapter 8 on attached diskette.

In this program:

1. Computation of capacitor associated models is performed using the procedures 'Reverse__Euler', 'Trapez', 'Gear__2' and 'Gear__3'.

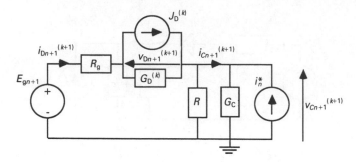

Figure 8.12   Linearized circuit of Fig. 8.11

2. Computation of the diode resistive companion model is performed using the procedure 'Comp_NR'.
3. The iterative algorithm is used to calculate the voltages of the circuit for every time step (capacitors have been replaced by their temporal companion models). The convergence method used in this procedure is method 4 presented in Section 6.5. The nodal method is used for the calculation of node voltages.
4. Two kinds of source are defined and used: a pulse signal (delay time, initial voltage, pulse voltage, pulse width and period, duration time) described by the function 'Pulse' and a sinusoidal source (average value, amplitude, delay time, frequency) described by the function 'Sinus'.

Results of the analysis. Let us consider the circuit of Fig. 8.10 with $R_g = 100\ \Omega$, $I_S = 10^{-13}$ A, $C = 1\ \mu F$ and $R = 1\ k\Omega$. The input source is a pulse source with the following parameters:

1. initial voltage: 0 V,
2. pulse voltage: 5 V,
3. delay time: 1 ms,
4. pulse width: 2 ms,
5. period: 4 ms.

This circuit has been simulated by a PSPICE simulator in the same condition with a maximum time step equal to 0.01 ms (PSPICE calculates the value of h automatically depending on the circuit and input signal; it uses trapezoidal and bear algorithms). In Table 8.2, the results of voltage $v_C(t)$ obtained using the program 'FILT' are shown for two time steps for backward Euler and trapezoidal algorithms. The results show the importance of the time step h.

The second example circuit shown in Fig. 8.13 analyses the voltage doubler. In order to calculate the voltages and currents of this circuit, the same procedure as in the previous circuit is used. So the linearized circuit at $t = (n + 1)h$ is shown in Fig. 8.14.

The program 'DOUBLER' shown in Program 8.4 permits the simulation of the

Table 8.2   Results of $v_C(t)$ computation for the circuit shown in Fig. 8.10 using the program 'FILT'; PSPICE results are given for comparison

t(ms)	Euler algorithm		Trapezoidal algorithm		PSPICE
	Results: $v_C(t)$				
	$h = 0.1$ ms	$h = 0.01$ ms	$h = 0.1$ ms	$h = 0.01$ ms	$h \leqslant 0.01$ ms
1.1	2.0573	2.5532	1.3915	2.6888	2.622
1.2	3.0438	3.5002	3.1948	3.5278	3.507
1.3	3.5192	3.7963	3.7313	3.8147	3.810
1.4	3.7494	3.9005	3.8946	3.9145	3.915
1.5	3.8614	3.9472	3.9453	3.9495	3.952
2	3.9161	3.9685	3.9686	3.9685	3.972
4	3.9671	3.9686	3.9686	3.9686	3.972
4.1	3.6078	3.5250	3.7796	3.5375	3.534
4.2	3.2799	3.1882	3.4197	3.2009	3.252
4.3	2.9817	2.8864	3.0940	2.8969	2.942
4.4	2.7106	2.6130	2.5327	2.3713	2.662
4.5	2.4642	2.3655	2.5327	2.3713	2.402
5	1.5301	1.4672	1.2570	1.4382	1.461

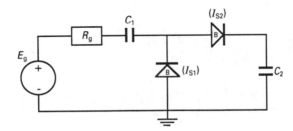

Figure 8.13   The voltage doubler

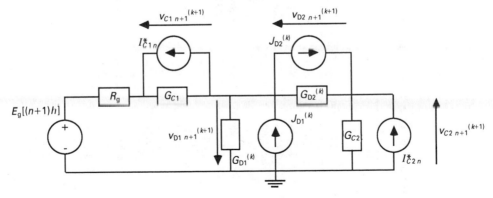

Figure 8.14   Linearized circuit of the voltage doubler of Fig. 8.13

Table 8.3 Results of $v_{C2}(t)$ computation for the circuit shown in Fig. 8.13 using the program 'DOUBLER'; PSPICE results are given for comparison

Results: $v_{C2}(t)$

t(ms)	Euler algorithm			Trapezoidal algorithm			Gear2 algorithm			PSPICE
	$h = 300\ \mu s$	$h = 50\ \mu s$	$h = 10\ \mu s$	$h = 300\ \mu s$	$h = 50\ \mu s$	$h = 10\ \mu s$	$h = 300\ \mu s$	$h = 50\ \mu s$	$h = 10\ \mu s$	
0.5	1.7617	1.9851	2.0559	3.0721	2.0838	2.0770	2.1877	2.1185	2.0788	2.079
1.5	3.6478	4.6501	4.8644	6.3344	4.9421	4.9239	4.6787	5.0177	4.9279	4.928
2.5	4.5409	6.0861	6.3393	6.9898	6.4328	6.4087	5.5816	6.5036	6.4131	6.414
3.5	5.9203	6.9197	7.1580	8.0238	7.2477	7.2240	7.0459	7.2999	7.2280	7.230
4.5	6.1869	7.4366	7.6424	8.0238	7.7219	7.7006	7.2092	7.7529	7.7039	7.707
5.5	6.1870	7.7656	7.9476	8.0238	8.0123	7.9976	7.2152	8.0269	8.0002	8.004
6.5	7.0263	7.9920	8.1508	8.3515	8.2057	8.1935	7.8129	8.2089	8.1956	8.200
7.5	7.1662	8.1537	8.2929	8.3516	8.3375	8.3294	7.8851	8.3335	8.3310	8.336
8.5	7.1662	8.2722	8.3962	8.3516	8.4312	8.4276	7.8878	8.4222	8.4289	8.434
9.5	7.6232	8.3613	8.4739	8.4075	8.5010	8.5012	8.1388	8.4891	8.5022	8.507
10.5	7.7073	8.4304	8.5341	8.4161	8.5560	8.5580	8.1753	8.5420	8.5588	8.564
11.5	7.7073	8.4851	8.5818	8.4161	8.6008	8.6030	8.1772	8.5857	8.6036	8.609
12.5	7.9584	8.5316	8.6205	8.4292	8.6382	8.6394	8.2604	8.6226	8.6399	8.646
13.5	8.0116	8.5707	8.6525	8.4423	8.6700	8.6694	8.3150	8.6542	8.6698	8.676
14.5	8.0116	8.6046	8.6792	8.4509	8.6973	8.6946	8.3150	8.6817	8.6949	8.701

circuit in similar conditions to those for the previous circuit, that is,

1. application of associated models for capacitors,
2. choice of time step h,
3. application of iterative companion models for diodes,
4. input source $E_g(t)$ is sinusoidal (average value, frequency, amplitude).

> **See the listing of the program DOUBLER Directory of Chapter 8 on attached diskette.**

The main difference in this circuit is its complexity: it contains two diodes and two capacitors. The procedures 'Calcul_mat' and 'Mat33' of the program 'DOUBLER' permit this computation.

Results of the analysis. Let us consider the circuit of Fig. 8.13 with the following parameters: $R_g = 100\ \Omega$, $I_S = 10^{-13}\ A$ (for the two diodes), $C_1 = C_2 = 1\ \mu F$. The input source is a sinusoidal source with the following parameters:

1. delay time: 0 s,
2. average value: 0 V,
3. magnitude: 5 V,
4. frequency: 1000 Hz.

This circuit has been simulated with a PSPICE simulator with maximum time step $h = 0.01$ ms. The results obtained for the voltage capacitor C_2 (for every maximum value of the input source) are presented in Table 8.3. This circuit is very interesting in testing the validity of algorithms after a 'long time' with time step h. A time step equal to 300 μs gives false results. A time step equal to 50 μs is sufficient for trapezoidal or Gear2 algorithms. The error is bigger for the Euler algorithm. A time step equal to 10 μs is correct for the three algorithms.

8.4.2 Circuits containing diodes and non-linear capacitors

The semiconductor elements are modelled in the dynamic state by equivalent circuits containing non-linear resistors and non-linear capacitors which are functions of the voltage. All these dynamic models are described in Chapter 1. In this section we recall only a simplified model of the diode, which is shown in Fig. 8.15 (model with one Boltzmann diode and $R_S = 0$).

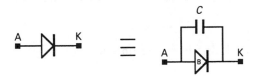

Figure 8.15 Simplified dynamic model of a semiconductor diode

The capacitance C of the diode is defined by the following expression: $C = C_D + C_J$, with

$$C_D = \tau \frac{I_S}{V_T} \exp\left[\frac{v}{V_T}\right] \tag{8.57}$$

$$C_J = C_{J0} \frac{1}{\sqrt{1 - \dfrac{v}{0.7}}} \qquad v < 0.35 \text{ V} \tag{8.58a}$$

$$C_J = 2.83 C_{J0} \left[0.25 + \frac{v}{1.4}\right] \qquad v > 0.35 \text{ V} \tag{8.58b}$$

In the case of a non-linear capacitor, the current is defined by the following expression (use of the charge concept):

$$i_C = \frac{dq}{dt} = \frac{dq}{dv_C} \frac{dv_C}{dt} = C(v_C) \frac{dv_C}{dt} \tag{8.59}$$

The previous algorithms developed for linear capacitors cannot be used in this case. So we have to define the discrete models of non-linear capacitors associated with the backward Euler, trapezoidal and Gear algorithms.

Backward Euler algorithm. Considering the equation

$$i_{Cn+1} = \left(\frac{dq}{dt}\right)_{n+1} = \left(\frac{dq}{dv_C}\right)_{n+1} \left(\frac{dv_C}{dt}\right)_{n+1} = C_{n+1} \left(\frac{dv_C}{dt}\right)_{n+1} \tag{8.60}$$

and applying the backward Euler algorithm, we obtain

$$i_{Cn+1} = C_{n+1} \frac{v_{Cn+1} - v_{Cn}}{h} \tag{8.61}$$

Thus, the associated model of the capacitor is represented by a non-linear resistor depending on the time step. Therefore, an iterative companion model is necessary to calculate the value of v_{Cn+1}.

Trapezoidal algorithm. Considering the equation

$$i_{Cn+1} = \left(\frac{dq}{dt}\right)_{n+1} = C_{n+1} \left(\frac{dv_C}{dt}\right)_{n+1} \tag{8.62}$$

and using the trapezoidal algorithm, we obtain

$$\left(\frac{dv_C}{dt}\right)_{n+1} + \left(\frac{dv_C}{dt}\right)_{n} = 2 \frac{v_{Cn+1} - v_{Cn}}{h} \tag{8.63}$$

or

$$\left(\frac{dv_C}{dt}\right)_{n+1} = 2\,\frac{v_{Cn+1} - v_{Cn}}{h} - \left(\frac{dv_C}{dt}\right)_n; \qquad \left(\frac{dv_C}{dt}\right)_n = \frac{i_{Cn}}{C_n} \tag{8.64}$$

So

$$i_{Cn+1} = C_{n+1}\,\frac{2}{h}\,(v_{Cn+1} - v_{Cn}) - \frac{i_{Cn}}{C_n}\,C_{n+1} = \frac{2C_{n+1}v_{Cn+1}}{h} - C_{n+1}\left(\frac{2v_{Cn}}{h} + \frac{i_{Cn}}{C_n}\right) \tag{8.65}$$

Gear2 algorithm. In the same way,

$$i_{Cn+1} = \left(\frac{dq}{dt}\right)_{n+1} = C_{n+1}\left(\frac{dv_C}{dt}\right)_{n+1} \tag{8.66}$$

Applying the Gear2 algorithm, we obtain

$$\left(\frac{dv_C}{dt}\right)_{n+1} = \frac{3}{2h}\,v_{Cn+1} - \frac{2}{h}\,v_{Cn} + \frac{1}{2h}\,v_{Cn-1} \tag{8.67}$$

or

$$i_{Cn+1} = C_{n+1}\left(\frac{3}{2h}\,v_{Cn+1} - \frac{2}{h}\,v_{Cn} + \frac{1}{2h}\,v_{Cn-1}\right) \tag{8.68}$$

Now we will apply these algorithms to the circuit shown in Fig. 8.16.

At time $t = (n+1)h$, the circuit of Fig. 8.16 is represented by the circuit of Fig. 8.17. This circuit contains two non-linear elements:

1. A Boltzmann diode represented by the equation

$$i_{Dn+1} = I_S\left[\exp\left(\frac{v_{Dn+1}}{V_T}\right) - 1\right]$$

2. A non-linear resistor modelling the capacitor of the diode. The relation $i_{Cn+1}(v_{Cn+1})$ has been developed previously (equations (8.61), (8.65) and (8.68)).

In order to determine the voltage $v_{n+1} = v_{Dn+1} = v_{Cn+1}$ of this circuit, the Newton–Raphson algorithm is used with an improved convergence method. At each

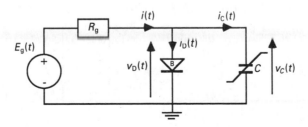

Figure 8.16 A dynamic circuit containing a semiconductor diode

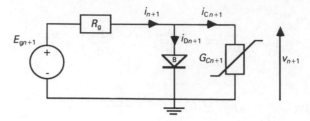

Figure 8.17   Discretized non-linear model of the dynamic circuit of Fig. 8.16

iteration (k), the iterative companion models of the diode and capacitor are recalculated. Thus, we obtain the linearized circuit represented in Fig. 8.18.

The elements of this circuit are described by the following expressions:

$$G^{(k)}_{Dn+1} = \left.\frac{d i_{Dn+1}}{d v_{n+1}}\right|_{v^{(k)}_{n+1}} = \frac{I_S + i^{(k)}_{Dn+1}}{V_T} \tag{8.69a}$$

$$J^{(k)}_{Dn+1} = i^{(k)}_{Dn+1} - G^{(k)}_{Dn+1} v^{(k)}_{n+1} \tag{8.69b}$$

$$G^{(k)}_{Cn+1} = \left.\frac{d i_{Cn+1}}{d v_{n+1}}\right|_{v^{(k)}_{n+1}} \tag{8.70}$$

The expression for $G^{(k)}_{Cn+1}$ depends on the algorithm used to define the associated model of the capacitor. For a different algorithm, we obtain the following relations.

Backward Euler algorithm:

$$G^{(k)}_{Cn+1} = \frac{C_{n+1}}{h} + \left.\frac{\delta C_{n+1}}{\delta v_{n+1}}\right|_{v^{(k)}_{n+1}} \frac{v^{(k)}_{n+1} - v_n}{h} \tag{8.71a}$$

Trapezoidal algorithm:

$$G^{(k)}_{Cn+1} = \frac{2C_{n+1}}{h} + \left.\frac{\delta C_{n+1}}{\delta v_{n+1}}\right|_{v^{(k)}_{n+1}} \left(\frac{2(v^{(k)}_{n+1} - v_n)}{h} - \frac{i_{Cn}}{C_n}\right) \tag{8.71b}$$

Gear2 algorithm:

$$G^{(k)}_{Cn+1} = \frac{3C_{n+1}}{2h} + \left.\frac{\delta C_{n+1}}{\delta v_{n+1}}\right|_{v^{(k)}_{n+1}} \left(\frac{3v^{(k)}_{n+1}}{2h} - \frac{2v_n}{h} + \frac{v_{n-1}}{2h}\right) \tag{8.71c}$$

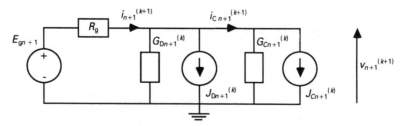

Figure 8.18   Linearized model (at iteration k) of the circuit of Fig. 8.17

The calculation of $J_{Cn+1}^{(k)}$ is based on the relation

$$J_{Cn+1}^{(k)} = i_{Cn+1}^{(k)} - G_{Cn+1}^{(k)} v_{n+1}^{(k)} \tag{8.72}$$

The program 'DIODE_DYN' shown in Program 8.5 permits us to analyze the circuit of Fig. 8.18. In this program, it is possible to choose

1. time step,
2. source generator (pulse or sinusoidal),
3. companion model of the capacitor.

> **See the listing of the program DIODE_DYN.pas Directory of Chapter 8 on attached diskette.**

Results of the analysis. Let us consider the circuit of Fig. 8.16 with $R_g = 1000\ \Omega$, $I_S = 10^{-13}$ A, $C_{J0} = 1$ pF and $\tau = 5$ ns. The input source is a pulse source with the

Table 8.4 The time responses $v(t)$ and $i(t)$ of Fig. 8.16 obtained by the 'DIODE_DYN' program with the Euler and trapezoidal algorithms; PSPICE results are given for comparison

			Results: v(t) and i(t)			
	Euler algorithm		*Trapezoidal algorithm*		*PSPICE*	
t(ns)	$v(t)$	$i(t)$	$v(t)$	$i(t)$	$v(t)$	$i(t)$
0	-5	-10^{-13}	-5	-10^{-13}	-5	-5×10^{-12}
0.2	-2.0527	7.0527×10^{-3}	-2.3224	7.3225×10^{-3}	-2.013	7.013×10^{-3}
0.4	-0.4830	5.4830×10^{-3}	-0.4148	5.4148×10^{-3}	-0.1731	5.174×10^{-3}
0.6	0.3711	4.6289×10^{-3}	0.4598	4.5489×10^{-3}	0.5023	4.498×10^{-3}
0.8	0.5230	4.4770×10^{-3}	0.5703	4.4375×10^{-3}	0.5518	4.449×10^{-3}
1	0.5538	4.4462×10^{-3}	0.5804	4.4261×10^{-3}	0.5680	4.432×10^{-3}
2	0.5953	4.4047×10^{-3}	0.6034	4.3977×10^{-3}	0.5974	4.403×10^{-3}
3	0.6091	4.3911×10^{-3}	0.6136	4.3914×10^{-3}	0.6088	4.391×10^{-3}
4	0.6167	4.3833×10^{-3}	0.6197	4.3819×10^{-3}	0.6155	4.385×10^{-3}
5	0.6217	4.3783×10^{-3}	0.6238	4.3768×10^{-3}	0.6199	4.380×10^{-3}
10	0.6323	4.3677×10^{-3}	0.6329	4.3671×10^{-3}	0.6294	4.372×10^{-3}
20	0.6364	4.3637×10^{-3}	0.6365	4.3636×10^{-3}	0.6331	4.368×10^{-3}
25	0.6367	4.3633×10^{-3}	0.6368	4.3632×10^{-3}	0.6335	4.368×10^{-3}
25.2	0.6342	-5.6346×10^{-3}	0.6362	-5.6412×10^{-3}	0.6327	-5.632×10^{-3}
26	0.6242	-5.6242×10^{-3}	0.6252	-5.6219×10^{-3}	0.6220	-5.624×10^{-3}
27	0.6017	-5.6017×10^{-3}	0.6047	-5.5905×10^{-3}	0.6017	-5.602×10^{-3}
28	-1.0601	-3.9399×10^{-3}	0.1770	5.1770×10^{-3}	0.5208	-5.521×10^{-3}
29	-4.3921	-6.0791×10^{-4}	-4.1156	-8.8434×10^{-4}	-4.036	-9.635×10^{-4}
30	-4.9294	-7.0568×10^{-5}	-4.9188	-8.1162×10^{-5}	-4.940	-6.046×10^{-5}
31	-4.9921	-7.9236×10^{-6}	-4.9930	-6.9657×10^{-6}	-4.997	-3.502×10^{-6}
32	-4.9991	-8.8636×10^{-7}	-4.9994	-5.9436×10^{-7}	-5	-2.019×10^{-7}
33	-4.9999	-9.9113×10^{-8}	-4.9999	-5.0698×10^{-8}	-5	-1.164×10^{-8}
34	-5	-1.1082×10^{-8}	-5	-4.3229×10^{-9}	-5	-6.758×10^{-10}
35	-5	-1.2392×10^{-9}	-5	-7.7162×10^{-10}	-5	-4.373×10^{-11}

following parameters:

1. initial voltage: -5 V,
2. pulse voltage: 5 V,
3. delay time: 0,
4. pulse width: 25 ns,
5. period: 50 ns.

This circuit has been simulated by the PSPICE simulator in the same conditions. The results obtained for the voltage $v_D(t)$ and current $i_{DT}(t)$ of the diode are presented in Table 8.4 with a step time equal to 0.1 ns for Euler and trapezoidal algorithms.

Exercises

Exercise 1. Analyze the circuits shown in Fig. E8.1(a, b, c) using the state variable method and write the equations in the normal form.

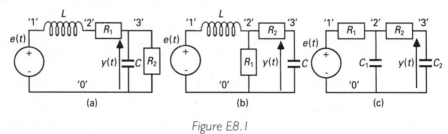

Figure E8.1

Exercise 2. Analyze the circuits shown in Fig. E8.2 (a, b, c) using the state variable method and write the equations in the normal form.

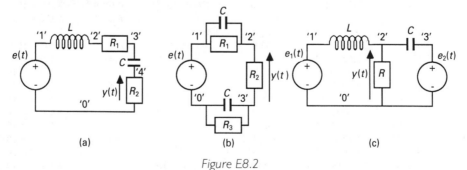

Figure E8.2

Exercise 3. Analyze the circuit shown in Fig. E8.3 using the state variable method and write the equations in the normal form (note: the circuit contains a capacitive loop).

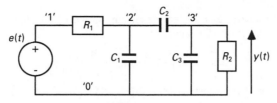

Figure E8.3

Exercise 4. Analyze the circuit shown in Fig. E8.4 using the state variable method and write the equations in the normal form (note: the circuit contains an inductive cut-set).

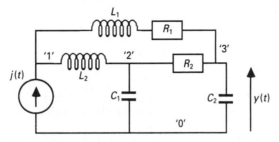

Figure E8.4

Exercise 5. Analyze the circuits shown in Fig. E8.5(a, b) using the state variable method on the Laplace plane (variable s) and write the equations in the normal form.

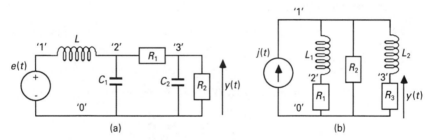

(a)   (b)

Figure E8.5

Exercise 6. Analyze the circuits shown in Fig. E8.6(a, b) (containing controlled sources) using the state variable method and write the equations in the normal form.

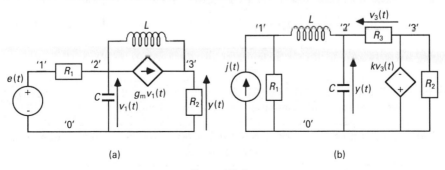

(a)   (b)

Figure E8.6

Exercise 7. The theoretical analysis of the circuit shown in Fig. 8.2 leads to the following expression for the voltage on the capacitor:

$$v(t) = \left\{ 2 - e^{-t/4} \left[\cos\left(\frac{\sqrt{15}}{4} t\right) + \frac{2}{\sqrt{15}} \sin\left(\frac{\sqrt{15}}{4} t\right) \right] \right\} u(t)$$

Modify the outputs of the program 'MULDIFSO' in order to compare the computation results with the exact values ($u(t)$: unit step function).

Exercise 8. The differential equations of order greater than one can be transformed in a first-order differential equations system. For example, the second-order differential equation $d^2x(t)/dt^2 + \omega^2 x(t) = 0$ is equivalent to the first-order differential equations system:

$$\frac{dx(t)}{dt} = z(t)$$

$$\frac{dz(t)}{dt} = -\omega^2 x(t)$$

Modify the program 'MULDIFSO' to solve this linear system and compare the results with the theoretical values (consider several couples of initial conditions).

Exercise 9. Apply the theory presented in exercise 8 to solve the second-order differential equation characterizing the motion of the pendulum:

$$\frac{d^2\alpha(t)}{dt^2} + k \sin(\alpha(t)) = 0; \qquad \alpha(0) = \frac{\pi}{4}; \qquad \left.\frac{d\alpha(t)}{dt}\right|_{t=0} = 0$$

The algorithms developed in Chapter 7 can be used.

Exercise 10. Apply the program 'MULDIFVA' to the circuit of Fig. 8.2 assuming that:

$$C = \frac{0.5}{1 + \alpha \sin t}; \qquad -1 < \alpha < 1$$

Exercise 11. Apply the program 'MULDIFVA' to the circuit shown in Fig. E8.7.

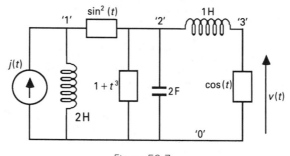

Figure E8.7

Exercise 12. The program 'MULDIFNL' is a modification of the program 'MULDIFSO'; use this program to solve the non-linear circuit shown in Fig. 8.8.

> **See the listing of the program MULDIFNL Directory of Chapter 8 on attached diskette.**

Exercise 13.   In Section 8.2, the RLC circuit shown in Fig. 8.2 is analyzed. In Section 8.3, using the program 'MULDIFSO', numerical results are presented. In this exercise, the resistance R is a non-linear element characterized by the equation $i = -k_1 v + k_3 v^3$:

1.   Analyze the form of this characteristic for different values of k_1 and k_3.
2.   Apply the program 'MULDIFSO' for the case where $i_s = 0$ assuming several couples of initial conditions.

Exercise 14.   The program 'HEUN' developed in Chapter 7 analyzes the same circuit (see Fig. 7.6) as the program 'FILT' (Program 8.3) in Chapter 8 (see Fig. 8.10), with the same assumptions on the diode model. Run the two programs for the same input data and compare the results.

Exercise 15.   The circuit shown in Fig. E8.8 contains a 'free wheel' diode. Develop a program similar to the program 'FILT' (Program 8.3) permitting analysis of the transient behaviour of this circuit.

For the voltage source $e(t)$, consider the following time functions (pulse sources described in Section 8.4):

$e_1(t)$: initial voltage = -5 V, pulse voltage = 5 V, delay time = 0, pulse width = 1 μs, period = 2 μs,

$e_2(t)$: initial voltage = -5 V, pulse voltage = 5 V, delay time = 0, pulse width = 10 μs, period = 20 μs.

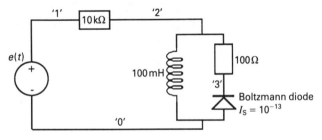

Figure E8.8

Exercise 16.   The circuit shown in Fig. E8.9 is called the 'clamping circuit'. Modifying the program 'FILT' (Program 8.3), determine the voltage on the resistor of 1 kΩ as a function of time when the voltage source $e(t)$ has the following form: $e(t) = 5$ V for $0 < t < T$; 0 V in the other case. Consider three values for T: 10 ms, 1 ms, 100 μs.

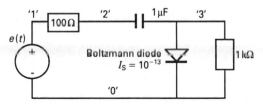

Figure E8.9

Exercise 17.   Develop a program similar to the program 'FILT' for the circuit shown in Fig. E8.10 assuming that the initial charge of the capacitor corresponds to the voltage $v_C(0)$

($v_C(0) = 0$ V, $v_C(0) = -5$ V, $v_C(0) = 5$ V). The voltage source $e(t)$ is equal to the unit step: $e(t) = 5$ V for $t > 0$. Determine $v_C(t)$ and $i(t)$: $L = 1$ mH and 100 mH; $C = 1\ \mu$F; $R = 1$ kΩ.

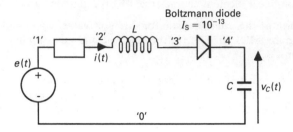

Figure E8.10

Exercise 18.   Develop a program similar to the program 'FILT' (Program 8.3) permitting analysis of the basic NMOS inverter shown in Fig. E8.11.

E_G is equal to the unit step: $E_G(t) = 5$ V for $t > 0$. The characteristics of the NMOS transistor are as follows: $V_{Th} = 1$ V; $K_P = 20\ \mu$A/V^2; $W = 25\ \mu$m and $L = 5\ \mu$m.

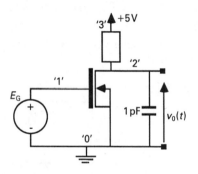

Figure E8.11

Exercise 19.   Develop a program similar to the program 'DIODE_DYN' (Program 8.5) permitting analysis of the influence of non-linear capacitors in the diode model for the circuit analyzed by the program 'DOUBLER' (see Fig. 8.13 and Program 8.4) and the circuit of exercise 16. The models of the two diodes are the same and are characterized by the following parameters:

$$C_{J0} = 1\ \text{pF};\ \phi = 0.7\ \text{V};\ M = 0.5;\ F_C = 0.5;\ \tau = 1\ \text{ns};\ I_S = 10^{-13}\ \text{A}$$

Exercise 20.   To consider the dynamic characteristics of a Zener diode, we use the circuit shown in Fig. E8.12. In exercise 8 of Chapter 6, the program 'COMP_ZENER' has been developed. Modify this program to take into account the non-linear capacitors in the Zener diode model (composed of two Boltzmann diodes as described in Section 1.2). The Zener diode model is characterized by the following parameters:

'direct' diode: $I_S = 10^{-13}$ A; $C_{J0} = 1$ pF; $\phi = 0.7$ V; $M = 0.5$; $F_C = 0.5$; $\tau = 1$ ns;

'Reverse' diode: $I_{bd} = 16.748 \times 10^{-13}$ A; $V_{bd} = 4.3$ V; $n_{bd} = 1.7936$

The input source $e(t)$ is a sinusoidal source: $e(t) = -3 + 10 \sin(10\pi \times 10^4 t)$ V. Plot the voltage $v_1(t)$ as a function of $v_2(t)$ and notice a hysteresis phenomenon.

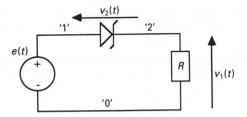

Figure E8.12

Exercise 21.   Develop a program similar to the program 'FILT' (see Program 8.3) permitting analysis of the NMOS inverter shown in Fig. E8.13.

E_G is equal to the unit step $E_G(t) = 5$ V for $t > 0$. The characteristics of the NMOS transistors are as follows: for T_1: $V_{Th} = -4$ V; $K_P = 20$ $\mu A/V^2$; $W = 5$ μm and $L = 5$ μm; for T_2: $V_{Th} = +1$ V; $K_P = 20$ $\mu A/V^2$; $W = 25$ μm and $L = 5$ μm.

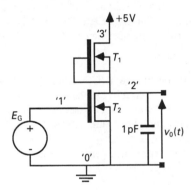

Figure E8.13

Exercise 22.   Develop a program similar to the program 'FILT' (see Program 8.3) permitting analysis of the CMOS inverter (with an NMOS and a PMOS transistor) shown in Fig. E8.14.

E_G is equal to the unit step $E_G(t) = 5$ V for $t > 0$. The characteristics of the MOS transistor are as follows: for T_1 (PMOS transistor): $V_{Th} = -1$ V; $K_P = 20$ $\mu A/V^2$; $W = 10$ μm and $L = 10$ μm; for T_2 (NMOS transistor): $V_{Th} = +1$ V; $K_P = 20$ $\mu A/V^2$; $W = 10$ μm and $L = 10$ μm.

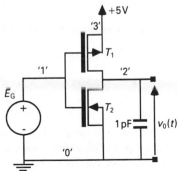

Figure E8.14

Exercise 23. Develop a program similar to the program 'DOUBLER' (see Program 8.4) permitting analysis of the 'tripler' circuit shown in Fig. E8.15. The models of the three diodes are the same and are characterized by $I_S = 10^{-13}$ A.

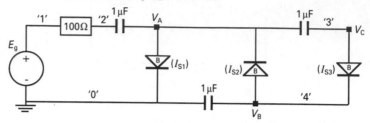

Figure E8.15

Exercise 24. Modify the program 'DOUBLER' (see Program 8.4) in order to take into account the non-linear capacitors of the diode models. Compare the computation results with those obtained with the program 'DOUBLER'. The models of the two diodes are the same and are characterized by the following parameters: $C_{J0} = 1$ pF; $\phi = 0.7$ V; $M = 0.5$; $F_C = 0.5$; $\tau = 1$ ns; $I_S = 10^{-13}$ A.

Exercise 25. In Chapter 6, the program 'COMP_TRAN' permits us to compute a DC analysis of a bipolar transistor circuit. Develop a program similar to the program 'DIODE_DYN' taking into account the base–emitter and base–collector non-linear capacitors of transistor models and permitting analysis of the circuit shown in Fig. E8.16(a) (the input voltage is shown in Fig. E8.16(b)).

The parameters of the transistor model are as follows: $C_{JE0} = C_{JC0} = 1$ pF; $\phi_E = \phi_C = 0.7$ V; $M_E = M_C = 0.5$; $F_C = 0.5$; $\tau_F = 1$ ns; $\tau_R = 20$ ns; $\beta_F = 100$; $\beta_R = 1$; $I_S = 10^{-13}$ A.

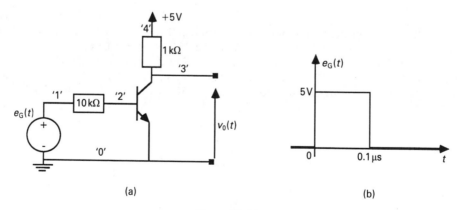

(a)

(b)

Figure E8.16

Exercise 26. In order to improve the commutation times of the circuit shown in exercise 25, the circuit is modified as shown in Fig. E8.17. Modify the program of exercise 25 assuming that the diode D_1 is modelled by a Boltzmann diode. Next, take into account the non-linear capacitors of the diode model. The model of the diode is characterized by the following parameters: $C_{J0} = 1$ pF; $\phi = 0.7$ V; $M = 0.5$; $F_C = 0.5$; $\tau = 1$ ns; $I_S = 10^{-10}$ A.

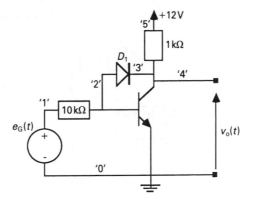

Figure E8.17

Exercise 27. Consider the circuit shown in Fig. E8.18 (voltage controlled oscillator). Develop a program permitting us to obtain the voltage $v_0(t)$. $E_1 = 6\, u(t)$ (V). The two transistors are identical and are characterized by the following parameters: $\beta_F = 100$; $\beta_R = 1$ and $I_S = 10^{-13}$ A. The initial conditions are $v_4(0) = 0$ V and $v_2(0) = 12$ V. Compute the variations of $v_0(t)$ for several values of $E_1(3$ V, 6 V, 9 V, 12 V$)$.

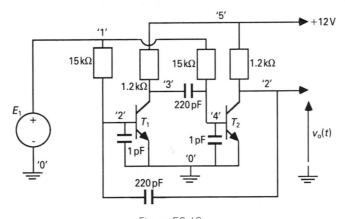

Figure E8.18

Exercise 28. Again, take exercise 27, taking into account the non-linear capacitors in the transistor models. The parameters of these models are as follows: $C_{JE0} = C_{JC0} = 1$ pF; $\phi_E = \phi_C = 0.7$ V; $M_E = M_C = 0.5$; $F_C = 0.5$; $\tau_F = 1$ ns; $\tau_R = 20$ ns; $\beta_F = 100$; $\beta_R = 1$; $I_S = 10^{-13}$ A.

Exercise 29. Consider the circuit shown in Fig. E8.19 in which for $t \geqslant 0$: $v_1(t) = V_1 \sin(\omega_1 t)$ and $v_2(t) = V_2 \sin(\omega_2 t)$. Develop a program permitting us to obtain the variations of $v_0(t)$ as a function of time. The transistor model used is characterized by the following parameters: $\beta_F = 100$; $\beta_R = 1$ and $I_S = 10^{-13}$ A; $\omega_1 = 10^{+4}$ rad/s; $\omega_2 = 1.1 \times 10^{+4}$ rad/s; $V_1 = V_2 = 0.6$ V.

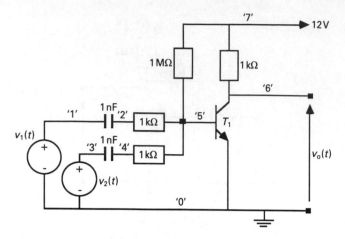

Figure E8.19

Exercise 30. Again, take exercise 29, taking into account the non-linear capacitors of the transistor models. The models are described by the following values: $C_{JE0} = C_{JC0} = 1$ pF; $\phi_E = \phi_C = 0.7$ V; $M_E = M_C = 0.5$; $F_C = 0.5$; $\tau_F = 1$ ns; $\tau_R = 20$ ns; $\beta_F = 100$; $\beta_R = 1$; $I_S = 10^{-13}$ A.

9 Computer Generation of State Equations and their Solutions in the Time Domain

9.1 Introduction

As was mentioned in Chapter 8 the state variables are usually chosen as capacitor voltages, inductor currents, and resistor voltages or currents. This leads to circuit description in terms of the set of the first-order differential equations. In this method all four types of controlled source can be used. The general normal form of the state and output equation is

$$\dot{\mathbf{x}} = \mathbf{A}\mathbf{x} + \mathbf{B}\mathbf{u} \tag{9.1}$$

$$\mathbf{y} = \mathbf{C}\mathbf{x} + \mathbf{D}\mathbf{u} + (\mathbf{D}_1\dot{\mathbf{u}} + \mathbf{D}_2\ddot{\mathbf{u}} + \cdots) \tag{9.2}$$

where

$\mathbf{x}$ = state vector consisting of n independent state variables,
$\dot{\mathbf{x}}$ = vector of state variable derivatives,
$\mathbf{u}$ = independent source vector (excitations) composed of m sources,
$\mathbf{y}$ = vector of p outputs (voltages and/or currents),
$\mathbf{A}, \mathbf{B}, \mathbf{C}, \mathbf{D}, \mathbf{D}_1, \ldots,$ = real matrices with dimensions $n \times n$, $n \times m$, $p \times n$, $p \times m$, $p \times m$, ..., respectively.

The constant n represents the order of the circuit (see Chapter 8):

$$0 \leqslant n \leqslant \Sigma C + \Sigma L - \Sigma MCE - \Sigma C_s LI \tag{9.3}$$

where

1. ΣC = number of capacitors,
2. ΣL = number of inductors,
3. ΣMCE – number of independent meshes (loops) composed of capacitors and independent voltage sources,
4. $\Sigma C_s LI$ = number of independent cut-sets composed of inductors and independent current sources.

To show that the order n can be smaller than the number of capacitors let us consider the circuits composed of two capacitors and one voltage source shown in Fig. 9.1(a, b, c). All these circuits have one CE mesh.

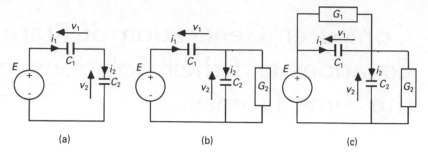

Figure 9.1 Simple circuits that contain two capacitors and one CE mesh

For these simple circuits we can write the initial state equations by inspection, or, more systematically, using the superposition principle. For this reason we substitute independent voltage sources for each capacitor and independent current sources for each inductor. However, when CE meshes exist in the circuit, we have to 'break' these meshes by replacing one of the capacitors by the independent current source. The circuits of Fig. 9.1 (a, b, c) redrawn for analysis by superposition are shown in Fig. 9.2 (a, b, c).

For the circuits of Fig. 9.2(a) we have directly

$$i_1 = i_2, \quad v_2 = -v_1 + E \tag{9.4a}$$

Since $i_1 = C_1\dot{v}_1$ and $i_2 = C_2\dot{v}_2$ we can rewrite relations (9.4a) in an initial state equation form

$$C_1\dot{v}_1 - C_2\dot{v}_2 = 0$$
$$0 = -v_1 - v_2 + E \tag{9.4b}$$

or, in matrix form,

$$\begin{bmatrix} C_1 & -C_2 \\ 0 & 0 \end{bmatrix} \begin{bmatrix} \dot{v}_1 \\ \dot{v}_2 \end{bmatrix} = \begin{bmatrix} 0 & 0 \\ -1 & -1 \end{bmatrix} \begin{bmatrix} v_1 \\ v_2 \end{bmatrix} + \begin{bmatrix} 0 \\ 1 \end{bmatrix} E \tag{9.4c}$$

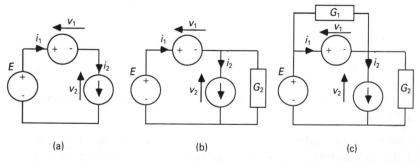

Figure 9.2 Circuits of Fig. 9.1 redrawn for analysis by superposition

Since the voltages v_1 and v_2 are not independent we can eliminate v_1 from equations (9.4) to obtain

$$v_2 = \frac{C_1}{C_1 + C_2} E \tag{9.4d}$$

which shows that the circuit of Fig. 9.1(a) is zero order.

For the circuit of Fig. 9.2(b) we obtain relations

$$i_1 = -G_2 v_1 + i_2 + G_2 E$$
$$v_2 = -v_1 + E \tag{9.5a}$$

or

$$C_1 \dot{v}_1 = -G_2 v_1 + C_2 \dot{v}_2 + G_2 E$$
$$v_2 = -v_1 + E \tag{9.5b}$$

or, in matrix form,

$$\begin{bmatrix} C_1 & -C_2 \\ 0 & 0 \end{bmatrix} \begin{bmatrix} \dot{v}_1 \\ \dot{v}_2 \end{bmatrix} = \begin{bmatrix} -G_2 & 0 \\ -1 & -1 \end{bmatrix} \begin{bmatrix} v_1 \\ v_2 \end{bmatrix} + \begin{bmatrix} G_2 \\ 1 \end{bmatrix} E \tag{9.5c}$$

Since voltages v_1 and v_2 are not independent, elimination of v_1 gives

$$\dot{v}_2 = -\frac{G_2}{C_1 + C_2} v_2 + \frac{C_1}{C_1 + C_2} \dot{E} \tag{9.5d}$$

which shows that this circuit is first order. Note that in equation (9.5d) the derivative of the source voltage $\dot{E}$ appears; therefore this equation is not in normal form.

For the circuit of Fig. 9.2(c) we obtain

$$i_1 = -(G_1 + G_2) v_1 + i_2 + G_2 E$$
$$v_2 = -v_1 + E \tag{9.6a}$$

or

$$C_1 \dot{v}_1 - C_2 \dot{v}_2 = -(G_1 + G_2) v_1 + G_2 E$$
$$0 = -v_1 - v_2 + E \tag{9.6b}$$

or, in matrix form,

$$\begin{bmatrix} C_1 & -C_2 \\ 0 & 0 \end{bmatrix} \begin{bmatrix} \dot{v}_1 \\ \dot{v}_2 \end{bmatrix} = \begin{bmatrix} -(G_1 + G_2) & 0 \\ -1 & -1 \end{bmatrix} \begin{bmatrix} v_1 \\ v_2 \end{bmatrix} + \begin{bmatrix} G_2 \\ 1 \end{bmatrix} E \tag{9.6c}$$

Elimination of the dependent voltage v_1 gives a single first-order differential equation

$$\dot{v}_2 = \frac{-(G_1 + G_2)}{C_1 + C_2} v_2 + \frac{G_1}{C_1 + C_2} E + \frac{C_1}{C_1 + C_2} \dot{E} \tag{9.6d}$$

Equation (9.6d) is also not in normal form. Elimination of the derivative $\dot{E}$ from equations (9.5d) and (9.6d) can be obtained by changing the variable. For equation (9.6d) we can substitute

$$v_2 = z + \frac{C_1}{C_1 + C_2}\, E, \quad \dot{v}_2 = \dot{z} + \frac{C_1}{C_1 + C_2}\, \dot{E} \tag{9.7}$$

to obtain

$$\dot{z} + \frac{C_1}{C_1 + C_2}\, \dot{E} = -\frac{(G_1 + G_2)}{C_1 + C_2}\, z - \frac{(G_1 + G_2)C_1}{(C_1 + C_2)^2}\, E + \frac{G_1}{C_1 + C_2}\, E + \frac{C_1}{C_1 + C_2}\, \dot{E}$$

or

$$\dot{z} = -\frac{(G_1 + G_2)}{C_1 + C_2}\, z - \frac{(G_2 C_1 - G_1 C_2)}{(C_1 + C_2)^2}\, E \tag{9.8}$$

Equation (9.8) is in normal form. Solution of this equation for z gives us the value of v_2. Equations (9.5c) and (9.6c) containing voltages on all capacitors as state variables are referred to as the initial state equations, while equations (9.5d) and (9.6d) containing the derivatives of the excitation sources are referred to as the intermediate state equations. Transformation of the initial and intermediate state equations to normal form is performed automatically by a computer program.

Generation of the state and output equations in normal form can be performed in three steps:

1. Step 1: generation of initial state and output equations in the forms

   $$\mathbf{M}^{(0)}\dot{\mathbf{x}}^{(0)} = \mathbf{A}^{(0)}\mathbf{x}^{(0)} + \mathbf{B}^{(0)}\mathbf{u} \tag{9.9}$$

   $$\mathbf{y} = \mathbf{N}^{(0)}\dot{\mathbf{x}}^{(0)} + \mathbf{C}^{(0)}\mathbf{x}^{(0)} + \mathbf{D}^{(0)}\mathbf{u} \tag{9.10}$$

   where the superscript (0) indicates that the state variable vector $[\mathbf{x}^{(0)}]$ is composed of all capacitor voltages and all inductor currents of the analyzed circuit. These variables can be linearily dependent. Elimination of dependent variables will be performed in the next step.

2. Step 2: transformation of equations (9.9) and (9.10) to the intermediate form:

   $$\dot{\mathbf{x}} = \mathbf{A}^{(r)}\mathbf{x} + \mathbf{B}^{(r)}\mathbf{u} + \mathbf{B}_1^{(r)}\dot{\mathbf{u}} + \mathbf{B}_2^{(r)}\ddot{\mathbf{u}} + \cdots \tag{9.11}$$

   $$\mathbf{y} = \mathbf{C}^{(r)}\mathbf{x} + \mathbf{D}^{(r)}\mathbf{u} + \mathbf{D}_1^{(r)}\dot{\mathbf{u}} + \mathbf{D}_2^{(r)}\ddot{\mathbf{u}} + \cdots \tag{9.12}$$

   where the superscript (r) indicates that (r) dependent variables have been eliminated, and $\mathbf{x}$ is the subset of the vector $\mathbf{x}^{(0)}$;

3. Step 3: reduction of equations (9.11) and (9.12) to the normal form described by equations (9.1) and (9.2).

9.2 Generation of hybrid matrix and initial state equations

Computer generation of a hybrid matrix and initial state equations for a given RLC circuit is a lengthy process described in detail in the book *Computer Aided Analysis of Electronic Circuits* by L.O Chua and L.P Lin, Prentice Hall, 1975. In this chapter only the main steps of the process are described and will be illustrated by numerical example. In the process of the generation of computer initial state equations the following concepts are used:

1. generation of the normal tree,
2. generation of topological matrices (incidence, cut-set),
3. generation of the hybrid matrix of a resistive circuit,
4. linear matrix transformations (echelon form, unity submatrix).

In the procedure of hybrid matrix generation the concept of a resistive circuit for the RLC circuit is important. The resistive circuit is formed in the following way:

1. All capacitors and inductors are extracted from the circuit and substituted by voltage or current ports (independent sources).
2. All selected output voltages, treated as voltages on zero-valued current ports, are extracted ($i_{OC} = 0$).
3. All selected output currents, treated as currents in zero-valued voltage ports, are extracted ($u_{SC} = 0$).
4. All independent voltage and current sources, treated as voltage and current ports, respectively, are extracted from the circuit.

Matrix hybrid equation for such a resistive n-port has the form

$$
\begin{bmatrix} i_{SC} \\ i_U \\ i_a \\ u_b \\ u_J \\ u_{OC} \end{bmatrix} =
\begin{bmatrix}
H_{11} & H_{12} & H_{13} & H_{14} & H_{15} & H_{16} \\
H_{21} & H_{22} & H_{23} & H_{24} & H_{25} & H_{26} \\
H_{31} & H_{32} & H_{33} & H_{34} & H_{35} & H_{36} \\
H_{41} & H_{42} & H_{43} & H_{44} & H_{45} & H_{46} \\
H_{51} & H_{52} & H_{53} & H_{54} & H_{55} & H_{56} \\
H_{61} & H_{62} & H_{63} & H_{64} & H_{65} & H_{66}
\end{bmatrix} =
\begin{bmatrix} u_{SC} \\ U \\ u_a \\ i_b \\ J \\ i_{OC} \end{bmatrix}
\tag{9.13}
$$

Where the subvectors:

1. u_a, i_a are associated with voltage ports of extracted L and C elements,
2. i_b, u_b are associated with current ports of extracted L and C elements,
3. i_{SC}, u_{SC} are associated with extracted zero-valued voltage ports (short-circuits, $u_{SC} = 0$).
4. u_{OC}, i_{OC} are associated with extracted zero-valued current ports (open-circuits, $i_{OC} = 0$),
5. U, i_U are associated with extracted independent voltage sources,
6. u_J, J are associated with extracted independent current sources.

Since $u_{SC} = 0$ and $i_{OC} = 0$, equation (9.13) can be rewritten in the form

$$i_U = H_{23}u_a + H_{24}i_b + H_{22}U + H_{25}J \tag{9.14}$$

$$
\begin{bmatrix} i_a \\ u_b \end{bmatrix} =
\begin{bmatrix} H_{33} & H_{34} \\ H_{43} & H_{44} \end{bmatrix}
\begin{bmatrix} u_a \\ i_b \end{bmatrix} +
\begin{bmatrix} H_{32} & H_{35} \\ H_{42} & H_{45} \end{bmatrix}
\begin{bmatrix} U \\ J \end{bmatrix}
\tag{9.15}
$$

$$
\begin{bmatrix} i_{SC} \\ u_{OC} \end{bmatrix} =
\begin{bmatrix} H_{13} & H_{14} \\ H_{63} & H_{64} \end{bmatrix}
\begin{bmatrix} u_a \\ i_b \end{bmatrix} +
\begin{bmatrix} H_{12} & H_{15} \\ H_{62} & H_{65} \end{bmatrix}
\begin{bmatrix} U \\ J \end{bmatrix}
\tag{9.16}
$$

Now we can relate the variables i_a and $\mathbf{u}_b$ with $\mathbf{u}_a$ and i_b by insertion of the branch C and L element relations

$$i_C = C\dot{\mathbf{u}}_C, \ v_L = L\dot{i}_L \tag{9.17}$$

Next, it is necessary to reorder equation (9.15) to obtain all the terms with the derivatives on the left-hand side; this leads to the initial state equation and the initial output equation in the form given by relations (9.9) and (9.10). Systematic generation of the hybrid matrix starts by reading in the circuit elements and creating an incidence matrix based on the circuit directed graph. For example, the directed graph for the circuit of Fig. 9.3(a) is shown in Fig. 9.3(b and c). The reduced incidence matrix with the number of rows equal to the number of independent nodes and the number of columns equal to the number of branches depends on the enumeration of the nodes and branches.

For branch enumeration as in Fig. 9.3(b), the incidence matrix is

$$\begin{array}{cccc} 1 & 2 & 3 & 4 \quad \text{branches} \end{array}$$
$$\mathbf{A} = \begin{array}{c} 1 \\ 2 \end{array}\left[\begin{array}{cccc} -1 & 1 & 1 & 0 \\ 0 & 0 & -1 & 1 \end{array}\right] \tag{9.18}$$
$$\text{nodes}$$

and for branch enumeration as in Fig. 9.3(c) the incidence matrix is

$$\begin{array}{cccc} 1 & 2 & 3 & 4 \end{array}$$
$$\mathbf{A} = \begin{array}{c} 1 \\ 2 \end{array}\left[\begin{array}{cccc} 1 & 1 & 0 & -1 \\ -1 & 0 & 1 & 0 \end{array}\right] \tag{9.19}$$

Next, it is necessary to form a tree and co-tree for the directed graph. The tree is created from the graph branches touching each independent node and not forming the cycles (closed loops). The rest of the graph forms the co-tree. The tree, referred to as a normal tree, is formed from graph branches selected in the following way:

1. all independent voltage sources,
2. all current outputs (short circuits),
3. as many as possible capacitors,
4. some resistors.

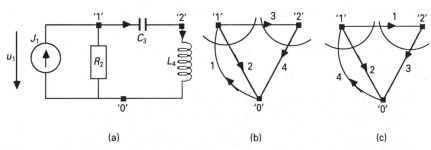

(a)   (b)   (c)

Figure 9.3   (a) Simple RLC circuit. (b and c) Directed graphs corresponding to the circuit

The co-tree should contain graph branches (links) of:

1. all independent current sources,
2. all voltage outputs (open circuits),
3. as many as possible inductors,
4. some resistors.

Based on this selection of the tree, the graph branches can be divided into four categories:

port branches in the tree, a,
port links in the co-tree, b,
tree branches not forming ports, z,
co-tree branches not forming ports. y

Tree branches can be formed from the incidence matrix $\mathbf{A}$ transformed to the echelon form $\mathbf{A}_{ech}$. The matrix $\mathbf{A}_{ech}$ is shown in Fig. 9.4.

In such a case, the leftmost 1 in each row indicates the columns corresponding to the tree branches; in this example, the tree branches are 1, 2, 5, 7. In our simple example, the matrix $\mathbf{A}$ given by (9.19) can be transformed into echelon form by elementary row equations, that is, by the addition of row 1 to row 2:

$$\begin{array}{cccc} 1 & 2 & 3 & 4 \text{ branches} \end{array}$$

$$\mathbf{A}_{ech} = [\mathbf{A}_T \mid \mathbf{A}_L] = \begin{bmatrix} 1 & 1 & 0 & -1 \\ 0 & 1 & 1 & -1 \end{bmatrix} \tag{9.20}$$

Thus, branches 1 and 2 form the tree, and links 3 and 4 form the co-tree. Next, it is necessary to find the cut-set matrix $\mathbf{D}$ which represents the Kirchhoff current law:

$$\mathbf{D}\,\mathbf{i} = 0 \tag{9.21}$$

The cut-set matrix $\mathbf{D}$ can be transformed into the form

(tree) (links)

$$[\mathbf{1}_T \mid \mathbf{D}_L]$$

$$\begin{array}{ccccccccc} & 1 & 2 & 3 & 4 & 5 & 6 & 7 & 8 & \text{Branches} \end{array}$$

$$\mathbf{A}_{ech} = \begin{bmatrix} \boxed{1} & 1 & 0 & 0 & 1 & 0 & 0 & 0 \\ 0 & \boxed{1} & -1 & 0 & 0 & 0 & 0 & 0 \\ 0 & 0 & 0 & 0 & \boxed{1} & -1 & 0 & 0 \\ 0 & 0 & 0 & 0 & 0 & 0 & \boxed{1} & -1 \end{bmatrix}$$

Figure 9.4 Matrix $\mathbf{A}_{ech}$ transformed to the echelon form

which can be obtained from $\mathbf{A}_{\text{ech}}$ by transformation of $\mathbf{A}_T$ to the unit submatrix. In our simple example, transformation of (9.20) gives

$$
\begin{array}{cccc}
1 & 2 & 3 & 4 \quad \text{branches}
\end{array}
$$

$$
\mathbf{D} = [\mathbf{1}_T \mid \mathbf{D}_L] = \begin{bmatrix} 1 & 0 & | & -1 & 0 \\ 0 & 1 & | & 1 & -1 \end{bmatrix} \begin{matrix} 1 \\ 2 \end{matrix}
\tag{9.22}
$$

$$
\text{cut-sets}
$$

According to the branch categories described earlier, the $\mathbf{D}$ matrix can be written as

$$
\text{(tree)} \qquad \text{(co-tree)}
$$

$$
\mathbf{D} = [\mathbf{1}_T \mid \mathbf{D}_L] = \begin{bmatrix} \mathbf{1}_{aa} & \mathbf{0}_{az} & | & \mathbf{D}_{ay} & \mathbf{D}_{ab} \\ \mathbf{0}_{za} & \mathbf{1}_{zz} & | & \mathbf{D}_{zy} & \mathbf{D}_{zb} \end{bmatrix}
\tag{9.23}
$$

and Kirchhoff's current law as

$$
\begin{bmatrix} \mathbf{1}_{aa} & \mathbf{0}_{az} & | & \mathbf{D}_{ay} & \mathbf{D}_{ab} \\ \mathbf{0}_{za} & \mathbf{1}_{zz} & | & \mathbf{D}_{zy} & \mathbf{D}_{zb} \end{bmatrix} \begin{bmatrix} \mathbf{i}_a \\ \mathbf{i}_z \\ \hline \mathbf{i}_y \\ \mathbf{i}_b \end{bmatrix} = 0
\tag{9.24}
$$

Kirchhoff's voltage law can be expressed as

$$
\mathbf{B}\,\mathbf{u} = \mathbf{B} \begin{bmatrix} \mathbf{u}_a \\ \mathbf{u}_z \\ \mathbf{u}_y \\ \mathbf{u}_b \end{bmatrix} = 0
\tag{9.25}
$$

where $\mathbf{B}$ is known as the loop matrix. The relation between the $\mathbf{B}$ and $\mathbf{D}$ matrices is as follows:

$$
\mathbf{B} = [\mathbf{B}_T \mid \mathbf{1}] = [-\mathbf{D}_L^t \mid \mathbf{1}]
\tag{9.26}
$$

Therefore, the Kirchhoff voltage law can be expressed as

$$
\begin{bmatrix} -\mathbf{D}_{ay}^t & -\mathbf{D}_{zy}^t & | & \mathbf{1}_{yy} & \mathbf{0}_{by} \\ -\mathbf{D}_{ab}^t & -\mathbf{D}_{zb}^t & | & \mathbf{0}_{yb} & \mathbf{1}_{bb} \end{bmatrix} \begin{bmatrix} \mathbf{u}_a \\ \mathbf{u}_z \\ \hline \mathbf{u}_y \\ \mathbf{u}_b \end{bmatrix} = [\mathbf{0}]
\tag{9.27}
$$

According to equations (9.24) and (9.27) the Kirchhoff current and voltage laws for the circuit of Fig. 9.3(a) with the directed graph of Fig. 9.3(c) are, respectively,

$$
i_a - i_y = 0
$$
$$
i_z + i_y - i_b = 0
\tag{9.28}
$$

or

$$i_C - i_L = 0$$
$$i_R + i_L - J_1 = 0$$

(9.29)

and

$$u_a - u_z + u_y = 0$$
$$u_z + u_b = 0$$

(9.30)

or

$$u_C - u_R + u_L = 0$$
$$u_R + u_1 = 0$$

(9.31)

The branches not forming the ports in the tree and co-tree can be described by the equation (branch relations)

$$\mathbf{F}_{iz}\mathbf{i}_z + \mathbf{F}_{uy}\mathbf{u}_y + \mathbf{F}_{iy}\mathbf{i}_y + \mathbf{F}_{uz}\mathbf{u}_z + \mathbf{F}_{ia}\mathbf{i}_a + \mathbf{F}_{ub}\mathbf{u}_b + \mathbf{F}_{ua}\mathbf{u}_a + \mathbf{F}_{ib}\mathbf{i}_b = \mathbf{0}$$

Rewriting equations (9.24), (9.27) and (9.32) in one matrix equation, we obtain the tableau description of the circuit. After the additional elimination of vectors $[\mathbf{i}_z]$ and $[\mathbf{u}_y]$, we obtain the following relation:

$$[\mathbf{F}_b \mid \mathbf{F}_p] \begin{bmatrix} \mathbf{i}_y \\ \mathbf{u}_z \\ \hline \mathbf{i}_a \\ \mathbf{u}_b \\ \mathbf{u}_a \\ \mathbf{i}_b \end{bmatrix} = \begin{bmatrix} \mathbf{D}_{ay} & \mathbf{0} & | & \mathbf{1} & \mathbf{0} & \mathbf{0} & \mathbf{D}_{ab} \\ \mathbf{0} & -\mathbf{D}_{zb}^t & | & \mathbf{0} & \mathbf{1} & -\mathbf{D}_{ab}^t & \mathbf{0} \\ \hat{\mathbf{F}}_{iy} & \hat{\mathbf{F}}_{uz} & | & \hat{\mathbf{F}}_{ia} & \hat{\mathbf{F}}_{ub} & \hat{\mathbf{F}}_{ua} & \hat{\mathbf{F}}_{ib} \end{bmatrix} \begin{bmatrix} \mathbf{i}_y \\ \mathbf{u}_z \\ \hline \mathbf{i}_a \\ \mathbf{u}_b \\ \mathbf{u}_a \\ \mathbf{i}_b \end{bmatrix} = [\mathbf{0}]$$

(9.33)

where submatrices $\mathbf{F}_b$ and $\mathbf{F}_p$ correspond to the branch and port variables, respectively, and

$$\hat{\mathbf{F}}_{iy} = \mathbf{F}_{iy} - \mathbf{F}_{iz}\mathbf{D}_{zy}$$
$$\hat{\mathbf{F}}_{uz} = \mathbf{F}_{uz} + \mathbf{F}_{uy}\mathbf{D}_{zy}^t$$
$$\hat{\mathbf{F}}_{ib} = \mathbf{F}_{ib} - \mathbf{F}_{iz}\mathbf{D}_{zb}$$
$$\hat{\mathbf{F}}_{ua} = \mathbf{F}_{ua} + \mathbf{F}_{uy}\mathbf{D}_{ay}^t$$
$$\hat{\mathbf{F}}_{ub} = \mathbf{F}_{ub}$$
$$\hat{\mathbf{F}}_{ia} = \mathbf{F}_{ia}$$

(9.34)

To obtain relations between the port variables $\mathbf{i}_a, \mathbf{u}_b, \mathbf{u}_a, \mathbf{i}_b$, it is necessary to eliminate the variable vectors $\mathbf{i}_y, \mathbf{u}_z$ from equation (9.33). This can be performed by reduction of the matrix $[\mathbf{F_b}\,|\,\mathbf{F_p}]$ to the echelon form. Then the right lower part of the matrix in the echelon form is the required port matrix $\mathbf{P}$:

$$
\begin{bmatrix}
\text{echelon form} & | & x.\,.\,.\,.\,.\,x \\
 & | & \cdots\cdots \\
0\,.\,.\,1\,x\,. & | & x\,.\,.\,.\,.\,.\,x \\
\text{-------} & | & \text{-------} \\
 & | & \\
\mathbf{0} & | & \mathbf{P}
\end{bmatrix}
\begin{bmatrix}
\mathbf{i}_y \\
\mathbf{u}_z \\
- \\
\mathbf{i}_a \\
\mathbf{u}_b \\
\mathbf{u}_a \\
\mathbf{i}_b
\end{bmatrix}
= \mathbf{0}
\tag{9.35}
$$

Then, we can write

$$
\mathbf{P}
\begin{bmatrix}
\mathbf{i}_a \\
\mathbf{u}_b \\
\mathbf{u}_a \\
\mathbf{i}_b
\end{bmatrix}
= \mathbf{0}
\tag{9.36}
$$

since the left lower part of the $[\mathbf{F_b}\,|\,\mathbf{F_p}]$ matrix is the $\mathbf{0}$ submatrix. After the port matrix $\mathbf{P}$ is created, we can split it into the left and right parts:

$$
[\mathbf{P_L}\,|\,\mathbf{P_R}]
\begin{bmatrix}
\mathbf{i}_a \\
\mathbf{u}_b \\
- \\
\mathbf{u}_a \\
\mathbf{i}_b
\end{bmatrix}
= \mathbf{0}
\tag{9.37}
$$

Transformation of the left part of this matrix to the unity matrix gives us the desired hybrid matrix $\mathbf{H}$:

$$
[\mathbf{P_L}\,|\,\mathbf{P_R}]
\begin{bmatrix}
\mathbf{i}_a \\
\mathbf{u}_b \\
- \\
\mathbf{u}_a \\
\mathbf{i}_b
\end{bmatrix}
= \mathbf{0} \Rightarrow [\mathbf{1}\,|\,-\mathbf{H}]
\begin{bmatrix}
\mathbf{i}_a \\
\mathbf{u}_b \\
- \\
\mathbf{u}_a \\
\mathbf{i}_b
\end{bmatrix}
= \mathbf{0}
\tag{9.38}
$$

and finally:

$$
\begin{bmatrix}
\mathbf{i}_a \\
\mathbf{u}_b
\end{bmatrix}
= \mathbf{H}
\begin{bmatrix}
\mathbf{u}_a \\
\mathbf{i}_b
\end{bmatrix}
\tag{9.39}
$$

Recall, that $\mathbf{i}_a$ and $\mathbf{u}_a$ correspond to all currents and voltages in the tree ports, and $\mathbf{u}_b$ and $\mathbf{i}_b$ correspond to all voltages and currents in the co-tree ports. Replacing the currents and voltages in $\mathbf{i}_a$ and $\mathbf{u}_b$ that correspond to capacitive and inductive ports by the branch relations

$$
i_C = C\dot{v}_C \text{ and } v_L = L\dot{i}_L
$$

we can transform equation (9.39) into the initial state and output equations, as will be shown in the next section.

9.3 Numerical example for generation of hybrid matrix and initial state and output equations

To illustrate all the steps in the processes of initial and output equation generation, let us consider the active RC circuit shown in Fig. 9.5. The nodes and branches of this circuit are denoted by arabic and roman digits, respectively; the branches are directed by arrows. The normal tree of this circuit is shown in Fig. 9.6 (dotted lines represent the links of the co-tree).

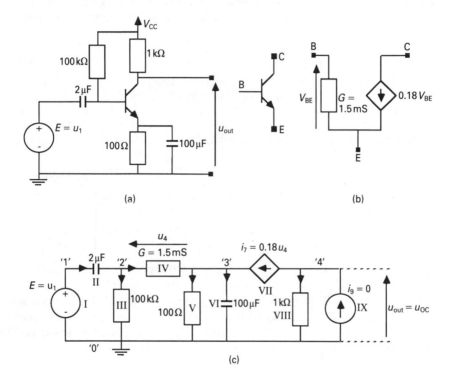

(a)   (b)

(c)

Figure 9.5 (a) Schematic diagram of RC amplifier. (b) Simplified transistor model. (c) Equivalent circuit of the RC amplifier

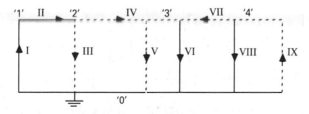

Figure 9.6 Tree and co-tree for the circuit of Fig. 9.5(c)

The incidence matrix $\mathbf{A}$ is

$$
\begin{array}{c}
\begin{array}{cccccccccc}
\text{Nodes} & \text{I} & \text{II} & \text{III} & \text{IV} & \text{V} & \text{VI} & \text{VII} & \text{VIII} & \text{IX} & \text{Branches}
\end{array} \\
\mathbf{A} = \begin{array}{c} 1 \\ 2 \\ 3 \\ 4 \end{array}
\left[\begin{array}{ccccccccc}
-1 & 1 & 0 & 0 & 0 & 0 & 0 & 0 & 0 \\
0 & -1 & 1 & 1 & 0 & 0 & 0 & 0 & 0 \\
0 & 0 & 0 & -1 & 1 & 1 & -1 & 0 & 0 \\
0 & 0 & 0 & 0 & 0 & 0 & 1 & 1 & -1
\end{array} \right]
\end{array}
$$

The normal tree is formed by branches: I, II, VI, VIII. According to the rules described in the preceding section, the branches can be reordered in the following way:

I	II	VI	VIII	V	III	IV	VII	IX
E	C	C	R	R	R	G	VCCS	U_{out}

The incidence matrix for reordered branches is

$$
[\mathbf{A}_T \,|\, \mathbf{A}_L] =
\begin{array}{c}
\begin{array}{ccccccccc}
\text{I} & \text{II} & \text{VI} & \text{VIII} & \text{V} & \text{III} & \text{IV} & \text{VII} & \text{IX} \quad \text{Branches}
\end{array} \\
\left[\begin{array}{cccc|ccccc}
-1 & 1 & 0 & 0 & 0 & 0 & 0 & 0 & 0 \\
0 & -1 & 0 & 0 & 0 & 1 & 1 & 0 & 0 \\
0 & 0 & 1 & 0 & 1 & 0 & -1 & -1 & 0 \\
0 & 0 & 0 & 1 & 0 & 0 & 0 & 1 & -1
\end{array} \right] \\
\begin{array}{cc} \text{Tree} & \qquad\qquad \text{Co-tree} \end{array}
\end{array}
$$

The branches I, II, VI form ports in the tree, while the branch IX forms a port in the co-tree. Transformation of $\mathbf{A}_T$ to a unity matrix $\mathbf{1}$ gives the cut-set matrix $\mathbf{D}$. This can be performed by multiplication of the first two rows by -1 and addition of the second row to the first row:

$$[\mathbf{A}_T \,|\, \mathbf{A}_L] \Rightarrow [\mathbf{1} \,|\, \mathbf{D}_L]$$

Then, the submatrices $\mathbf{D}_{ay}$, $\mathbf{D}_{ab}$, $\mathbf{D}_{zy}$ and $\mathbf{D}_{zb}$ are (see equation (9.24) and Fig. 9.7)

$$
\mathbf{D}_{ay} = \begin{bmatrix} 0 & -1 & -1 & 0 \\ 0 & -1 & -1 & 0 \\ 1 & 0 & -1 & -1 \end{bmatrix}, \quad
\mathbf{D}_{ab} = \begin{bmatrix} 0 \\ 0 \\ 0 \end{bmatrix}
$$

$$\mathbf{D}_{zy} = [0 \ 0 \ 0 \ 1], \quad \mathbf{D}_{zb} = [-1]$$

The branches which are not in the tree and co-tree ports can be described by the

$$\mathbf{D} = [\mathbf{1} \mid \mathbf{D_l}] = \begin{bmatrix} 1 & 0 & 0 & 0 & 0 & -1 & -1 & 0 & 0 \\ 0 & 1 & 0 & 0 & 0 & -1 & -1 & 0 & 0 \\ 0 & 0 & 1 & 0 & 1 & 0 & -1 & -1 & 0 \\ 0 & 0 & 0 & 1 & 0 & 0 & 0 & 1 & -1 \end{bmatrix}$$

Figure 9.7   Decomposition of the cut-set matrix $\mathbf{D}$ in submatrices

relations

$$\begin{aligned} -10^3\, i_8 + u_8 &= 0 \\ u_5 - 10^2\, i_5 &= 0 \\ u_3 - 10^5\, i_3 &= 0 \\ -1.5 \times 10^{-3}\, u_4 + i_4 &= 0 \\ -0.18\, u_4 + i_7 &= 0 \end{aligned}$$

or, in matrix form (see equation (9.32)),

$$
\begin{array}{c}\mathbf{F}_{iz}\\ \begin{bmatrix} -10^3 \\ 0 \\ 0 \\ 0 \\ 0 \end{bmatrix}\end{array} [i_8] +
\begin{array}{c}\mathbf{F}_{uy}\\ \begin{bmatrix} 0 & 0 & 0 & 0 \\ 1 & 0 & 0 & 0 \\ 0 & 1 & 0 & 0 \\ 0 & 0 & -1.5\times 10^{-3} & 0 \\ 0 & 0 & -0.18 & 0 \end{bmatrix}\end{array}
\begin{bmatrix} u_5 \\ u_3 \\ u_4 \\ u_7 \end{bmatrix} +
\begin{array}{c}\mathbf{F}_{iy}\\ \begin{bmatrix} 0 & 0 & 0 & 0 \\ -10^2 & 0 & 0 & 0 \\ 0 & -10^5 & 0 & 0 \\ 0 & 0 & 1 & 0 \\ 0 & 0 & 0 & 1 \end{bmatrix}\end{array}
\begin{bmatrix} i_5 \\ i_3 \\ i_4 \\ i_7 \end{bmatrix} +
\begin{array}{c}\mathbf{F}_{uz}\\ \begin{bmatrix} 1 \\ 0 \\ 0 \\ 0 \\ 0 \end{bmatrix}\end{array} [u_8]
$$

$$
+ \begin{array}{c}\mathbf{F}_{ia}\\ [0]\end{array}\begin{bmatrix} i_1 \\ i_2 \\ i_6 \end{bmatrix} +
\begin{array}{c}\mathbf{F}_{ub}\\ [0]\end{array}[u_9] +
\begin{array}{c}\mathbf{F}_{ua}\\ [0]\end{array}\begin{bmatrix} u_1 \\ u_2 \\ u_6 \end{bmatrix} +
\begin{array}{c}\mathbf{F}_{ib}\\ [0]\end{array}[i_9] = \mathbf{0}
$$

Now we have to find submatrices $\hat{\mathbf{F}}_{iy}$, $\hat{\mathbf{F}}_{uz}$, $\hat{\mathbf{F}}_{ia}$, $\hat{\mathbf{F}}_{ub}$, $\hat{\mathbf{F}}_{ua}$ and $\hat{\mathbf{F}}_{ib}$ to obtain the matrix $[\mathbf{F_b} \mid \mathbf{F_p}]$ (see equation (9.33)). Using the matrix equation 9.34 we obtain

$$
\hat{\mathbf{F}}_{iy} = \begin{bmatrix} 0 & 0 & 0 & 10^3 \\ -10^2 & 0 & 0 & 0 \\ 0 & -10^5 & 0 & 0 \\ 0 & 0 & 1 & 0 \\ 0 & 0 & 0 & 1 \end{bmatrix}, \quad
\hat{\mathbf{F}}_{uz} = \begin{bmatrix} 1 \\ 0 \\ 0 \\ 0 \\ 0 \end{bmatrix}, \quad
\hat{\mathbf{F}}_{ia} = \begin{bmatrix} 0 & 0 & 0 \\ 0 & 0 & 0 \\ 0 & 0 & 0 \\ 0 & 0 & 0 \\ 0 & 0 & 0 \end{bmatrix},
$$

$$
\hat{\mathbf{F}}_{ub} = \begin{bmatrix} 0 \\ 0 \\ 0 \\ 0 \\ 0 \end{bmatrix}, \quad
\hat{\mathbf{F}}_{ua} = \begin{bmatrix} 0 & 0 & 0 \\ 0 & 0 & 1 \\ -1 & -1 & 0 \\ 1.5\times 10^{-3} & 1.5\times 10^{-3} & 1.5\times 10^{-3} \\ 0.18 & 0.18 & 0.18 \end{bmatrix}, \quad
\hat{\mathbf{F}}_{ib} = \begin{bmatrix} -10^3 \\ 0 \\ 0 \\ 0 \\ 0 \end{bmatrix}
$$

Now, we can form equation (9.33), which in our case will be

$$
0 = [\mathbf{F_b} \mid \mathbf{F_p}]
\begin{bmatrix}
i_5 \\
i_3 \\
i_4 \\
i_7 \\
\hline
u_8 \\
\hline
i_1 \\
i_2 \\
i_6 \\
u_9 \\
u_1 \\
u_2 \\
u_6 \\
i_9
\end{bmatrix}
$$

$\left.\begin{array}{c} \\ \\ \\ \\ \end{array}\right\}$ $\mathbf{i}_y$ = currents in the co-tree branches not forming ports

$\mathbf{u}_z$ = voltage on the tree branch not forming ports

$\left.\begin{array}{c} \\ \\ \\ \end{array}\right\}$ $\mathbf{i}_a$ = currents in the tree port branches

$\mathbf{u}_b$ = voltage on the co-tree port branch (also $\mathbf{u}_{OC}$)

$\left.\begin{array}{c} \\ \\ \\ \end{array}\right\}$ $\mathbf{u}_a$ = voltages on the tree port branches

$\mathbf{i}_b$ = current on the co-tree port branch (also $\mathbf{i}_{OC}$)

After transformation of the matrix $[\mathbf{F_b} \mid \mathbf{F_p}]$ to the form given by equation (9.35), that is, after elimination of the $\mathbf{i}_y$ and $\mathbf{u}_z$ variables, we obtain the port matrix $\mathbf{P}$ (see equation (9.37))

$$
[\mathbf{P_L} \mid \mathbf{P_R}]
\begin{bmatrix}
\mathbf{i}_a \\
\mathbf{u}_b \\
\hline
\mathbf{u}_a \\
\mathbf{i}_b
\end{bmatrix}
=
\begin{bmatrix}
1 & 0 & -1 & 10^{-3} & \mid & 10^{-5} & 10^{-5} & -10^{-2} & 1 \\
0 & 1 & -1 & 10^{-3} & \mid & 10^{-5} & 10^{-5} & -10^{-2} & 1 \\
\hline
0 & 0 & 1 & -10^3 & \mid & 1.5\times10^{-3} & 1.5\times10^{-3} & 1.15\times10^{-2} & -1 \\
0 & 0 & 0 & 1 & \mid & 180 & 180 & 180 & 10^3
\end{bmatrix}
\begin{bmatrix}
i_1 \\
i_2 \\
i_6 \\
u_9 \\
\hline
u_1 \\
u_2 \\
u_6 \\
i_9
\end{bmatrix}
= 0
$$

To obtain the relations between the port currents and voltages it is necessary to transform the left-hand submatrix $\mathbf{P_L}$ to a unity matrix. This can be achieved in three elementary row operations:

1. addition of the third row to the first one,
2. addition of the third row to the second one,
3. multiplication of the fourth row by 10^{-3} and addition to the third one.

This yields

$$
[\mathbf{1} \mid -\mathbf{H}] =
\begin{bmatrix}
1 & 0 & 0 & 0 & \mid & 1.51\times10^{-3} & 1.51\times10^{-3} & 1.5\times10^{-3} & 0 \\
0 & 1 & 0 & 0 & \mid & 1.51\times10^{-3} & 1.51\times10^{-3} & 1.5\times10^{-3} & 0 \\
0 & 0 & 1 & 0 & \mid & 0.1815 & 0.1815 & 0.1915 & 0 \\
0 & 0 & 0 & 1 & \mid & 180 & 180 & 180 & 10^3
\end{bmatrix}
$$

Hence, the hybrid equation is

$$
\begin{bmatrix} i_1 \\ i_2 \\ i_6 \\ u_9 \end{bmatrix} = \begin{bmatrix} -1.51 \times 10^{-3} & -1.51 \times 10^{-3} & -1.5 \times 10^{-3} & 0 \\ -1.51 \times 10^{-3} & -1.51 \times 10^{-3} & -1.5 \times 10^{-3} & 0 \\ -0.1815 & -0.1815 & -0.1915 & 0 \\ -180 & -180 & -180 & -10^3 \end{bmatrix} \begin{bmatrix} u_1 \\ u_2 \\ u_6 \\ i_9 \end{bmatrix}
$$

Comparison of this hybrid equation with equations (9.14), (915) and (9.16), with

$$
\mathbf{i}_U = \mathbf{i}_1, \ \mathbf{i}_a = \begin{bmatrix} i_2 \\ i_6 \end{bmatrix}, \ \mathbf{u}_{OC} = u_9, \ \mathbf{U} = u_1, \ \mathbf{u}_a = \begin{bmatrix} u_2 \\ u_6 \end{bmatrix}, \ \mathbf{i}_{OC} = i_9
$$

gives

$$
\begin{bmatrix} i_1 \\ \hline i_2 \\ i_6 \\ \hline u_9 \end{bmatrix} = \begin{bmatrix} \mathbf{H}_{22} & | & \mathbf{H}_{23} & | & \mathbf{H}_{26} \\ \hline \mathbf{H}_{32} & | & \mathbf{H}_{33} & | & \mathbf{H}_{36} \\ \hline \mathbf{H}_{62} & | & \mathbf{H}_{63} & | & \mathbf{H}_{66} \end{bmatrix} \begin{bmatrix} u_1 \\ \hline u_2 \\ u_6 \\ \hline i_9 \end{bmatrix}
$$

This matrix equation corresponds to the situation where branches I, II, VI are voltage ports and branch IX is a current port. Since i_2 and i_6 are currents in the tree voltage ports obtained by the extraction of capacitors C_2 and C_6, we can therefore substitute

$$i_2 = C_2 \dot{u}_2 = 2 \times 10^{-6} \dot{u}_2$$

$$i_6 = C_6 \dot{u}_6 = 1 \times 10^{-4} \dot{u}_6$$

Then, using second, third and fourth hybrid equations we obtain the initial state and output equations:

$$2 \times 10^{-6} \dot{u}_2 = -1.51 \times 10^{-3} u_2 - 1.5 \times 10^{-3} u_6 - 1.51 \times 10^{-3} u_1$$

$$10^{-6} \dot{u}_6 = -0.1815 u_2 - 0.1915 u_6 - 0.1815 u_1$$

$$u_{OC} = u_9 = -180 u_2 - 180 u_6 - 180 u_1$$

or, in a matrix form (see equations (9.9) and (9.10)),

$$
\begin{bmatrix} 2 \times 10^{-6} & 0 \\ 0 & 10^{-4} \end{bmatrix} \begin{bmatrix} \dot{u}_2 \\ \dot{u}_6 \end{bmatrix} = \begin{bmatrix} -1.51 \times 10^{-3} & -1.5 \times 10^{-3} \\ -0.1815 & -0.1915 \end{bmatrix} \begin{bmatrix} u_2 \\ u_6 \end{bmatrix} + \begin{bmatrix} -1.51 \times 10^{-3} \\ -0.1815 \end{bmatrix} [u_1]
$$

$$
[u_9] = [0 \ \ 0] \begin{bmatrix} \dot{u}_2 \\ \dot{u}_6 \end{bmatrix} + [-180 \ \ -180] \begin{bmatrix} u_2 \\ u_0 \end{bmatrix} + [-180] [u_1]
$$

9.4 Description of programs for generation of the initial state equation

The programs for the generation of initial state and output equations consist of two files 'INPUT' and 'HYBRID' shown in Program 9.1 and Program 9.2.

> **See the listing of the program INPUT Directory of Chapter 9 on attached diskette.**

> **See the listing of the program HYBRID Directory of Chapter 9 on attached diskette.**

Program 'INPUT' is used for the interactive description of the analyzed circuit. The program prints out on the screen each type of element and its meaning (see Table 9.1).

Then the program asks questions to enter digital data in the following order:

```
Type of the element - ?
Number of branches - ?
From node - ?
To node - ?
Number of control branch - ? [only for controlled sources]
Value - ? [in the case of input and output, enter given
                                                    number 1 or 2]
```

After entering the last branch of the circuit, press the key ⟨ENTER⟩ for the question concerning the type of element. Then the circuit data are automatically saved in the disk-file 'NAME.NET'. For example, for the circuit of Fig. 9.8, the

Table 9.1   Print-out for the program 'INPUT'

Type of elements		Meaning
Digital code	Symbol	
0	OutJ	Output current port
1	U	Independent voltage source and voltage input
2	C	Capacitance
3	Uu	Voltage controlled voltage source
4	Uj	Current controlled voltage source
5	R	Resistance
6	G	Conductance
7	Ju	Voltage controlled current source
8	Jj	Current controlled current source
9	L	Inductance
10	J	Independent current source and current input
11	OutU	Output voltage port

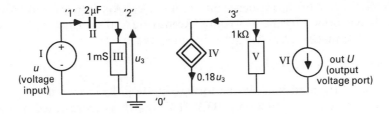

Figure 9.8 Example circuit for entering data in the program 'INPUT'

circuit data are

Type of element	Number of branch	From node	To node	Number of cont. branch	Value
1	1	0	1	–	1
2	2	1	2	–	2E-6
6	3	2	0	–	1E-3
7	4	3	0	3	0.18
5	5	3	0	–	1E3
11	6	0	3	–	2

<ENTER>

Program 'HYBRID' reads in the computer memory the circuit data from the disk-file 'NAME.NET' (created by the program 'INPUT') and generates the hybrid matrix and initial state and output equations. Then the matrices corresponding to the initial state and output equations are saved automatically in the disk-file 'NAME.INR' for further use. In the program, the number of inputs and outputs of the circuit is restricted to two. Other circuit restrictions (numbers of elements, nodes, ports, etc.) are entered in the program as 'Const' (which can be found in the listing of the program).[1] The constant 'dysk:string[20] = 'a:\';' indicates the disk drive (in this case a:\) at which the program 'HYBRID' and the disk-files 'NAME.NET' and 'NAME.INR' are located (saved).

The program 'HYBRID' consists of several procedures. The process of the generation of the initial state and output equations can be divided into eight steps:

1. Reading in the circuit description from the file 'NAME.NET' and writing into the variable 'list' composed of records of the same type as records on the file 'NAME.NET'.
2. Generation of incidence matrix **A**.
3. Selection of the tree branches: this is performed using the procedure 'echelon-form', which converts the matrix **A** into the echelon form, and procedure 'Tree,' which detects the leftmost '1' in each row of the matrix $\mathbf{A}_{ech}$, and writes the numbers of the corresponding columns into the vector 'treecol'. These numbers indicate the numbers of the tree branches. Then the incidence matrix **A** is reordered to obtain the form $[\mathbf{A}_T \mid \mathbf{A}_L]$.

4. Transformation of the incidence matrix $\mathbf{A} = [\mathbf{A}_T \mid \mathbf{A}_L]$ into the cut-set matrix $\mathbf{D}$. This is performed by the procedure 'ReedukcjaAinD', which converts the left part of the matrix $[\mathbf{A}_T \mid \mathbf{A}_L]$ into the unity matrix $\mathbf{1}$:

$$[\mathbf{A}_T \mid \mathbf{A}_L] \Rightarrow [\mathbf{1}_T \mid \mathbf{D}_L]$$

5. Creation of the matrix in tableau form (including the branch description of the resistive circuits (see equation (9.33))). This matrix is represented by the matrix variable '**Ans**'.

6. Reduction of the matrix **Ans** to the form given by equation (9.35) to obtain the port matrix **P**. This is performed using the procedure 'Stairway', which is similar to the procedure 'Echelform'. The matrix **P** has dimensions $p \times 2p$, where p is the number of extracted ports.

7. Transformation of the port matrix $\mathbf{P} = [\mathbf{P}_L \mid \mathbf{P}_R]$ according to equation (9.38):

$$[\mathbf{P}_L \mid \mathbf{P}_R] \Rightarrow [\mathbf{1} \mid -\mathbf{H}]$$

   to the hybrid matrix. This is performed using elementary row operations.

8. Creation of the initial state and output equations by replacing capacitor currents and inductor voltages in the hybrid equation by their branch relations (9.17). This leads to the initial state and output equations given by relations (9.9) and (9.10), which, according to the program matrix variable notation, can be written as

$$\mathbf{M}\dot{\mathbf{x}} = [p\mathbf{A}]\mathbf{x} + [p\mathbf{B}]\mathbf{u}$$

$$\mathbf{y} = [p\mathbf{C}p]\dot{\mathbf{x}} + [p\mathbf{C}]\mathbf{x} + [p\mathbf{D}]\mathbf{u}$$

The last operation of the program is saving the initial state and output equations in the disk-file 'NAME.INR'. This is performed by the procedure 'Save_on_disk'. The file 'NAME.INR' contains

1. the number of state variables 'nvar',
2. the number of input ports 'nin',
3. the number of output ports 'nout',
4. the description of inputs and outputs (number and type given by the user in the description of the circuit),
5. the matrices of the initial equations: $\mathbf{M}$, $[p\mathbf{A}]$, $[p\mathbf{B}]$, $[p\mathbf{C}p]$, $[p\mathbf{C}]$, $[p\mathbf{D}]$.

The file 'NAME.INR' is used by the next program 'REDUCT', which converts the initial state and output equations to the normal form given by relations (9.1) and (9.2). Running the program 'HYBRID', after reading the elements of the circuit of Fig. 9.5 using the program 'INPUT', gives us the intermediate expositions on the screen (only more important results are presented):

Branch order after entering the elements

```
Branch number :  1  2  3  4  5  6 7   8 9
Element type  :  U  C  R  G  R  C Ju  R OutU
```

Branch order after ordering the branches according to elements type

```
Branch number : 1 2 6 5 3 8 4   7  9
Element type :   U C C R R R G   Ju OutU
```

Incidence matrix A

```
-1  1 0 0 0 0  0  0  0
 0 -1 0 0 1 0  1  0  0
 0  0 1 1 0 0 -1 -1  0
 0  0 0 0 0 1  0  1 -1
```

Incidence matrix A in a stairway form

```
1 -1 0 0  0 0  0  0  0
0  1 0 0 -1 0 -1  0  0
0  0 1 1  0 0 -1 -1  0
0  0 0 0  0 1  0  1 -1
```

The first four positions with '1' form the tree:

```
1 2 6 8 5 3 4   7  9
U C C R R R G   Ju OutU
```

Incidence matrix A for ordered branches

```
-1  1 0 0 0 0  0  0  0
 0 -1 0 0 0 1  1  0  0
 0  0 1 0 1 0 -1 -1  0
 0  0 0 1 0 0  0  1 -1
```

trport	trnoport	lnknoport	lnkport	trport: ports inside the tree
				trnoport: tree branches not forming ports
				lnknoport: links not forming ports
3	1	4	1	lnkports: ports inside the co-tree

Cut-set matrix D for ordered branches

```
1 0 0 0 0 -1 -1  0 0
0 1 0 0 0 -1 -1  0 0
0 0 1 0 1  0 -1 -1 0
0 0 0 1 0  0  0  1 1
```

```
    Matrix [Fiz]        Matrix [Fuy]

      -1.0E+3      0.0 0.0   0.0   0.0
       0.0E+0      1.0 0.0   0.0   0.0
       0.0E+0      0.0 1.0   0.0   0.0
       0.0E+0      0.0 0.0 -1.5E-3 0.0
       0.0E+0      0.0 0.0 -1.8E-1 0.0
```

Dimension of the port matrix: 4X8.

The port matrix

```
1.0 0.0 -1.0  1.0E-3 1.0E-5 1.0E-5 -1.0E-2   1.0
0.0 1.0 -1.0  1.0E-3 1.0E-5 1.0E-5 -1.0E-2   1.0
0.0 0.0  1.0 -1.0E-3 1.5E-3 1.5E-3  1.15E-2 -1.0
0.0 0.0  0.0  1.0    1.8E+2 1.8E+2  1.8E+2   1.0E+3
```

The number of ports in the tree TP and co-tree LP

TP = 3   LP = 1

Branches forming the ports

1 2 6 9

Hybrid matrix obtained after reduction of the first four columns of the port matrix to the unity matrix:

```
-1.510E-3 -1.510E-3 -1.50E-3   0.0
-1.510E-3 -1.510E-3 -1.50E-3   0.0
-1.815E-1 -1.815E-1 -1.915E-1  0.0
-1.80E+2  -1.80E+2  -1.80E+2  -1.00E+3
```

Initial state equations

Matrix [M]

```
    2.0E-6   0.0
     0.0    1.0E-4
```

State matrix [A]

```
    -1.510E-3 -1.500E-3

    -1.815E-1 -1.915E-1
```

Matrix [B]

```
    -1.510E-3

    -1.815E-1
```

Matrix [*N*]

    0.0 0.0

Matrix [*C*]

     -1.80E+2 -1.80E+2

Matrix [*D*]

     -1.80E+2

Now, you can use the program 'Reduct.pas'
  The End

9.5 Transformation of the initial state and output equations to the normal form

9.5.1 Elimination of dependent variables from initial equations

In equation (9.9) the vector $\mathbf{x}^{(0)}$ is composed of all capacitor voltages and all inductor currents in the circuit considered. When some of these variables are dependent it is necessary to detect and eliminate them. To do this, we first create a coefficient matrix composed of the initial state equation:

$$[\mathbf{M}^{(0)} \mid \mathbf{A}^{(0)} \mid \mathbf{B}^{(0)}] \tag{9.40}$$

and transform this matrix into echelon form:

$$[\mathbf{M}_e^{(0)} \mid \mathbf{A}_e^{(0)} \mid \mathbf{B}_e^{(0)}] \tag{9.41}$$

When $\mathbf{M}^{(0)}$ is not singular, that is, the determinant $\det[\mathbf{M}^{(0)}] \neq \mathbf{0}$, we can multiply both sides of the initial state equation by $[\mathbf{M}^{(0)}]^{-1}$ to obtain the normal state equation. Actually, instead of matrix inversion and multiplication we reduce $\mathbf{M}^{(0)}$ to the unity matrix $\mathbf{1}$:

$$[\mathbf{M}_e^{(0)} \mid \mathbf{A}_e^{(0)} \mid \mathbf{B}_e^{(0)}] \Rightarrow [\mathbf{1} \mid \mathbf{A} \mid \mathbf{B}] \tag{9.42}$$

Then the matrices $\mathbf{A}$ and $\mathbf{B}$ are of normal state form as in equation (9.1).

However, when $\mathbf{M}^{(0)}$ is singular, then in the transformation to echelon form at least one of the rows of $\mathbf{M}_e^{(0)}$ (let it be the jth row) is composed of zeros. Then three cases are possible:

1. The jth row in the second and third blocks of the matrix (9.41) are also posed of zeros. Then we have no unique solution.
2. The jth row in the $\mathbf{A}_e^{(0)}$ matrix is composed of zeros and the jth row in the matrix is non-zero. Then we obtain the jth equation:

$$\mathbf{0}_j \dot{\mathbf{x}}^{(0)} = \mathbf{0}_j \mathbf{x}^{(0)} + \mathbf{B}_j \mathbf{u}$$

which is contradictory; there is no solution.

3. The jth row in the $\mathbf{A}_e^{(0)}$ and $\mathbf{B}_e^{(0)}$ matrices are non-zero. Then we have the equation:

$$0_j \dot{\mathbf{x}}^{(0)} = \mathbf{A}_j \mathbf{x}^{(0)} + \mathbf{B}_j \mathbf{u} \qquad (9.44)$$

where $\mathbf{A}_j$ and $\mathbf{B}_j$ represent the jth row of matrices $\mathbf{A}_e^{(0)}$ and $\mathbf{B}_e^{(0)}$:

$$\mathbf{A}_j = [a_{j1} \ a_{j2} \dots a_{jn}]$$
$$\qquad (9.45)$$
$$\mathbf{B}_j = [b_{j1} \ b_{j2} \dots b_{jm}]$$

Since at least one of the elements of $\mathbf{A}_j$ is non-zero, let us say a_{jk}, we can find the variable x_k:

$$x_k = -\frac{1}{a_{jk}} \left(\sum_{\substack{i=1 \\ i \neq k}}^{n} a_{ji} x_i + \sum_{i=1}^{m} b_{ji} u_i \right) \qquad (9.46a)$$

and

$$\dot{x}_k = -\frac{1}{a_{jk}} \left(\sum_{\substack{i=1 \\ i \neq k}}^{n} a_{ji} \dot{x}_i + \sum_{i=1}^{m} b_{ji} \dot{u}_i \right) \qquad (9.46b)$$

Substitution of equations (9.46) into the initial state and output equations eliminates one of the dependent variables. Since in (9.46b) the derivative of excitation $\dot{\mathbf{u}}$ exists, then after dependent variable elimination we obtain

$$\mathbf{M}_e^{(0)} \dot{\mathbf{x}}^{(1)} = \mathbf{A}^{(1)} \mathbf{x}^{(1)} + \mathbf{B}^{(1)} \mathbf{u} + \mathbf{B}_1^{(1)} \dot{\mathbf{u}} \qquad (9.47)$$

$$\mathbf{y} = \mathbf{N}^{(1)} \dot{\mathbf{x}}^{(1)} + \mathbf{C}^{(1)} \mathbf{x}^{(1)} + \mathbf{D}^{(1)} \mathbf{u} + \mathbf{D}_1^{(1)} \dot{\mathbf{u}} \qquad (9.48)$$

This operation is continued until in the rth step we obtain $[\mathbf{M}_e^{(r)}]$, which is non-singular upper triangular; this corresponds to the equations

$$\mathbf{M}_e^{(r)} \dot{\mathbf{x}}^{(r)} = \mathbf{A}^{(r)} \mathbf{x}^{(r)} + \mathbf{B}^{(r)} \mathbf{u} + \mathbf{B}_1^{(r)} \dot{\mathbf{u}} + \cdots + \mathbf{B}_r^{(r)} \frac{d^r}{dt^r} \mathbf{u} \qquad (9.49)$$

$$\mathbf{y} = \mathbf{N}^{(r)} \dot{\mathbf{x}}^{(r)} + \mathbf{C}^{(r)} \mathbf{x}^{(r)} + \mathbf{D}^{(r)} \mathbf{u} + \mathbf{D}_1^{(r)} \dot{\mathbf{u}} + \cdots + \mathbf{D}_r^{(r)} \frac{d^r}{dt^r} \mathbf{u} \qquad (9.50)$$

Multiplying both sides of equation (9.49) by $[\mathbf{M}_e^{(r)}]^{-1}$, we obtain an intermediate form of the state equation as in (9.11); then, substitution of equation (9.11) into (9.50) eliminates $\dot{\mathbf{x}}^{(r)}$ and gives the intermediate output equation as in (9.12) or

$$\dot{\mathbf{x}} = \mathbf{A}\mathbf{x} + \mathbf{B}\mathbf{u} + \mathbf{B}_1 \frac{d}{dt} \mathbf{u} + \cdots + \mathbf{B}_r \frac{d^r}{dt^r} \mathbf{u} \qquad (9.51)$$

$$\mathbf{y} = \mathbf{C}\mathbf{x} + \mathbf{D}\mathbf{u} + \mathbf{D}_1 \frac{d}{dt} \mathbf{u} + \cdots + \mathbf{D}_r \frac{d^r}{dt^r} \mathbf{u} \qquad (9.52)$$

where the vector [x] is a subset of vector [$x^{(0)}$] with r dependent variables eliminated.

9.5.2 Reduction of the intermediate state and output equations to the normal form

The derivatives of the excitation (independent source) vector $d^i u / dt^i$ which exist in the intermediate state equation (9.51) can be eliminated by the substitution

$$\mathbf{x} = \mathbf{z} + \mathbf{B}_r \frac{d^{r-1}}{dt^{r-1}} \mathbf{u} \tag{9.53a}$$

and

$$\dot{\mathbf{x}} = \dot{\mathbf{z}} + \mathbf{B}_r \frac{d^r}{dt^r} \mathbf{u} \tag{9.53b}$$

Substitution of equations (9.53) into (9.51) and (9.52) yields

$$\dot{\mathbf{z}} = \mathbf{A}\mathbf{z} + \mathbf{B}\mathbf{u} + \mathbf{B}_1 \frac{d}{dt} \mathbf{u} + \cdots + (\mathbf{A}\mathbf{B}_r + \mathbf{B}_{r-1}) \frac{d^{r-1}}{dt^{r-1}} \mathbf{u} \tag{9.54}$$

$$\mathbf{y} = \mathbf{C}\mathbf{z} + \mathbf{D}\mathbf{u} + \mathbf{D}_1 \frac{d}{dt} \mathbf{u} + \cdots + (\mathbf{C}\mathbf{B}_r + \mathbf{D}_{r-1}) \frac{d^{r-1}}{dt^{r-1}} \mathbf{u} + \mathbf{D}_r \frac{d^r}{dt^r} \mathbf{u} \tag{9.55}$$

With the new state variable $\mathbf{z}$ the highest order of the excitation vector $\mathbf{u}$ derivative in equation (9.54) is reduced by one. Repeating r-times the substitution given by (9.53) we can eliminate all derivatives of the vector $\mathbf{u}$ and obtain the normal form. However, the highest order of the vector $\mathbf{u}$ derivative in the output equation remains unchanged. For example, for the intermediate equations

$$\dot{\mathbf{x}} = \mathbf{A}\mathbf{x} + \mathbf{B}\mathbf{u} + \mathbf{B}_1\dot{\mathbf{u}} + \mathbf{B}_2\ddot{\mathbf{u}}$$

$$\mathbf{y} = \mathbf{C}\mathbf{x} + \mathbf{D}\mathbf{u} + \mathbf{D}_1\dot{\mathbf{u}} + \mathbf{D}_2\ddot{\mathbf{u}}$$

the substitution

$$\mathbf{x} = \mathbf{z} + \mathbf{B}_2\dot{\mathbf{u}}$$

$$\dot{\mathbf{x}} = \dot{\mathbf{z}} + \mathbf{B}_2\ddot{\mathbf{u}}$$

gives as a result

$$\dot{\mathbf{z}} = \mathbf{A}\mathbf{z} + \mathbf{B}\mathbf{u} + (\mathbf{A}\mathbf{B}_2 + \mathbf{B}_1)\dot{\mathbf{u}}$$

$$\mathbf{y} = \mathbf{C}\mathbf{z} + \mathbf{D}\mathbf{u} + (\mathbf{C}\mathbf{B}_2 + \mathbf{D}_1)\dot{\mathbf{u}} + \mathbf{D}_2\ddot{\mathbf{u}}$$

The second substitution:

$$\mathbf{z} = \mathbf{v} + (\mathbf{AB}_2 + \mathbf{B}_1)\mathbf{u}$$

eliminates the derivative $\dot{\mathbf{u}}$ from the state equation. The new variables are linear combinations of $\mathbf{x}^{(0)}$ and $\mathbf{u}$. The derivatives of the excitation vector $\mathbf{u}$ appear in the intermediate equation for circuits containing CE meshes (as for the circuit of Fig. 9.1) and/or *LI* cut-sets.

9.5.3 Description of the program for reduction of the initial state and output equations to the normal form

The program 'REDUCT' is shown in Program 9.3. This program can be divided into the following four steps:

1. Reading into the computer memory the initial state and output equations from the disk-file 'NAME.INR'. This is performed by the procedure 'Read_init_state_eqns'.
2. Elimination of *r* dependent state variables, as is described in Section 9.5.1, to obtain the intermediate state and output equations.
3. Change of variables and reduction of the intermediate equations to the normal form.
4. Saving of the matrix coefficients of the normal state and output equations to the disk-file 'NAME.MAT' for further use. This is performed by the procedure 'Save_on_disk'.

See the listing of the program REDUCT Directory of Chapter 9 on attached diskette.

The program 'REDUCT' uses the output data of the program 'HYBRID' which are stored in the disk-file 'NAME.INR'. Running the program 'REDUCT' for the circuit of Fig. 9.5 we obtain the following exposition on the screen:

```
Matrix [M]

      2.00E-6      0.00
      0.00         1.00E-4

Matrix [N]

      0.00         0.00

Matrix [A]

     -1.510E-3    -1.500E-3
     -1.815E-1    -1.915E-1
```

Matrix [B]

    -1.510E-3
    -1.815E-1

Matric [C]

    -1.80E+2   -1.80E+2

Matrix [D]

    -1.80E+2

Matrix [S] = [M | A | B] after reading in

    2.00E-6   0.00   -1.510E-3   -1.500E-3 -1.510E-3

    0.00   1.00E-4   -1.815E-1   -1.915E-1 -1.815E-1

Matrix [S] after "stairwaying"

    1.00   0.00   -7.550E+2   -7.500E+2 -7.550E+2

    0.00   1.00   -1.815E+3   -1.915E+3 -1.815E+3

Matrix [A] (normal form)

    -7.550E+2   -7.49999E+2

    -1.815E+3   -1.9150E+3

Matrix [B] (normal form)

    -7.550E+2

    -1.815E+3

Matrix [C]

    -1.80E+2   -1.80E+2

Matrix [D]

    -1.80E+2

Now you can use the program 'timeresp.pas'. *The end*

9.6 Solution of the state and output equations in the time domain

Solution of the state equation given by equation (9.1) in the continuous time domain is given by the relation

$$\mathbf{x}(t) = e^{[\mathbf{A}]t} \int_{t_0}^{t} e^{-[\mathbf{A}]\tau}\mathbf{Bu}(\tau)\ d\tau + e^{[\mathbf{A}](t-t_0)}\mathbf{x}(t_0) \tag{9.56}$$

Substitution of (9.56) into output equation (9.2) gives the continuous time response of the circuit considered:

$$y(t) = \mathbf{C} \, e^{[\mathbf{A}](t-t_0)}\mathbf{x}(t_0) + \mathbf{C} \, e^{[\mathbf{A}]t} \int_{t_0}^{t} e^{-[\mathbf{A}]\tau}\mathbf{B}\mathbf{u}(\tau) \, d\tau + \mathbf{D}\mathbf{u}(t) + \mathbf{D}_1\dot{\mathbf{u}} + \cdots$$

$$(9.57)$$

The first term of relation (9.57) is referred to as the transient response, and the second term as the forced, or steady state, response. The matrix $e^{[\mathbf{A}]t}$, existing in equations (9.56) and (9.57), defined by the infinite series

$$e^{[\mathbf{A}]t} = 1 + \mathbf{A}t + \frac{1}{2!}(\mathbf{A}t)^2 + \frac{1}{3!}(\mathbf{A}t)^3 + \cdots \tag{9.58}$$

is referred to as the transition matrix.

9.6.1 Conversion of the state equation into the difference form

To compute the time response given by (9.56) numerically it is necessary to convert this equation into a proper difference equation in which time is discretized. Then using the computer, we can compute the response $\mathbf{x}(t)$ at the discrete time points $t = kT$, $k = 0, 1, 2, \ldots$, where T is the time step. Since the input signals $\mathbf{u}(kT)$ are known for all k, it is necessary to find a relation between $\mathbf{x}((k+1)T)$, $\mathbf{x}(kT)$ and $\mathbf{u}(kT)$. Such an equation is a particular form of a difference equation.

Let us assume that in equation (9.56), $t_0 = kT$ and $t = (k+1)T$. Then we obtain

$$\mathbf{x}((k+1)T) = e^{[\mathbf{A}]T}\mathbf{x}(kT) + e^{[\mathbf{A}](k+1)T}\int_{kT}^{(k+1)T} e^{-[\mathbf{A}]\tau}\mathbf{B}\mathbf{u}(\tau) \, d\tau \tag{9.59}$$

The integral in equation (9.59) can be computed exactly for the following cases:

1. The input function $\mathbf{u}(t)$ is piecewise constant, such that

$$\mathbf{u}(t) = \mathbf{u}(kT), \text{ for } kT \leqslant t \leqslant (k+1)T, \quad k = 0, 1, 2, \ldots \tag{9.60}$$

Then equation (9.59) takes the form

$$\mathbf{x}((k+1)T) = e^{[\mathbf{A}]T}\mathbf{x}(kT) + (e^{[\mathbf{A}]T} - 1)\mathbf{A}^{-1}\mathbf{B}\mathbf{u}(kT) \tag{9.61}$$

The expression $(e^{[\mathbf{A}]T} - 1)\mathbf{A}^{-1}$ can be computed without inversion of the matrix $\mathbf{A}$, since

$$(e^{[\mathbf{A}]T} - 1)\mathbf{A}^{-1} = \left\{ \left(1 + \mathbf{A}T + \frac{1}{2!}(\mathbf{A}T)^2 + \cdots \right) - 1 \right\}\mathbf{A}^{-1}$$

$$= \sum_{n=0}^{\infty} \frac{1}{(n+1)!}(\mathbf{A}T)^n \, T \tag{9.62}$$

2. The input function $\mathbf{u}(t)$ is piecewise linear, such that

$$\mathbf{u}(t) = \mathbf{u}(kT) + \frac{\{\mathbf{u}((k+1)T)) - \mathbf{u}(kT)\}}{T}(t - kT) \qquad (9.63)$$

for $kT \leqslant t \leqslant (k+1)T, \quad k = 0, 1, 2, \ldots$

Then equation (9.59) takes the form

$$\mathbf{x}((k+1)T) = \mathbf{F}\mathbf{x}(kT) + \mathbf{F}1\mathbf{u}(kT) + \mathbf{F}2\mathbf{u}((k+1)T) \qquad (9.64)$$

where

$$\mathbf{F} = e^{[\mathbf{A}]T} = \sum_{n=0}^{\infty} \frac{1}{n!}(\mathbf{A}T)^n$$

$$\mathbf{F}1 = \{e^{[\mathbf{A}]T}(-1 + \mathbf{A}T) + 1\}(\mathbf{A}^2T^2)^{-1}\mathbf{B}T$$

$$= \sum_{n=0}^{\infty} \frac{1}{n!(n+2)}(\mathbf{A}T)^n\mathbf{B}T$$

$$\mathbf{F}2 = \{e^{[\mathbf{A}]T} - 1 - \mathbf{A}T\}(\mathbf{A}^2T^2)^{-1}\mathbf{B}T$$

$$= \sum_{n=0}^{\infty} \frac{1}{(n+2)!}(\mathbf{A}T)^n\mathbf{B}T$$

However, when the input function $\mathbf{u}(t)$ is continuous and is approximated by function (9.60) or function (9.63), as is shown in Fig. 9.9, then equations (9.61) and (9.64) give approximate solutions for $\mathbf{x}(kT)$. Solution errors in this case can be reduced by choosing a smaller time step T. Other formulas for approximate solutions can be obtained using different assumptions. If we assume that derivatives of

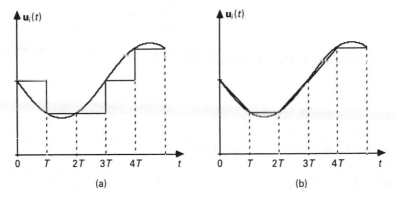

Figure 9.9 Approximation of continuous function by (a) piecewise constant function, (b) piecewise linear function

the state equation (9.1) are piecewise constant, then we can write

$$\dot{\mathbf{x}} = \frac{\mathbf{x}((k+1)T) - \mathbf{x}(kT)}{T} = \mathbf{A}\mathbf{x}(kT) + \mathbf{B}\mathbf{u}(kT) \tag{9.65}$$

or we obtain a formula

$$\mathbf{x}((k+1)T) = (1 + \mathbf{A}T)\mathbf{x}(kT) + T\mathbf{B}\mathbf{u}(kT) \tag{9.66}$$

which is known as the Euler extrapolation algorithm.

If we assume, on the other hand, that the integral function

$$\mathbf{f}(t) = e^{-[\mathbf{A}]t}\mathbf{B}\mathbf{u}(t) \tag{9.67}$$

in formula (9.59) is a piecewise constant function (as in Fig. 9.9(a)), then we obtain

$$\mathbf{x}((k+1)T) = e^{[\mathbf{A}]T}\mathbf{x}(kT) + e^{[\mathbf{A}]T}T\mathbf{B}\mathbf{u}(kT) \tag{9.68}$$

A very useful formula for approximate solutions of state equations is obtained when the integral function (9.67) is approximated by a second-degree polynomial in each time step T. Then we obtain a relation known as the Simpson formula:

$$\mathbf{x}((k+1)T) = e^{[\mathbf{A}]T}\mathbf{x}(kT) + e^{[\mathbf{A}]T}\frac{T}{6}\mathbf{B}\mathbf{u}(kT)$$

$$+ e^{[\mathbf{A}]T/2}\frac{4}{6}T\mathbf{B}\mathbf{u}((k+1/2)T) + \frac{T}{6}\mathbf{B}\mathbf{u}((k+1)T) \tag{9.69}$$

In the Simpson formula, the values of the input function $\mathbf{u}(t)$ should be known every half of the time step, that is, kT, $(k+1/2)T$, $(k+1)T$, etc.

9.6.2 Computation of the transition matrix $e^{[\mathbf{A}]T}$

To use the relations for time response given in Section 9.6.1, it is necessary to compute the transition matrix $e^{[\mathbf{A}]T}$ for given $[\mathbf{A}]$ and T. Since the transition matrix $e^{[\mathbf{A}]T}$ is defined by an infinite series, in practical computer calculations we approximate this matrix by a finite series composed of $K+1$ terms:

$$e^{[\mathbf{A}]T} \approx \mathbf{F} = 1 + \mathbf{A}T + \frac{1}{2!}(\mathbf{A}T)^2 + \cdots + \frac{1}{K!}(\mathbf{A}T)^K = f_{ij} \tag{9.70}$$

The error matrix $\mathbf{R}$, defined as the difference of the right-hand sides of relations (9.58) and (9.70), is given by the equation

$$\mathbf{R} = \frac{1}{(k+1)!}(\mathbf{A}T)^{(K+1)} + \frac{1}{(K+2)!}(\mathbf{A}T)^{(K+2)} + \cdots = \sum_{k=K+1}^{\infty}\frac{(\mathbf{A}T)^k}{k!} \tag{9.71}$$

Then it is possible to compute the upper bound U for elements $|r_{ij}|$ of the error $\mathbf{R}$, which is given by the relation

$$|r_{ij}| \leqslant \frac{\|AT\|^{(K+1)}}{(K+1)!} \frac{1}{1-\|AT\|} = U, \qquad \text{for } \|AT\| < 1 \tag{9.72}$$

where

r_{ij} is an arbitrary element of the error matrix $\mathbf{R}$,

$\|AT\|$ is a norm of the matrix AT, which is defined as follows:

$$\|AT\| = \max_i \sum_{j=1}^{n} |a_{ij}| T \tag{9.73}$$

or

$$\|AT\| = \max_j \sum_{i=1}^{n} |a_{ij}| T \tag{9.74}$$

that is, the norm of a quadratic matrix $n \times n$ corresponds to the maximal sum of the absolute values of the elements $|a_{ij}| T$ in the row (relation (9.73)) or in the column (relation 9.74)).

The upper bound U of the error in (9.72) can be minimized, or the accuracy of the $e^{[A]T}$ computation increased, by taking larger values of K (larger number of terms in (9.70)), since $\|AT\| < 1$. The largest relative error RE of $\mathbf{F} = [f_{ij}]$ in relation (9.70) is obtained by the division of U by the absolute value of the smallest element $f_{ij}|_{\min}$:

$$RE = \frac{U}{|f_{ij}|_{\min}} \tag{9.75}$$

Generally, if the value of the relative error RE of any number d is smaller than $0.5 \, 10^{-ND}$, then the number d is accurate to the ND digits (there are ND accurate significant digits in the number d). Thus, to obtain larger ND, it is necessary to take into account a larger number of terms in (9.70). As an example let us compute the transition matrix $\mathbf{F} = e^{[A]T}$ for the system matrix $\mathbf{A}$ considered in Section 9.5.3:

$$\mathbf{A} = \begin{bmatrix} -7.550 \times 10^2 & -7.4999 \times 10^2 \\ -1.815 \times 10^3 & -1.9150 \times 10^3 \end{bmatrix}$$

Let us assume the time step $T = 2 \times 10^{-5}$ s. Then the row norm given by relation (9.73) is

$$\|AT\|_r = 5.33 \times 10^{-2}$$

and the column norm given by relation (9.74) is $\|AT\|_c = 7.46 \times 10^{-2}$.

As the norm of the matrix AT we take the smaller value, that is, 5.33×10^{-2}. Then from relation (9.70), with $K = 6$, we obtain

$$\mathbf{F} = e^{[A]T} = \begin{bmatrix} 0.98487 & -0.0150 \\ -0.0363 & 0.9617 \end{bmatrix}$$

The upper bound of the error computed from relation (9.72) is

$$U = \frac{(5.33 \times 10^{-2})^7}{7!(1 - 5.33 \times 10^{-2})} = 2.096 \times 10^{-13}$$

The absolute value of the smallest element of $\mathbf{F}$, $|f_{ij}|_{min}| = 0.015$; thus the relative error RE is

$$RE = \frac{U}{|f_{ij}|_{min}|} = \frac{2.096 \times 10^{-13}}{1.5 \times 10^{-2}} = 1.3973 \times 10^{-11} = 0.13973 \times 10^{-10}$$

Thus, the elements of the computed transition matrix are accurate to $ND = 10$ significant digits.

9.7 Description of the program for solution of the state equation in the time domain

The program 'TIMERESP', shown in Program 9.4, performs the computation of the time response of linear circuits for a given excitation (input) function, based on the circuit state and output equations in normal form.

> **See the listing of the program TIMERESP Directory of Chapter 9 on attached diskette.**

The input data for this program, in the form of the matrix coefficients of the normal state and output equations, are read-in from the disk file 'NAME.MAT', which has been created by the program 'REDUCT'. The time response of the analyzed circuit is computed using the following three methods of approximation:

1. The input function $\mathbf{u}(t)]$ is of piecewise constant type (see equation (9.61)).
2. The input function $[\mathbf{u}(t)]$ is of piecewise linear type (see equation (9.64)).
3. The integral function (9.67) is approximated by a second-degree polynomial (Simpson's method (9.69)).

The program 'TIMERESP' can be divided into several parts:

1. Reading in the computer memory the state and output equation (in the normal form) coefficients from the disk file 'NAME.MAT' created by the previous program 'REDUCT'. This is performed by the procedure 'Read-disk'.
2. Computation of auxiliary matrices:

$$\mathbf{F} = e^{[A]T}, \ \mathbf{P} = (e^{[A]T} - 1)\mathbf{A}^{-1}, \ \mathbf{Q} = e^{[A]T/2}$$

$$\mathbf{G} = [e^{[A]T}(-1 + \mathbf{A}T) + 1] \, (\mathbf{A}^2 T^2)^{-1},$$

$$\mathbf{H} = [e^{[A]T} - 1 - \mathbf{A}T] \, (\mathbf{A}^2 T^2)^{-1}$$

This is performed by the procedure 'Aux__matrices'. To compute the transition matrix $e^{[A]T}$ with given accuracy, it is necessary to assume that the norm of the matrix AT is smaller than 1; thus it is necessary to enter the time step T:

$$T < \frac{1}{\|A\|}, \text{ where } \|A\| \text{ is the norm of } A$$

Additionally, to obtain accuracy of all terms of the transition matrix up to ND significant digits, it is necessary to sum up the subsequent terms of the series (9.70) and to check the relative error RE given by (9.75). The summing up in the procedure loop is stopped when the relative error is smaller than $RE < 0.5 \times 10^{-ND}$. The value of ND is introduced in the run-time by the user.

3. Description of input (excitation) functions. This is performed by the procedure 'Input__functions' with subprocedure 'Menu' for function selection. In the program, it is possible to select ten functions:

(1) $u(t) = 0 =$ no excitation
(2) rectangular pulse with amplitude 'amplit' and duration equal to half the analysis time
(3) function $u(t) = \text{amplit} \times 1$
(4) linear function $u(t) = \text{slope} \times t$
(5) sinusoidal function $u(t) = \text{amplit} \times \sin(2\pi ft)$
(6,7,8,9) functions 2, 3, 4, 5, respectively, with negative sign
(10) user function defined in the function 'funuzyt'; this function can be changed easily before running the program.
   In the program we have the function

$$\text{funuzyt} = 50(1 - e^{-t})\ 1(t) - 50(1 - e^{-(t-1)})1(t-1)$$

   After selection of the input function (menu is exposed on the screen), the selected function is tabulated for discrete time points $t = kT$ and $t = kT + T/2$ using the procedure 'Tabularization' for further use in the time response computation.

4. Computation of time responses using formulas (9.61), (9.64) and (9.69). This is performed by the procedure 'State__out'. For piecewise constant approximation of $u(t)$ the values of the time response are saved automatically in the file 'NAME.OTC', for piecewise linear approximation in the file 'NAME.OTL', and for Simpson's formula in the file 'NAME.OTS', for eventual further use (plots characteristics).

   Running the program 'TIMERESP' for the circuit of Fig. 9.5 we obtain the following expositions on the screen:

The Time Domain Response

n_variables	n_inputs	n_outputs	n_derivatives
2	1	1	0

Enter time-step T (T lower than 3.56E-4) = ?   2E-5

```
Enter the number ND of significant digits in
                                exp([A]T), f. ex. ND=8    8

Matrix [A]

    -7.550E+2   -7.500E+2
    -1.815E+3   -1.915E+3

Matrix [F]

     9.849E-1   -1.500E-2
    -3.630E-2    9.617E-1

Enter the number of analysis points (<=2000) - ?  20
```

If you would like to change the user defined function (U9) or the parameters AMPLIT, SLOPE, FREQ, break the program (Ctrl-C) or as a response enter the letter Q; then change the function 'funuzyt' in the source program or the parameter values in the 'const' declaration.

```
Accessible functions types

^^^^^^^^^^^^^^^^^^^^^^^^^^
0 - Zero function              U0(t) = 0

1 - rectangular pulse with amplitude = AMPLIT, duration
                                   time : 1/2 of analysis time

2 - unit step function         U2(t) = AMPLIT*1(t)

3 - linear function            U3(t) = SLOPE*t

4 - sinusoidal function        U4(t) =
                                   AMPLIT*sin(2*Pi*FREQ*t)

5 - function reciprocal to U1  U5(t) = -U1(t)

6 - function reciprocal to U2  U6(t) = -U2(t)

7 - function reciprocal to U3  U7(t) = -U3(t)

8 - function reciprocal to U4  U8(t) = -U4(t)

9 - User defined function      U9(t) = funuzyt(t)

AMPLIT, SLOPE, FREQ - constant program parameters

Enter the excitation function U1 - ?  3 {linear function}


Piecewise constant approximation

{t(s)          u(t)         y(t)}

t = 0.0        0.0          0.0
t = 2.0E-5     2.0E-5       -3.600E-3
```

```
t = 4.0E-5    4.0E-5    -7.020E-3
  .             .          .
t = 3.8E-4    3.8E-4    -4.513E-2
t = 4.0E-4    4.0E-4    -4.653E-2
```

Piecewise linear approximation

```
t = 0.0       0.0        0.0
t = 2.0E-5    2.0E-5    -3.509E-3
t = 4.0E-5    4.0E-5    -6.843E-3
  .             .          .
t = 3.8E-4    3.8E-4    -4.402E-2
t = 4.0E-4    4.0E-4    -4.539E-2
```

Simpson's method

```
t = 0.0       0.0        0.0
t = 2.0E-5    2.0E-5    -3.509E-3
t = 4.0E-5    4.0E-5    -6.843E-3
  .             .          .
t = 3.8E-4    3.8E-4    -4.402E-2
t = 4.0E-4    4.0E-4    -4.539E-2
```

```
* The End *
```

Since the input function $\mathbf{u}(t)$ in this example is linear in time, the output $\mathbf{y}(t)$ is identical for piecewise linear approximation and the Simpson method.

Exercises

Exercise 1.   Verify that for the linear circuits shown in Fig. E9.1 (a, b, c), the complexity order corresponds to that given by equation (9.3).

Exercise 2.   Verify that for the linear circuit shown in Fig. E9.2 the complexity order is given by the equality or the inequality in equation (9.3) according to the value of G_a and G_b. The two outputs are $v_1(t)$ and $v_2(t)$.

Exercise 3.   Verify that for the linear circuits shown in Fig. E9.3(a, b), the complexity order is given by the equality or the inequality in equation (9.3) according to the value of k. The circuits contain an ideal voltage amplifier of gain k. Its model is shown in Fig. E9.3(c).

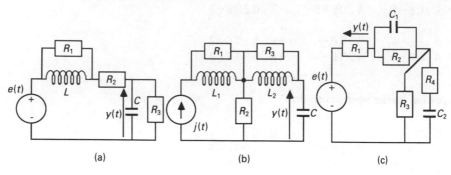

Figure E9.1

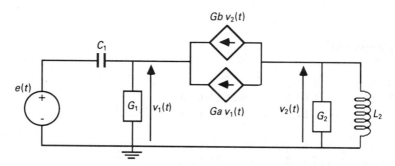

Figure E9.2

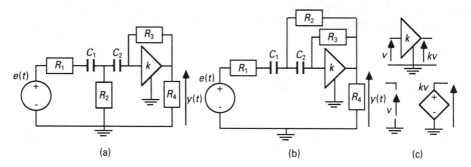

Figure E9.3

Exercise 4. Draw the normal trees and incidence matrices for the circuits shown in Fig. E9.1 (a, b, c).

Exercise 5. Determine the complexity orders for the circuits shown in Fig. E9.4(a, b).

Exercise 6. Run the program 'INPUT' for the circuit shown in Fig. E9.5(a). The model of the two transistors is shown in Fig. E9.5(b). Assume $e(t) = 1$ mv.

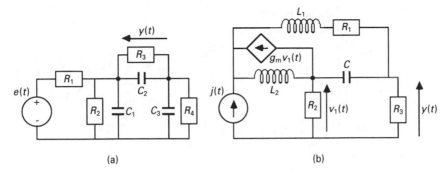

(a)

(b)

Figure E9.4

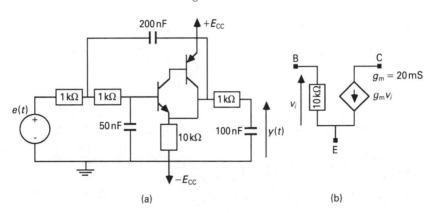

(a)

(b)

Figure E9.5

Exercise 7. Run the program 'INPUT' for the three circuits of exercise 1. All the element values are equal to unity in appropriate units.

Exercise 8. Run the program 'INPUT' for the circuit of exercise 2; $e(t) = 1$ V; $G_1 = 2$ S; $G_2 = 1$ S; choose different values for G_a and G_b.

Exercise 9. Run the program 'INPUT' for the two circuits of exercise 3; $e(t) = 1$ V; $R_1 = R_2 = R_3 = R_4 = 1$ Ω; $C_1 = C_2 = 1$ F; choose two values of k: $k = 1$ and $k = 1/2$.

Exercise 10 Run the program 'INPUT' for the circuits shown in Fig. 9.4(a, b) and E9.5(a, b).

Exercise 11. Run the program 'INPUT' for the circuit shown in Fig. E9.6(a) (consider the 'small signal' analysis; $e(t)$ is a sinusoidal source of frequency equal to 30 kHz). The two transistors are identical and are represented by the model shown in Fig. E9.6(b).

Exercise 12. Run the program 'INPUT' for the two circuits shown in Fig. E9.7(a, b). The two circuits contain ideal voltage amplifiers. Their models are the same as those shown in Fig. E9.3(c).

Exercise 13. Run the program 'INPUT' for the circuit shown in Fig. E9.8(a). The two circuits contain an INIC circuit with the following relations (with notations of Fig. E9.8(c)) $V_1 = V_2$; $I_1 = I_2$.

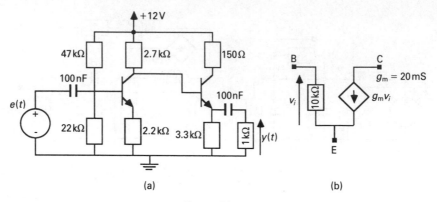

(a)   (b)

Figure E9.6

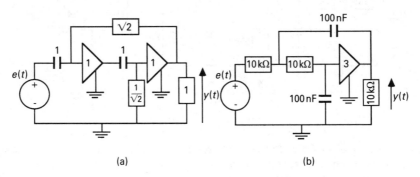

(a)   (b)

Figure E9.7

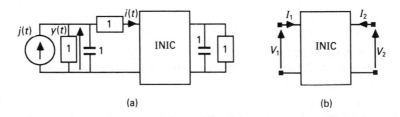

(a)   (b)

Figure E9.8

Exercise 14.   Run the program 'INPUT' for the two circuits shown in Fig. E9.9(a, b). The two circuits contain a gyrator with the following constitutive equations : $I_1 = V_2$; $I_2 = -V_1$, with the notation of Fig. E9.8(b).

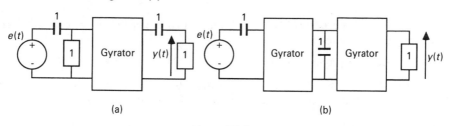

(a)   (b)

Figure E9.9

Exercise 15.   Run the program 'INPUT' for the two circuits shown in Fig. E9.10(a, b). The two circuits contain a transconductance amplifier. Its model is shown in Fig. E9.10(c).

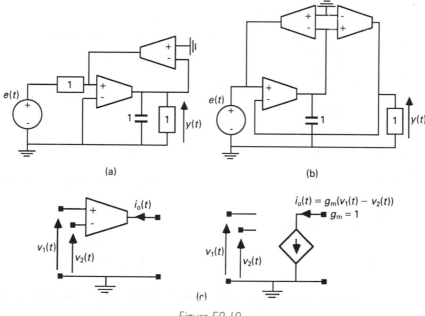

(a)   (b)

(c)

Figure E9.10

Exercise 16.   Using program 'INPUT', propose a solution to analyze an ideal transformer defined by the following constitutive equations $V_1 = V_2/n$; $I_1 = -nI_2$, according to the notations of Fig. E9.8(b).

Exercise 17.   Running the program 'HYBRID', determine the tree and the cut-set matrix for the examples of exercises 1, 2 and 3 using the results of exercises 7, 8 and 9.

Exercise 18.   Run the program 'HYBRID' for circuits of Fig. E9.4 and E9.5.

Exercise 19.   Run the program 'HYBRID' using the data obtained in exercise 11.

Exercise 20.   Run the program 'HYBRID' using the data obtained in exercises 12, 13, 14 and 15.

Exercise 21.   Using the program 'REDUCT', verify if some dependent variables of exercise 11 and exercise 19 are eliminated.

Exercise 22.   Run the program 'REDUCT' using the data obtained in exercise 20.

Exercise 23.   Modifying the program 'TIMERESP', develop a program named 'TIMERESP1' to input directly the matrix elements defined by equations (9.1) and (9.2).

Exercise 24.   Explain the advantages and the disadvantages of the state variable formulation.

Exercise 25.   Using relation (9.58), the properties of the nth power of a matrix and the Taylor series expansion, give the compact expression of the transition matrix $e^{\mathbf{A}t}$ in the two following cases:

$$\mathbf{A} = \begin{bmatrix} 1 & 0 \\ 0 & 2 \end{bmatrix} \text{ and } \mathbf{A} = \begin{bmatrix} 0 & 1 \\ -4 & 0 \end{bmatrix}.$$

Assume that the expansion converges.

Exercise 26.   Verify the following properties, satisfied by the transition matrix:

$$e^{[\mathbf{A}]t} \, e^{[\mathbf{B}]t} = e^{[\mathbf{A} + \mathbf{B}]t} \text{ if and only if } [\mathbf{A}][\mathbf{B}] = [\mathbf{B}][\mathbf{A}]$$

$$[e^{[\mathbf{A}]t}]^{-1} = e^{-[\mathbf{A}]t}$$

$$\frac{d}{dt} \{e^{[\mathbf{A}]t}\} = \mathbf{A}[e^{[\mathbf{A}]t}] = [e^{[\mathbf{A}]t}]\mathbf{A}$$

$$\int_0^a e^{[\mathbf{A}]t} \, dt = [\mathbf{A}]^{-1}\{e^{[\mathbf{A}]a} - \mathbf{1}\} = \{e^{[\mathbf{A}]a} - \mathbf{1}\}\mathbf{A}^{-1} \text{ if } \mathbf{A}^{-1} \text{ exists}$$

Exercise 27.   The functions $v_1(t) = e^{at} \cos(bt)$ and $v_2(t) = e^{at} \sin(bt)$ satisfy the following first-order differential equation system:

$$\frac{dv_1(t)}{dt} = av_1(t) - bv_2(t); \qquad v_1(0) = 1$$

$$\frac{dv_2(t)}{dt} = bv_1(t) + av_2(t); \qquad v_2(0) = 0$$

Transform this system into the form of (9.1) and (9.2). In this special case, write equation (9.57) again. Determine the approximate solutions, calculated with N terms in equation (9.58). Compare with exact values.

Exercise 28.   Run the program 'TIMERESP' for the examples of exercises 14 and 15 when $e(t)$ is equal to unit step.

Exercise 29.   Run the program 'TIMERESP' for the examples of exercises 12 and 13 when $e(t)$ is equal to the unit step.

Exercise 30.   Run the program 'TIMERESP' for the examples of exercise 11.

Exercise 31.   Run the program 'TIMERESP' for the examples of exercise 3 using all possible source functions.

Note 1. The programs 'INPUT', 'HYBRID', 'REDUCT' and 'TIMERESP' are modified versions of programs proposed in the Master Thesis by J. Pomirski, Technical University of Gdansk, Gdansk, Poland, 1988.

10 Computation of Transient Response Using the Inverse Laplace Transform

10.1 Introduction

Direct computation of transient response for a given circuit requires solutions of differential equations in the time domain, as was discussed in Chapters 7, 8 and 9. In these cases the variables were physical quantities such as time dependent voltages on capacitors or/and currents through inductors. However, in many cases an analysis is performed in the frequency domain, where the variables are treated as Laplace transforms of voltages and currents; in this case the variables are functions of a complex frequency $s = \sigma + j\omega$.

The Laplace transform of the time dependent function $f(t)$:

$$\mathcal{L}[f(t)] = F(s) = \int_0^\infty f(t)\, e^{-st}\, dt \qquad (10.1)$$

allows us to convert integro-differential equations describing the given circuit into algebraic equations. Relation (10.1) forms pairs of corresponding functions $f(t)$ and $F(s)$ and allows for fast transposition from the frequency domain function $F(s)$ into time domain function $f(t)$ and vice versa. For example, the Laplace transform of the unit step function $1(t)$ can be computed directly:

$$\mathcal{L}[1(t)] = \int_0^\infty 1\, e^{-st}\, dt = -\frac{1}{s}\, e^{-st}\Big|_0^\infty = -\frac{1}{s}\, e^{-\infty} + \frac{1}{s}\, e^{-0} = \frac{1}{s}$$

Examples of pairs of popular functions and their transforms are given in Table 10.1.

When circuit properties are described in the frequency domain an output–input relation is determined:

$$Y(s) = T(s)U(s) \qquad (10.2)$$

where $Y(s)$ is the transform of the output signal $y(t)$, $U(s)$ is the transform of the input signal $u(t)$ and $T(s)$ is the transmittance describing the frequency properties of the circuit; the procedure leading to computation of the time domain response $y(t)$ is composed of the following steps:

1. decomposition of the product $T(s)U(s)$ into simple quotients ($T(s)$ is a rational function, that is, the quotient of two polynomials in s),

Table 10.1 Examples of popular functions $f(t)$
and their Laplace transforms

$f(t)$	$F(s)$
$\delta(t)$ (Dirac impulse)	1
$1(t)$ (unit step)	$\dfrac{1}{s}$
t	$\dfrac{1}{s^2}$
e^{-at}	$\dfrac{1}{s+a}$
$\sin(\omega t)$	$\dfrac{\omega}{s^2+\omega^2}$
$\cos(\omega t)$	$\dfrac{s}{s^2+\omega^2}$
$e^{-at}\sin(\omega t)$	$\dfrac{\omega}{(s+a)^2+\omega^2}$
$e^{-at}\cos(\omega t)$	$\dfrac{s+a}{(s+a)^2+\omega^2}$
$\dfrac{d}{dt}f(t)$	$sF(s)-f(0^-)$
$\dfrac{d^2}{dt^2}f(t)$	$s^2F(s)-sf(0^-)-\dfrac{d}{dt}f(0^-)$
$\displaystyle\int_0^t f(\tau)\,d\tau$	$\dfrac{F(s)}{s}$

? application of the inverse Laplace transform to each simple quotient to obtain all terms of the time response $y(t)$.

To decompose the product $T(s)U(s)$ into simple quotients it is first necessary to compute the poles and, corresponding to these, the residues of the product $T(s)U(s)$. Then, to compute $y(t)$ we can use the formula

$$y(t) = \sum_{l=1}^{n} (\exp(s_{rl}t)\{k_{rl}\cos(s_{il}t) - k_{il}\sin(s_{il}t)\}) \tag{10.3}$$

where n is the number of poles (degree of denominator) of $T(s)U(s)$, s_{rl}, s_{il} are real and imaginary parts of the poles and k_{rl}, k_{il} are real and imaginary parts of the residues in these poles.

10.2 Determination of the response function $Y(s) = T(s)U(s)$, its decomposition into simple quotients and time response $y(t)$

Circuit diagrams in the time and frequency domains for a simple RLC circuit are shown in Fig. 10.1. Note that in the frequency domain impedances should be indicated for each circuit element. However, for simplicity, we denote the circuit elements using the R, L, C symbols.

The time domain output–input relation for the circuit of Fig. 10.1(a), with $y(t) = v_2(t) = v_C(t)$ and with initial conditions (that is $v_C(0^-) = 0$, $i_L(0^-) = 0$) is given by the integro-differential equation

$$Ri(t) + L\frac{di(t)}{dt} + \frac{1}{C}\int_0^t i(t)\,dt = v_1(t) \tag{10.4a}$$

which corresponds to Kirchhoff's voltage law

$$v_R(t) + v_L(t) + v_C(t) = v_1(t) \tag{10.4b}$$

Insertion of relations

$$v_2(t) = v_C(t) = \frac{1}{C}\int_0^t i(t)\,dt \text{ and } i(t) = C\frac{dv_C(t)}{dt}$$

into equation (10.4a) gives a second-order differential equation:

$$RC\frac{dv_2(t)}{dt} + LC\frac{d^2v_2(t)}{dt^2} + v_2(t) = v_1(t) \tag{10.5a}$$

which, after rearrangement, takes the form

$$\frac{d^2v_2(t)}{dt^2} + \frac{R}{L}\frac{dv_2(t)}{dt} + \frac{1}{LC}v_2(t) = \frac{1}{LC}v_1(t) \tag{10.5b}$$

Taking the Laplace transforms of all the terms of (10.5b) we obtain

$$s^2V_2(s) + \frac{R}{L}sV_2(s) + \frac{1}{LC}V_2(s) = \frac{1}{LC}V_1(s) \tag{10.6a}$$

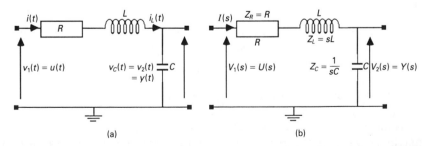

(a)   (b)

Figure 10.1 RLC circuit diagram. (a) In the time domain. (b) In the frequency domain

or

$$V_2(s) = \frac{\dfrac{1}{LC}}{s^2 + \dfrac{R}{L}s + \dfrac{1}{LC}}\, V_1(s) = T(s)\,V_1(s) \qquad (10.6b)$$

The same result is obtained directly from Fig. 10.1(b) in the frequency domain

$$I(s) = \frac{V_1(s)}{R + sL + \dfrac{1}{sC}}$$

$$V_2(s) = I(s)\frac{1}{sC} = \frac{\dfrac{1}{LC}}{s^2 + \dfrac{R}{L}s + \dfrac{1}{LC}}\, V_1(s) = T(s)\,V_1(s)$$

As an example let us find the time response $v_2(t)$ for the circuit of Fig. 10.1 with element values $R = 2.5\ \Omega$, $L = 0.5\ \text{H}$, $C = 1/3\ \text{F}$, and unit step input voltage $v_1(t) = 1(t)$, or $V_1(s) = 1/s$. The transfer function $T(s)$ is

$$\frac{6}{s^2 + 5s + 6} = \frac{6}{(s+2)(s+3)}$$

and

$$Y(s) = T(s)\,V_1(s) = \frac{6}{s(s+2)(s+3)}$$

with poles $s_1 = 0$; $s_2 = -2$; $s_3 = -3$.

To find the decomposition of $T(s)V_1(s)$ into the simple quotients of the form

$$Y(s) = T(s)\,V_1(s) = \sum_{i=1}^{nr} \frac{k_i}{s - s_i} + \sum_{ic=1}^{nc} \frac{k_{ic}}{s - s_{ic}} + \sum_{ic=1}^{nc} \frac{k_{ic}^*}{s - s_{ic}^*} \qquad (10.7)$$

where nr is the number of real poles, nc the number of complex poles pairs, s_i the single real poles, s_{ic} the single complex poles, s_{ic}^* the single complex pole coupled with s_{ic}, k_i the residue corresponding to real poles and k_{ic}, k_{ic}^* the complex residues corresponding to complex poles s_{ic} and s_{ic}^*, it is necessary to compute the poles s_i of $Y(s)$ and the residues for these poles using the formula (valid for single poles)

$$k_i = \frac{N(s)}{\dfrac{\mathrm{d}}{\mathrm{d}s}\{D(s)\}}\Bigg|_{s = s_i} \qquad (10.8)$$

where $N(s)$ and $D(s)$ are the numerator and denominator of the rational response function

$$Y(s) = T(s)\,U(s) = \frac{N(s)}{D(s)} \qquad (10.9)$$

Using formula (10.8) in our numerical example we obtain the values for the residues:

$$k_1 = \frac{6}{3s^2 + 10s + 6}\bigg|_{s=s_1=0} = \frac{6}{6} = 1$$

$$k_2 = \frac{6}{3s^2 + 10s + 6}\bigg|_{s=s_2=-2} = \frac{6}{-2} = -3$$

$$k_3 = \frac{6}{3s^2 + 10s + 6}\bigg|_{s=s_3=-3} = \frac{6}{3} = 2$$

Thus, the decomposition into simple quotients is

$$Y(s) = T(s)U(s) = \frac{1}{s} - \frac{3}{s+2} + \frac{2}{s+3}$$

Using the inverse Laplace transform (from Table 10.1) we obtain the time response $y(t)$:

$$y(t) = 1 - 3\,e^{-2t} + 2\,e^{-3t}$$

The same result is obtained from formula (10.3) (here $k_{il} = 0$, $s_{il} = 0$):

$$y(t) = \sum_{l=1}^{3} \{\exp(s_{rl}t)k_{rl}\} = 1 + \exp(-2t)(-3) + \exp(-3t)(2) = 1 - 3\,e^{-2t} + 2\,e^{-3t}$$

For the same circuit with element values $R = 1\,\Omega$, $C = 1$ F, $L = 0.5$ H, we obtain the transfer function with complex poles

$$T(s) = \frac{2}{s^2 + 2s + 2} = \frac{2}{[s - (-1 - j1)]\,[s - (-1 + j1)]} = \frac{2}{(s - s_2)(s - s_3)}$$

Assuming unit step input voltage $U(s) = 1/s$ we obtain

$$Y(s) = \frac{T(s)}{s} = \frac{2}{s^3 + 2s^2 + 2s} = \frac{N(s)}{D(s)} = \frac{2}{(s - s_1)(s - s_2)(s - s_3)}$$

and

$$\frac{d}{ds}D(s) = 3s^2 + 4s + 2$$

Now the residues are

$$k_1 = 1,$$

$$k_2 = -0.5 - j0.5, \qquad \text{or } k_{r2} = -0.5,\ k_{i2} = -0.5$$

$$k_3 = k_2^* = -0.5 + j0.5, \quad \text{or } k_{r3} = -0.5,\ k_{i3} = 0.5$$

Insertion of the real and imaginary parts of the poles and corresponding residues into relation (10.3) yields the result

$$y(t) = 1 - \exp(-1 \times t)\{\cos(1 \times t) + \sin(1 \times t)\}$$

Generally, any rational function $F(s)$:

$$F(s) = \frac{A(s)}{B(s)} = \frac{a_0 + a_1 s + \cdots + a_m s^m}{b_0 + b_1 s + \cdots + b_n s^n} \qquad (10.10)$$

can be decomposed into simple quotients when $m < n$. Thus, when $m \geqslant n$, it is necessary to first divide the numerator $A(s)$ by the denominator $B(s)$ to obtain the modified rational functions with $m = n - 1$.

In physical (real) electronic systems the degree m of the numerator of the transmittance cannot be larger than the degree n of the denominator. In the case where $m = n$, it is necessary to divide the numerator by the denominator to obtain

$$F(s) = \frac{A(s)}{B(s)} = d_0 + \frac{C(s)}{B(s)} = d_0 + \frac{c_0 + c_1 s + \cdots + c_{n-1} s^{n-1}}{b_0 + b_1 s + \cdots + b_n s^n} = d_0 + W(s) \quad (10.11)$$

Thus, to find the time response $y(t)$ based on the frequency domain function $Y(s) = T(s)U(s)$ it is necessary to proceed with several simple operations on complex variable polynomials:

1. division of polynomials – to obtain the rational function $Y(s)$ shown in relation (10.11),
2. computation of polynomial roots – to obtain the values of the poles of the rational function,
3. computation of a derivative of a polynomial – necessary in relation (10.8) to compute the residues in the poles,
4. computation of polynomial values for a given value of the complex variable s_j necessary to find the value of the residue k_j at given pole s_j; in this case it is necessary to compute the value s_j taken to the power n, that is, s_j^n.

10.3 Operations on complex variable polynomials

Complex variable polynomials used in the analysis of electronic circuits have real coefficients. Thus all operations on these polynomials are performed on real numbers.

10.3.1 Division of polynomials

To divide the polynomial $A(s)$ of degree m by the polynomial $B(s)$ of degree n, with $m \geqslant n$, we order the terms of both polynomials starting from the highest power of s:

$$A(s) = a_0 s^m + a_1 s^{m-1} + \cdots + a_m s^0$$

$$B(s) = b_0 s^n + b_1 s^{n-1} + \cdots + b_n s^0, \quad m \geqslant n$$

We start the process by division of the coefficients of the terms of the highest powers of s, that is, $d_0 = a_0/b_0$; then we multiply all the coefficients b_i by d_0 and subtract, respectively, from the coefficients a_i. The degree of the resultant polynomial is lowered by at least one. The process is repeated until degree m' of the resultant polynomial equals $m' = n - 1$. The algorithm is organized using two loops:

```
for i := 0 to m - n do
    begin
    d(i) := a(i)/b(0)
    for j := 1 to n do
        a(i+j) := a(i+j) - b(j)*d(i)
end;
```

For example, to divide the polynomial

$$A(s) = s^4 + 2s^3 + 3s^2 + 3s + 1$$

by:

$$B(s) = s^2 + s + 1$$

$$
\begin{array}{ccc}
d_0 & d_1 & d_2 \\
\downarrow & \downarrow & \downarrow \\
s^2 & s & 1
\end{array}
$$

$$\rule{4cm}{0.4pt}$$

$$s^4 + 2s^3 + 3s^2 + 3s + 1 : s^2 + s + 1$$

$(-)$ $\quad s^4 + s^3 + s^2$

$$\rule{3cm}{0.4pt}$$

$$0 + s^3 + 2s^2 + 3s + 1$$

$(-)$ $\qquad s^3 + s^2 + s$

$$\rule{3cm}{0.4pt}$$

$$0 + s^2 + 2s + 1$$

$(-)$ $\qquad\quad s^2 + s + 1$

$$\rule{2.5cm}{0.4pt}$$

remainder $\rightarrow \quad 0 + s + 0$

The result of the division is

$$\frac{s^4 + 2s^3 + 3s^2 + 3s + 1}{s^2 + s + 1} = s^2 + s + 1 + \frac{s + 0}{s^2 + s + 1}$$

Or, in general form,

$$\frac{A(s)}{B(s)} = D(s) + \frac{R(s)}{B(s)}$$

The process of polynomial division is shown in the program 'POLYDIV'.

See the listing of the program POLYDIV Directory of Chapter 10 on attached diskette.

Running the program 'POLYDIV' for polynomials

$$A(s) = s^4 + 2s^3 + 3s^2 + 4s + 4$$
$$B(s) = s^2 + 2s + 2$$

we obtain the following results:

The polynomial $D(s)$:

      1.000   0.000   1.000

The numerator of the remainder, $R(s)$:

      2.000   2.000

which corresponds to

$$\frac{A(s)}{B(s)} = s^2 + 1 + \frac{2s+2}{s^2 + 2s + 2}$$

Similarly, running the program for polynomials

$$A(s) = 2s^2 + 5s + 8; \ B(s) = s^2 + 2s + 3$$

we obtain the following results:

      $D(s)$: 2.000; $R(s)$: 1.000   2.000

which corresponds to

$$\frac{A(s)}{B(s)} = 2 + \frac{s+2}{s^2 + 2s + 3}$$

10.3.2 Differentiation of a polynomial

Differentiation of a polynomial $P(s)$ of degree n is very simple and is performed term-by-term. The resultant polynomial $Q(s)$ has degree $n - 1$ and its coefficients are products of former coefficients and powers of individual terms.
    For example

$$P(s) = 2s^3 + 4s^2 + 3s + 2$$

$$Q(s) = \frac{d}{ds} P(s) = 6s^2 + 8s + 3$$

10.3.3 Raising a complex number to the power N

Raising a complex number to the power N is necessary during computation of the polynomial value for a given value of a complex variable s_j. This is performed by multiple multiplication of the complex number by itself:

$$(a + jb)^N = (a + jb)(a + jb)...(a + jb) = \prod_{i=1}^{N} (a + jb)_i = e + jf \qquad (10.12)$$

The procedure for such a computation can be as follows:

```
set :   e := 1,   f = 0 ; (which corresponds to N = 0),
for i := 1 to n
  {compute : (a + jb)(e + jf) = (a*e - b*f) +
                                  j(a*f + b*e) = t + jf
    here temporary value is required : t = a*e - b*f, since
          previous value of e is needed for f computation}
  begin
    t := a*e - b*f
    f := a*f + b*e
    e := t
  end ;
```

For example, in computation

$$(1 + j1)^3 = -2 + j2$$

we obtain the intermediate results

$$i = 1: \qquad t_1 = 1, \qquad f_1 = 1, \qquad e_1 = 1$$

$$i = 2: \qquad t_2 = 0, \qquad f_2 = 2, \qquad e_2 = 0$$

$$i = 3: \qquad t_3 = -2, \qquad f_3 = 2, \qquad e_3 = -2$$

10.3.4 Computation of the roots of a polynomial

Computation of the roots of the first- and second-degree polynomials is straightforward. However, for higher than second-degree polynomials, the root computation requires more complicated procedures. One of them is based on the determination of a second-degree polynomial of the type

$$s^2 + b_1 s + b_0 = B(s)$$

of a given polynomial $A(s)$, and then extraction of $B(s)$ from $A(s)$ by division:

$$\frac{A(s)}{B(s)} = A_r(s)$$

Thus, the degree of the resulting polynomial $A_r(s)$ is lower by 2 in comparison with the degree of the initial polynomial $A(s)$. If the degree of $A_r(s)$ equals 1 or 2, then the roots of $A_r(s)$ are computed directly. If, however, the degree of $A_r(s)$ is higher than 2, the process is repeated and a new second-degree polynomial $B(s)$ is determined and extracted from $A_r(s)$ until all the roots are computed.

The first step in the procedure is the normalization of the coefficients of $A(s)$ to have the form

$$A(s) = s^n + a_{n-1}s^{n-1} + \cdots + a_1s + a_0$$

Then the degree n of the polynomial is tested. If $n = 1$, or $n = 2$, then the roots are computed directly from known formulas. If, however, $n > 2$ then the second-degree polynomial $B(s)$ is created:

$$B(s) = s^2 + b_1s + b_0$$

with initial values of coefficients

$$b_0 = \frac{a_0}{a_2}, \quad b_1 = \frac{a_1}{a_2}$$

Then, after the introduction of some auxiliary quantities, an iteration process is started in which division of $A(s)/B(s)$ is performed and correcting values Δb_0 and Δb_1 are computed to improve the values of b_0 and b_1, respectively, according to the following relations:

$$b_0 := b_0 - \Delta b_0$$
$$b_1 := b_1 - \Delta b_1$$

The values of b_0 and b_1 are modified until the increments Δb_0 and Δb_1 are smaller than some prescribed accuracy, for example 10^{-6}. If this condition is met, this means that b_0 and b_1 are the correct values and the roots of $B(s)$ are computed. Then, the degree of $A_r(s)$, which now equals $n - 2$, is tested and the process is repeated if $n - 2$ is greater than 2. The program 'POLYROOT' shown in Program 10.2, performs the polynomial root computation.

See the listing of the program POLYROOT Directory of Chapter 10 on attached diskette.

Here, the coefficients of polynomial $A(s)$ are read in during the run-time using the procedure 'Read-coefficients'. For example, for the polynomial

$$A(s) = 2s^2 + 6s + 4$$

we obtain the following results:

```
The normalized values of the coefficients are
       1.000   3.000   2.000
The roots of the polynomial are
     s[1] = -2.000 + j0.000
     s[2] = -1.000 + j0.000
```

For the polynomial

$$A(s) = s^3 + 3s^2 + 2s + 0$$

we obtain the following results:

The normalized values of the coefficients are

    1.000   3.000   2.000

The roots of the polynomial are

    s[1] = -2.000 + j0.000

    s[2] = -1.000 + j0.000

    s[3] = 0.000 + j0.000

For the polynomial

$$A(s) = 2s^4 + 4s^3 + 10s^2 + 10s + 9$$

we obtain the following results:

The normalized values of the coefficients are

    1.000   2.000   5.000   5.000   4.500

The roots of the polynomial are

    s[1] = -0.803478 + j(1.009116)

    s[2] = -0.803478 + j(-1.009116)

    s[3] = -0.196522 + j(1.632753)

    s[4] = -0.196522 + j(-1.632753)

10.3.5 Computation of the value of a polynomial $P(s_j)$ for a given value of a complex variable $s_j = \sigma_j + j\omega_j$

Any polynomial $P(s)$ of a complex variable s:

$$P(s) = p_0 + p_1 s + p_2 s^2 + \cdots + p_n s^n \tag{10.13}$$

after substitution of $s_j = \sigma_j + j\omega_j$, has the form

$$P(s_j) = p_0(\sigma_j + j\omega_j)^0 + p_1(\sigma_j + j\omega_j)^1 + p_2(\sigma_j + j\omega_j)^2 + \cdots + p_n(\sigma_j + j\omega_j)^n \tag{10.14}$$

To compute the values of $s_j^i = (\sigma_j + j\omega_j)^i$ we will use the procedure described in Section 10.3.3.

Now, we are interested in the computation of the real and imaginary parts of the polynomial $P(s)|_{s=sj} = P(s_j)$:

$$P(s_j) = P_r(s_j) + j P_i(s_j) \tag{10.15}$$

where the real and imaginary parts are

$$P_r(s_j) = p_0 \times 1 + p_1\sigma_j + p_2 \, \text{Re}(\sigma_j + j\omega_j)^2 + \cdots + p_n \, \text{Re}(\sigma_j + j\omega_j)^n \qquad (10.16)$$

$$P_i(s_j) = p_0 \times 0 + p_1\omega_j + p_2 \, \text{Im}(\sigma_j + j\omega_j)^2 + \cdots + p_n \, \text{Im}(\sigma_j + j\omega_j)^n \qquad (10.17)$$

The summations in relations (10.16) and (10.17) can be performed in a loop: 'for i: = 0 to n' according to the relations

```
pr:= pr + p[i]*e
```
$$\qquad (10.18)$$
```
pi:= pi + p[i]*f
```

where e and f represent the real and imaginary parts, respectively, of s_j taken to the ith power:

$$e + jf = (\sigma_j + j\omega_j)^i \qquad (10.19)$$

An example program 'POLYVAL' shown in Program 10.3 performs the computation of the polynomial value.

See the listing of the program POLYVAL Directory of Chapter 10 on attached diskette.

In this case the values of the polynomial coefficients and the s_j values are read in during the run-time; the degree of the polynomial is declared as a constant.

For example, for $P(s) = s^3 + 2s^2 + 5s + 2$ and value of $s_j = s_r + js_i = 1 + j1$, we obtain the following results:

$$P(1 + j1) = 5 + j11$$

10.4 Computation of residues in poles of rational function $Y(s) = T(s)U(s)$

As was discussed in Section 10.2, to find the jth residue for a single real or complex pole s_j it is necessary according to formula (10.8) to know the coefficients of the numerator $N(s)$ and the denominator $D(s)$ of the rational response function

$$Y(s) = T(s)U(s) = \frac{N(s)}{D(s)}$$

to compute the roots s_j of $D(s)$, that is, the poles s_j of $Y(s)$, to compute the derivative of $D(s)$, that is, $dD(s)/ds = D'(s)$, to compute the values of polynomials $N(s_j)$ and $D'(s_j)$ at s_j, and then to find the residue k_j: $k_j = N(s_j)/(D'(s_j))$.

To do this we can use programs for root computation, computation of the derivative of polynomials, and computation of the value of the polynomial for a given s_j, described in Section 10.3.

To make the program for residue calculation simpler let us assume that the degree m of the numerator $N(s)$ of the response function $Y(s) = N(s)/(D(s))$ is at least one less than the degree n of the denominator $D(s)$, that is, $m \leqslant n - 1$, which is usually fulfilled for electronic circuits excited by the step function $1(t)$ or more complex functions.

Also, we assume that the poles of $Y(s)$ are simple, which is fulfilled by the majority of electronic circuits; this assumption is necessary in order to use relations (10.7) and (10.8). An example program 'RESIDUES', shown in Program 10.4, performs residue computation.

See the listing of the program RESIDUES Directory of Chapter 10 on attached diskette.

After running the program, it is necessary to enter the coefficients of the numerator and the denominator of the response function $Y(s)$. Then, the roots of the denominator $D(s)$ are computed in the main part of the program; then, after differentiation of the denominator the real and imaginary parts of the residues, $k_j = k_{rj} + jk_{ij}$, are computed. As an example let us enter the $Y(s)$ function into the program 'RESIDUES':

$$Y(s) = \frac{5}{s(s+1)(s+2)} = \frac{5}{s^3 + 3s^2 + 2s}$$

The results of the program will be

The roots of the denominator are

$$s[1] = -2.000 + j(0.000)$$

$$s[2] = -1.000 + j(0.000)$$

$$s[3] = 0.000 + j(0.000)$$

The residues, $k = k_r + jk_i$, are

$$k_r(1) = 2.500 \; ; \; k_i(1) = 0.000$$

$$k_r(2) = -5.000 \; ; \; k_i(2) = 0.000$$

$$k_r(3) = 2.500 \; ; \; k_i(1) = 0.000$$

which corresponds to the simple quotient decomposition of $Y(s)$:

$$Y(s) = \frac{2.5}{(s+2)} - \frac{5}{(s+1)} + \frac{2.5}{s}$$

For another example response function

$$Y(s) = \frac{25}{2s^3 + 2s^2 + 2s + 1}$$

the results of the program will be

The roots of the denominator are

```
      s[1] = -0.176101 + j( 0.860717)
      s[2] = -0.176101 + j(-0.860717)
      s[3] = -0.647799 + j( 0.000000)
```

The residues, $k = k_r + jk_i$, are

```
      kr(1) =  -6.4879 ; ki(1) = -3.5556
      kr(2) =  -6.4879 ; ki(2) =  3.5556
      kr(3) =  12.9758 ; ki(3) =  0.0000
```

which corresponds to the simple quotient decomposition of $Y(s)$:

$$Y(s) = \frac{-6.4879 - j3.5556}{s + 0.176101 - j0.860717} + \frac{-6.4879 + j3.5556}{s + 0.176101 + j0.860717} + \frac{12.9758}{s + 0.647799}$$

10.5 Computation of the time response $y(t)$ based on pole and residue knowledge of the response function $Y(s)$

After pole and residue computation, which allows us to decompose the response function into simple quotients, we can find the values of the time response $y(t)$ using formula (10.3), that is,

$$y(t) = \sum_{l=1}^{n} (\exp(s_{rl}t)\{k_{rl}\cos(s_{il}t) - k_{il}\sin(s_{il}t)\})$$

To compute the time response $y(t)$ at discrete time steps

$$t = 0,\ h,\ 2h,\ 3h, \ldots, nh = tk$$

in the time range $(0, tk)$, it is necessary to form a 'time loop' (from $t = 0$, to tk) in which formula [10.3] is inserted in the inner loop (from $i = 0$, to $i = n$; $n =$ degree of the denominator). These two loops can be added at the end of the program 'RESIDUES' in the form

```
write("Enter the time range :"); readln(tk);
write("Enter the time step:"); readln(h);
    t:= 0;
    repeat
    yt:= 0
        for i:= 1 to n do
        yt = yt + exp(sr[i]*t)*(kr[i]*cos(si[i]*t) ·
                                        ki[i]*sin(s
        writeln('t = ',t:2:2,' y(t) ', yt:4:4);t:
        until t > tk;
    end.
```

The example program 'INLARESP' (time response based on inverse Laplace transform) based on the program 'RESIDUES' for the time response computation is shown in Program 10.5.

See the listing of the program INLARESP Directory of Chapter 10 on attached diskette.

It is necessary to note that the accuracy of the computations using this method depends on the accuracy of the computation of the poles and residues. As an example let us compute the time response for the RLC circuit of Fig. 10.1 with element values

$$R = 2.5 \ \Omega, \ L = 0.5 \ \text{H}, \ C = \frac{1}{3} \text{F}$$

The transfer function is then

$$T(s) = \frac{6}{(s+2)(s+3)}$$

Application of the input step function $1(t)$, gives the response function $Y(s)$:

$$Y(s) = T(s)U(s) = \frac{6}{s(s+2)(s+3)} = \frac{6}{s^3 + 5s^2 + 6s}$$

Running the program 'INLARESP' and entering this function we will obtain the following results:

```
The poles of the response function are

    s[1] = -3.000 + j(0.000)

    s[2] = -2.000 + j(0.000)

    s[3] =  0.000 + j(0.000)

The residues, k = kr + jki, are

    kr(1) =  2.000 ; ki(1) = 0.000

    kr(2) = -3.000 ; ki(2) = 0.000

    kr(3) =  1.000 ; ki(3) = 0.000
```

Entering the time range $tk = 5$, and the time step $h = 0.25$, we obtain the values of the time response $y(t)$ (Table 10.2).

When the poles and residues are computed exactly, the computed time response $y(t)$ is accurate independent of the time step h. This can be proved by tabulation of the exact time response $y(t)$, which is, for this example,

$$y(t) = 1 - 3 \ \exp(-2t) + 2 \ \exp(-3t)$$

Table 10.2

$t =$	$y(t) =$	$t =$	$y(t) =$
0.00	0.0000	3.00	0.9928
0.50	0.3426	3.50	0.9973
1.00	0.6936	4.00	0.9990
1.50	0.8729	4.50	0.9996
2.00	0.9500	5.00	0.9999
2.50	0.9809		

As a second example let us investigate the properties of an active RC filter with the ideal operational amplifiers shown in Fig. 10.2, using the program 'INLARESP'.
Direct analysis of this active RC circuit gives the relations

$$Z_1 = \frac{R_1}{1 + sR_1C}$$

$$V_2 = -\frac{R_2}{R} V_o$$

$$(10.20)$$

$$V_1 = -\frac{Z_1}{R} V_i - \frac{Z_1}{R} V_2$$

$$V_o = -\frac{1}{sRC} V_1$$

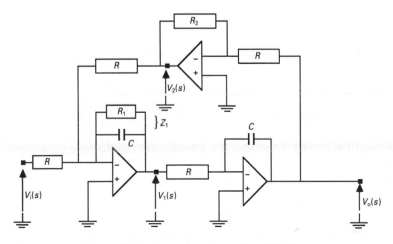

Figure 10.2 Second-order active RC filter

Combining relations (10.20) we obtain the transfer function $T(s)$:

$$T(s) = \frac{V_o(s)}{V_i(s)} = \frac{\dfrac{1}{(RC)^2}}{s^2 + \dfrac{R}{R_1}\dfrac{1}{RC}s + \dfrac{R_2}{R}\dfrac{1}{(RC)^2}} \tag{10.21}$$

which corresponds to the standard second-order low pass filter

$$T(s) = \frac{V_o(s)}{V_i(s)} = \frac{a\omega_0^2}{s^2 + \dfrac{\omega_0}{Q}s + \omega_0^2} \tag{10.22}$$

We can see that in this type of filter the Q value and the ω_0 value can be adjusted independently: Q by changing R_1 and ω_0 by changing R_2.

Let us assume that $\omega_0 = $ const, with $(R_2/R) = 1$, and investigate the residue and pole positions, and also the time response $y(t)$ for different $Q = R_1/R$ values. To do this let us assign the following values for R and C elements:

$$R = R_2 = 10^4\ \Omega = 10\ \text{k}\Omega, \quad C = 10^{-7}\ \text{F} = 0.1\ \mu\text{F}$$

Then, the frequency $\omega_0 = 10^3$ rad/s = 1 k rad/s and the time constant $\tau = RC = 10^{-3}$ s = 1 ms. The transfer function $T(s)$ then becomes

$$T(s) = \frac{10^6}{1 \times s^2 + \dfrac{R}{R_1}10^3 s + 10^6}$$

To lower the spread of the coefficients of the transfer function, let us normalize the values of the R and C elements, and the time constant and frequency ω_0:

$$R_n = \frac{R}{10^3} \qquad\qquad \text{expressed in kiloohms}$$

$$C_n = C \times 10^6 \qquad\qquad \text{expressed in microfarads}$$

$$\tau_n = R_n C_n = RC \times 10^3 \qquad \text{expressed in milliseconds}$$

$$\omega_{0n} = \frac{\omega_0}{10^3} = \frac{1}{RC \times 10^3} \qquad \text{expressed in kiloradians per second}$$

Using this normalization we obtain the transfer function

$$T(s) = \frac{1}{s^2 + \dfrac{R}{R_1}s + 1} = \frac{1}{s^2 + \dfrac{s}{Q} + 1}$$

which is convenient for further discussion and computations.

Let us use the unit step function $v_i(t) = 1(t)$ as an excitation. Then our output response will be

$$Y(s) = T(s)U(s) = \frac{T(s)}{s} = \frac{1}{s^3 + \dfrac{s^2}{Q} + s}$$

After running the program 'INLARESP' and subsequently entering the values of $Q = 0.25, 0.4, 0.5001, 1.0$ and 2.0, we obtain the values of the pole and residue positions in the complex plane and the values of the response $y(t) = v_0(t)$ in volts for the time values in the range $[0, 20]$ in milliseconds. The results of the computations are shown in Table 10.3. Graphical representations of the data of Table 10.3 are shown in Fig. 10.3.

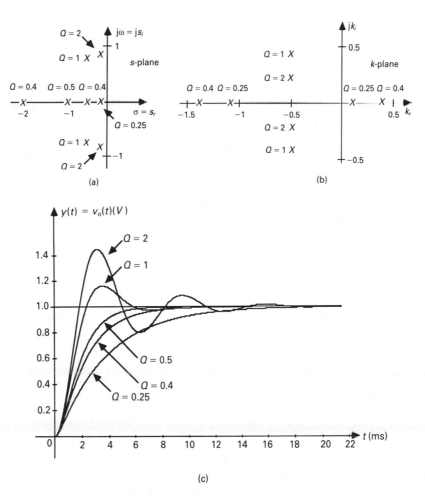

Figure 10.3 Graphical representation of the properties of the circuit of Fig. 10.2, excited by the unit step function $v_i(t) = 1(t)$, for different Q values. (a) Pole positions. (b) Residue positions. (c) $y(t)$ response

Table 10.3 Results of computation for the circuit of Fig. 10.2 with $v_i(t) = 1(t)\,V$

Q	0.25	0.4	0.5001	1	2
$sr(1) + jsi(1)$	$-3.7320 + j0$	$-2.0 + j0$	$-0.9998 + j0.01999$	$-0.5 + j0.866$	$-0.25 + j0.9682$
$sr(2) + jsi(2)$	$-0.2679 + j0$	$-0.5 + j0$	$-0.9998 - j0.01999$	$-0.5 - j0.866$	$-0.25 - j0.9682$
$sr(3) + jsi(3)$	$0 + j0$	$0 + j0$	$0 + j0$	$0 + j0$	$0 + j0$
$kr(1) + jki(1)$	$0.0774 + j0$	$0.3333 + j0$	$-0.5 + j24.9962$	$-0.5 + j0.2887$	$-0.5 + j0.1291$
$kr(2) + jki(2)$	$-1.0774 + j0$	$-1.3333 + j0$	$-0.5 - j24.9962$	$-0.5 - j0.2887$	$-0.5 - j0.1291$
$kr(3) + jki(3)$	$1.0 + j0$	$1.0 + j0$	$1 + j0$	$1 + j0$	$1 + j0$
$tk = 20$ ms $h = 2$ ms t(ms)	$v_o(t)(V)$	$v_o(t)(V)$	$v_o(t)(V$	$v_o(t)(V)$	$v_o(t)(V)$
0	0	0	0	0	0
2	0.3696	0.5156	0.5941	0.8494	1.0706
4	0.6311	0.8197	0.9085	1.1531	1.3372
6	0.7842	0.9336	0.9827	1.0023	0.8277
8	0.8737	0.9756	0.9970	0.9790	0.9507
10	0.9261	0.9910	0.9995	1.0022	1.0848
12	0.9568	0.9967	0.9999	1.0026	0.9814
14	0.9747	0.9988	1.0000	0.9994	0.9769
16	0.9852	0.9996	1.0000	0.9997	1.0169
18	0.9913	0.9998	1.0000	1.0001	1.0012
20	0.9949	0.9999	1.0000	1.0000	0.9933

To use formula (10.8) for the computation of residues it is necessary to avoid double (or multiple) poles which occur in this example for $Q = 0.5$ (then $s_r = s_i = -1.0$). In such a case we can use the Q value close to, but different from, 0.5, for example $Q = 0.5001$. It can be seen that the response $y(t)$ for the second-degree low pass circuit is oscillatory for $Q > 0.5$.

10.6 Computation of the time response using the convolution integral

The method of the computation of the time response $y(t)$ based on the decomposition of the response function $Y(s) = T(s)U(s)$ into simple quotients is easy when the input (excitation) function $U(s)$ takes the simple form, as, for example, 1, $1/s$, etc. However, for more complicated input functions the task may be much more difficult; the method cannot be used for input $u(t)$ functions not having the Laplace transform.

In such situations it is more convenient to compute the time response $y(t)$ directly in the time domain. If we assume that the input function $U(s) = 1$, which corresponds to the Dirac impulse $\delta(t)$ in the time domain, then the output function is

$$Y(s) = T(s) \times 1 = T(s) \tag{10.23}$$

Taking the inverse Laplace transform we obtain the impulse response $h(t)$:

$$y(t) = \mathscr{L}^{-1}[T(s)] = h(t) \tag{10.24}$$

Generally, the inverse Laplace transform of a product of two functions of a complex variable s gives the convolution integral

$$y(t) = \mathscr{L}^{-1}[T(s)U(s)] = \int_0^t h(t - \tau)u(\tau)\,d\tau = \int_0^t h(\tau)u(t - \tau)\,d\tau \tag{10.25}$$

with the assumption that $u(t)$ is defined for $t \geqslant 0$, that is, $u(t)1(t)$.

From relation (10.25) we see that if the impulse response $h(t)$ is known for a given circuit, then we can find the time response $y(t)$ for an arbitrary input function $u(t) \cdot 1(t)$ by computing the convolution integral (10.25) in the limits $[0, t]$. Since the upper limit of integration t is changing, the numerical computation technique is similar to the computation of the indefinite integral described in Chapter 2. Numerical integration is usually performed using the trapezoidal method or Simpson's method. As a simple example let us consider the second-degree transfer function

$$T(s) = \frac{2}{s^2 + 3s + 2} = \frac{2}{(s + 1)(s + 2)} = \frac{2}{(s + 1)} - \frac{2}{(s + 2)}$$

Taking the inverse Laplace transform $\mathscr{L}^{-1}[T(s)]$ we obtain the impulse response $h(t)$:

$$h(t) = 2\,e^{-t} - 2\,e^{-2t} = 2\{e^{-t} - e^{-2t}\}$$

Now, let us compute the time response $y(t)$ for the unit step input function $u(t) = 1(t)$ analytically and using the convolution integral, and compare the results.

For $u(t) = 1(t)$, that is, $U(s) = 1/s$, we have the output function $Y(s)$:

$$Y(s) = T(s)U(s) = \frac{2}{(s + 1)(s + 2)}\frac{1}{s} = \frac{1}{s} - \frac{2}{(s + 1)} + \frac{1}{(s + 2)}$$

and the exact time response function $y_{ex}(t)$ is

$$y_{ex}(t) = \mathscr{L}^{-1}[Y(s)] = 1 - 2\,e^{-t} + e^{-2t}$$

This $y_{ex}(t)$ function will be used for comparison with the numerical computations. To compute the time response $y(t)$ using the convolution integral the program 'CONVOLUT', shown in Program 10.6, has been prepared.

See the listing of the program CONVOLUT Directory of Chapter 10 on attached diskette.

In this program Simpson's method of integration is used, with p (p must be even) discrete points for the computation of the integrand function, and the nl number of the upper limit t of the integral changes. In the program the functions $u(\tau)$ and

$h(t) = ht(t - \tau)$ are defined:

```
u(τ) := 1;
ht(t-τ) := 2*(exp(-(t-τ)) - exp(-2*(t-τ)));
```

and the procedure for the convolution integral:

$$\int_0^t ht(t - \tau)u(\tau)\,d\tau$$

is written using Simpson's method; the variables *ta*, *tb*, *tc* represent the variable τ during Simpson integration. The process of the computation of the convolution integral for the $ht(t - \tau) \cdot u(\tau)$ integrand function is explained graphically in Fig. 10.4(a).

The same computation results are obtained if we define the function $u(t - \tau)$ and $h(\tau) = ht(\tau)$:

```
u(t-τ) := 1;
ht(τ) := 2*(exp(-τ) - exp(-2*τ));
```

This situation is explained graphically in Fig. 10.4(b). In the program of numerical integration, the time step or time increment *h* should be selected to be much smaller than the smallest time constant in the analyzed circuit, and smaller than the time constant and/or the time period of the input function $u(t)$. Running the program 'CONVOLUT' with input data *p* (even for Simpson's integration) and $nl = 20$ (number of changes of integral upper limit), we obtain the results of the computed $y(t)$ and the exactly evaluated $y_{ex}(t)$.

It can be seen that the accuracy of computation improves as *p* increases, because of the decrease of the time step $h = t/p$ (Table 10.4).

As a second example let us consider the simple first-degree transfer function

$$T(s) = \frac{4}{s + 1}$$

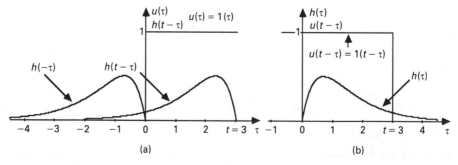

Figure 10.4 Graphical explanation of the convolution integration. (a) For $h(t - \tau)u(\tau)$. (b) For $h(\tau) \cdot u(t - \tau)$

Table 10.4

	$nl = 20$		
	$p = 8$	$p = 20$	
t	$y(t)$	$y(t)$	exact: $y_{ex}(t)$
0.2	0.03286	0.03286	0.03286
0.4	0.10869	0.10869	0.10869
1.0	0.39956	0.39958	0.39958
2.0	0.74735	0.74764	0.74765
3.0	0.90147	0.90278	0.90290
4.0	0.95941	0.96358	0.96370

with impulse response $h(t) = 4\,e^{-t}$ and the more complicated input function $u(t)$

$$u(t) = 3\,\sin^2(\omega t)\left\{1(t) - 1\left[t - \frac{T}{2}\right]\right\}$$

where the time period $T = 2\pi/\omega$, or $T/2 = \pi/\omega$ (for simplicity let us assume that $\omega = 1$ rad/s). The function $u(t)$ is shown in Fig. 10.5(a), and a graphical explanation of the convolution integral computation process is shown in Fig. 10.5(b). The program 'CONVOL__1' for the computation of the convolution integral for $u(t)$ defined in the finite time range is shown in Program 10.7.

See the listing of the program CONVOL__1 Directory of Chapter 10 on attached diskette.

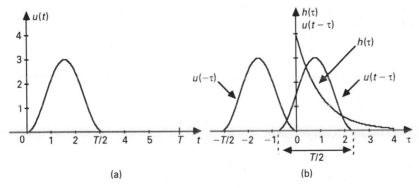

Figure 10.5 Graphical presentation of (a) $u(t) = 3\,\sin^2(\omega t)\left\{1(t) - 1\left[t - \frac{T}{2}\right]\right\}$,

(b) convolution integration process

Table 10.5

	$nl = 20$	$p = 8$	$p = 20$
j	t	$y(t)$	$y(t)$
1	0.1571	0.0148	0.0148
.	.	.	.
5	0.7854	1.4115	1.4115
.	.	.	.
10	1.5708	6.2015	6.2022
.	.	.	.
13	2.0420	8.0233	8.0241
14	2.1991	8.1206	8.1210
15	2.3562	7.9454	7.9451
.	.	.	.
$j = nl$	3.1416	4.6022	4.5928

Here again the Simpson integration method is used. It is necessary to note that for $t \leqslant T/2 = \pi/\omega$, the lower limit t_0 equals zero, and for $t > T/2 = \pi/\omega$, the lower limit of the integral equals $t_0 = t - T/2 = t - \pi/\omega$; this results from defining the input function $u(t)$ in the time range $[0, T/2]$.

This situation is described in the program 'CONVOL__1' using the instruction 'if ... then ...'.

Running the program 'CONVOL__1' with input data p and nl we obtain the results shown in Table 10.5.

Slight differences can be observed for the results for different p values; in this case the time increment is $h = (t - t_0)/p$, with $t_0 = 0$, for $t \leqslant \pi/\omega$, and $t_0 = t - \pi/\omega$, for $t > \pi/\omega$. The programs presented in this section can easily be modified for other $u(t)$ and $h(t)$ functions.

Exercises

Exercise 1.   Using Table 10.1, determine the Laplace transforms of the following functions:

$$e^{-2t}; \quad \sin(2t); \quad \cos\left\{\frac{t}{3} + \frac{\pi}{4}\right\}; \quad t\cos(t); \quad \frac{t^2}{2}$$

Note: use equation (10.1), which defines the Laplace transform $F(s)$ of a time function $f(t)$ defined for $t \geqslant 0$.

Exercise 2. Consider the function $f(t) = t(t-2)$ shown in Fig. E10.1. Construct the following functions: $f_1(t) = f(t) \times 1(t)$; $f_2(t) = f(t-2) \times 1(t-2)$; $f_3(t) = f(t) \times 1(t-2)$; $f_4(t) = f(t-2) \times 1(t)$. Determine the Laplace transforms of these four functions. Deduce how the Laplace transform of the function $f(t-t_0) \times 1(t-t_0)$ can be determined knowing the Laplace transform of the function $f(t) \times 1(t)$.

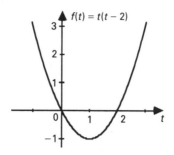

Figure E10.1

Exercise 3. Consider the function $f_1(t)$ shown in Fig. E10.2(a), equal to zero for $t < 0$ and for $t > T$ and the corresponding Laplace transform $F_1(s)$. The function $f_2(t)$ shown in Fig. E10.2(b) is obtained from $f_1(t)$ by the time shift T. Determine the Laplace transform $F_2(s)$ of $f_2(t)$ as a function of $F_1(s)$. Deduce the Laplace transform of the periodic function $f(t)$ with time period T for $t > 0$, shown in Fig. E10.2(c) (the elementary function $f_1(t)$ for one period is shown in Fig. E10.2(a)).

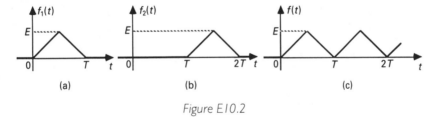

Figure E10.2

Exercise 4. Using equation (10.1), determine the Laplace transform of the functions shown in Fig. E10.3(a, b, c, d, e). Determine the limits of these Laplace transforms when Δ tends to zero.

Exercise 5. Consider the function $f(t)$ and its corresponding Laplace transform $F(s)$. We want to consider the influence of a time scale modification (t becomes (at), $a > 0$). Using equation (10.1), determine the Laplace transform of the function $f(at)$.

Exercise 6. Consider the circuits shown in Fig. E10.4(a, b). All the initial conditions are equal to zero. Determine the differential equation in terms of the current flowing through the resistor R_1; develop the expression for the transfer function. Assuming that the sources $e(t)$ and $j(t)$ are step functions, determine the two expressions for the current $i(t)$. Verify the results using the program 'INLARESP'.

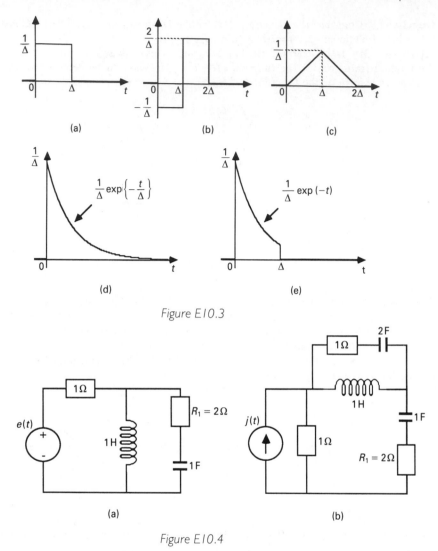

Figure E10.3

Figure E10.4

Exercise 7. The inductor shown in Fig. E10.5(a) has an initial condition $i_L(0)$. Demonstrate that in the complex frequency plane, the constitutive equation of the inductor can be represented by the equivalent circuits shown in Fig. E10.5(b, c). The inductors shown in Fig. E10.5(b, c) have zero initial conditions. Determine the values of these sources. Deduce, in the time domain, the equivalent circuit of an inductor with initial conditions.

Exercise 8. The capacitor shown in Fig. E10.6(a) has an initial condition $v_C(0)$. Demonstrate that in the complex frequency plane the constitutive equation of the capacitor can be represented by the equivalent circuits shown in Fig. E10.6(b, c). The capacitors shown in Fig. E10.6(b, c) have zero initial conditions. Determine the values of these sources. Deduce, in the time domain, the equivalent circuit of a capacitor with initial conditions.

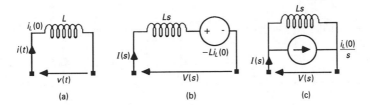

Figure E10.5

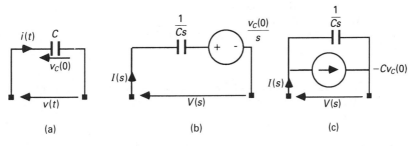

Figure E10.6

Exercise 9.   Run the program 'POLYDIV', assuming the following cases:

1.   The degree of the denominator is larger than the degree of the numerator.
2.   The degree of the numerator is equal to zero.
3.   The degree of the denominator is equal to zero.
4.   The numerator and the denominator have some identical roots.

Exercise 10.   Demonstrate that, for the two-terminal networks shown in Fig. E10.7(a, b), the impedance or admittance function has the following form:

$$Z(s) = L_1 s + \cfrac{1}{C_2 s + \cfrac{1}{L_3 s + \cfrac{1}{C_4 s + \cfrac{1}{L_5 s + \cfrac{1}{C_6 s}}}}}$$

$$Y(s) = C_1 s + \cfrac{1}{L_2 s + \cfrac{1}{C_3 s + \cfrac{1}{L_4 s + \cfrac{1}{C_5 s}}}}$$

Modifying the program 'POLYDIV', determine the value of the elements for the following two functions:

$$Z(s) = \frac{(s^2 + 1)(s^2 + 4)(s^2 + 9)}{s(s^2 + 2)(s^2 + 5)}; \quad Y(s) = \frac{s(s^2 + 2)(s^2 + 4)}{(s^2 + 1)(s^2 + 3)}$$

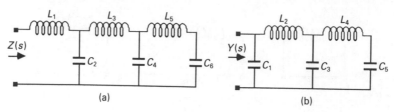

Figure E10.7

Exercise 11. The synthesis of Butterworth filters leads, for the normalized transfer function $F(s)$, to the following equality: $H(s) = F(s)F(-s) = 1/(1 + (-1)^n s^{2n})$. To determine the transfer function $F(s)$, we have to calculate the $H(s)$ poles and assign to $F(s)$ the poles of the left half-plane in order to obtain a stable filter. Construct the program 'BUTTWO__1' giving the poles of the transfer function as a function of n. Verify using the program 'POLYROOT'.

Exercise 12. Modify the program 'BUTTWO__1' in order to obtain the coefficients a_i of the transfer function $F(s)$ described by the following form:

$$F(s) = \frac{1}{a_0 + a_1 s + a_2 s^2 + a_3 s^3 + \ldots + a_{n-1}s^{n-1} + a_n s^n}$$

Verify that the coefficients have the values shown in Table E10.1.

Table E10.1   Values of coefficients for Butterworth transfer function

n	a_0	a_1	a_2	a_3	a_4	a_5	a_6	a_7	a_8	a_9	a_{10}
1	1	1									
2	1	1.4142	1								
3	1	2	2	1							
4	1	2.6131	3.4142	2.6131	1						
5	1	3.2360	5.2360	5.2360	3.2360	1					
6	1	3.8637	7.4641	9.1416	7.4641	3.8637	1				
7	1	4.4939	10.0978	14.5917	14.5917	10.0978	4.4939	1			
8	1	5.1258	13.1370	21.8461	25.6883	21.8461	13.1370	5.1258	1		
9	1	5.7587	16.5817	31.1634	41.9863	41.9863	31.1634	16.5817	5.7587	1	
10	1	6.3924	20.4317	42.8020	64.8823	74.2324	64.8823	42.8020	20.4317	6.3924	1

Exercise 13. The transfer function $F(s)$ of a linear phase filter can be put in the form $F(s) = K/B_n(s)$, where $B_n(s)$ is an nth degree polynomial satisfying the following recursive equation:

$$B_n(s) = (2n - 1)B_{n-1}(s) + s^2 B_{n-2}(s) \quad \text{with } B_0(s) = 1 \text{ and } B_1(s) = s + 1$$

Modify the program 'POLYROOT' in order to obtain the poles of the 'n order' function $F(s)$.

Exercise 14. We want to analyze the behaviour of the program 'POLYROOT' when two real roots are different but very close together; for this, we choose the following fourth-order test polynomial: $s^4 + (3 + a)s^3 + (2 + 3a)s^2 + 2as$. Verify that factorization leads to $s(s + 1)(s + 2)(s + a)$. Run the program 'POLYROOT' for values of a equal to and close to 0, 1 and 2. Give some conclusions.

Exercise 15. We want to analyze the behaviour of the program 'POLYROOT' when two imaginary roots are different but very close together; for this, we choose the following fourth-order test polynomial: $s^4 + (1 + a^2)s^2 + a^2$. Verify that factorization leads to $(s^2 + 1)(s^2 + a^2)$. Run the program 'POLYROOT' for values of a equal to and close to 1. Give some conclusions.

Exercise 16. We want to analyze the behaviour of the program 'POLYROOT' when two real roots are very different; for this, we choose the following fourth-order test polynomial: $s^4 + (a + b + 3)s^3 + (ab + 3a + 3b + 2)\ s^2 + (3ab + 2a + 2b)s + 2ab$. Verify that factorization leads to $(s + 1)(s + 2)(s + a)(s + b)$. Run the program 'POLYROOT' for the following values of a and b: $a = 10^{-n}$ and $b = 10^{+n}$, and $n = 100$. Give some conclusions.

Exercise 17. For the lumped networks, a transfer function $F(s)$ is the ratio of two polynomials of the complex variable s; use the program 'POLYVAL' to construct a program to obtain the value of the magnitude and phase of $F(s)$ for $s = j\omega$, for a given ω. Apply this program to plotting Bode and Nyquist diagrams.

Exercise 18. For the polynomials $D(s)$ defined in exercises 14 and 15, apply the program 'RESIDUES' to $(1/D(s))$ and observe the evolution of the value of each residue when the values of the roots are very close.

Exercise 19. In the second example of Section 10.5, the program 'INLARESP' is used to determine the response $v_0(t)$ to a unit step signal. Proceed in the same way but for the output $v_1(t)$ shown in Fig. 10.2.

Exercise 20. We want to compare the responses to a unit step for the Butterworth filters. Modify the program 'INLARESP' using the results of exercises 11 and 12.

Exercise 21. We want to verify that the inverse Laplace transform of the function $1/(s + 1)^2$ is $t\,e^{-t}$ for $t \geqslant 0$. To do so, run the program 'INLARESP' for the function $1/(s^2 + s(2 + \varepsilon) + 1 + \varepsilon)$; plot the relative error as a function of time for different values of ε.

Exercise 22. Apply the program 'INLARESP' to the function $F(s) = 2s/D(s)$, where $D(s)$ is the polynomial defined in exercise 15: $(s^2 + 1)(s^2 + a^2)$. Verify that, if a tends to 1, the inverse Laplace transform of $2s/(s^2 + 1)^2$ is $t\,\sin(t)$ for $t \geqslant 0$.

Exercise 23. Relation (10.25) shows that the inverse Laplace transform of the product of two functions $F_1(s)F_2(s)$ is equal to the convolution integral $\int_0^t f_1(\tau)f_2(t - \tau)\,d\tau$. Apply the program 'CONVOLUT' in order to verify that the inverse Laplace transform of the function $1/(s + 1)^2$ is $t\,e^{-t}$ for $t \geqslant 0$.

Exercise 24. Consider the same question as in Exercise 23 with the functions $1/s^2$ and $1/(s^2 + 1)^2$.

Exercise 25. For the circuit shown in Fig. E10.8, determine the voltage $v(t)$ when the voltage source $e(t)$ is represented by the following functions: $atan(t)$; $th(t)$; $1/ch(t)$; $e^{\sin(t)}$; $t/(t^2 + 1)$; $\sin(t)/t$ for $t \geqslant 0$ and zero for $t < 0$. All reactive elements have zero initial conditions.

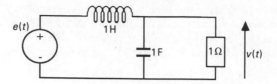

Figure E10.8

Exercise 26. Modify the program 'CONVOL_1' in order to study the response of a second-order system defined by $1/(s^2 + 1)$ when the inputs are as shown in Fig. E10.9(a, b, c).

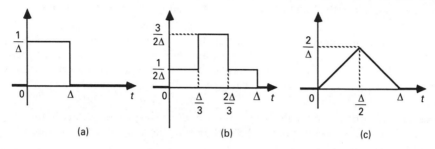

Figure E10.9

11 Sinusoidal Steady State Analysis in the Frequency Domain

11.1 Introduction

One of the most popular methods for the investigation of linear electronic circuit properties in the frequency domain is the sinusoidal steady state analysis. The results of this analysis are usually the frequency characteristics of magnitude (gain) $|T(\omega)|$ and phase $\phi(\omega)$ of the transfer function, or transmittance:

$$T(j\omega) = \frac{V_{out}(j\omega)}{V_{in}(j\omega)} = |T(\omega)|\,e^{j\phi(\omega)} \tag{11.1}$$

where V_{out} and V_{in} are output and input complex voltage amplitudes (phasors).

The frequency characteristics of magnitude, expressed in decibels:

$$|T(\omega)|_{dB} = 20\,\log[\,|T(\omega)|\,] \tag{11.2}$$

and of phase, expressed in degrees or radians, and drawn in rectangular coordinates, are known as Bode plots, while, when drawn in polar coordinates, they are known as Nyquist plots. Examples of Bode and Nyquist plots for low pass characteristics are shown in Fig. 11.1.

Generally, an analysis in the frequency domain is performed using the complex frequency $s = \sigma + j\omega$, and then the results are computed and drawn for $s = j\omega$. When the input and output ports of the analyzed circuit are determined, then the circuit can be treated as a two-port network loaded at the output by the load admittance Y_L (or impedance $Z_L = 1/Y_L$) and loaded at the input by the source admittance Y_S associated with the independent current source J_S, or the source impedance Z_S associated with the independent voltage source E_S, as shown in Fig. 11.2.

For the two-port of Fig. 11.2, we can define the following eight transmittances:

1. voltage transmittance: $T_v = \dfrac{V_L}{V_i}$, or voltage gain,

2. transducer voltage transmittance: $T_{vt} = \dfrac{V_L}{E_S}$, or transducer voltage gain,

3. current transmittance: $T_i = \dfrac{I_L}{I_i} = -\dfrac{I_o}{I_i}$, or current gain,

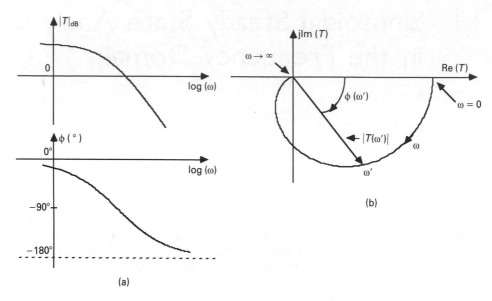

Figure 11.1 Graphical representation of transmittance. (a) Bode plots. (b) Nyquist plot

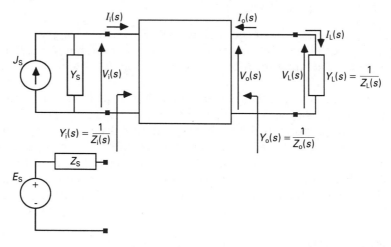

Figure 11.2 Linear two-port network with transmittance $T(s)$, excited by the current source J_S with source admittance Y_S, or by the voltage source E_S with source impedance

$$Z_S, \text{ and loaded by the admittance } Y_L \left(\text{impedance } Z_L = \frac{1}{Y_L}\right)$$

4. transducer current transmittance: $T_{\mathrm{it}} = \dfrac{I_{\mathrm{L}}}{J_{\mathrm{S}}}$, or transducer current gain,

5. voltage–current V–C transmittance: $T_z = \dfrac{V_{\mathrm{L}}}{I_{\mathrm{i}}}$, or transimpedance,

6. transducer V–C transmittance: $T_{\mathrm{zt}} = \dfrac{V_{\mathrm{L}}}{J_{\mathrm{S}}}$, or transducer transimpedance,

7. current–voltage transmittance: $T_y = \dfrac{I_{\mathrm{L}}}{V_{\mathrm{i}}}$, or transadmittance,

8. transducer C–V transmittance: $T_{\mathrm{yt}} = \dfrac{I_{\mathrm{L}}}{E_{\mathrm{S}}}$, or transducer transadmittance,

and also one-port input and output immittances (that is impedances, or admittances)

$$Y_{\mathrm{i}} = \frac{1}{Z_{\mathrm{i}}} = \frac{I_{\mathrm{i}}}{V_{\mathrm{i}}}$$

and

$$Y_{\mathrm{o}} = \frac{1}{Z_{\mathrm{o}}} = \frac{I_{\mathrm{o}}}{V_{\mathrm{o}}}$$

All the two-port functions defined here are rational functions, that is, they are ratios of the polynomials of the complex variable $s = \sigma + \mathrm{j}\omega$.

To find the frequency characteristics of any transmittance or immittance it is necessary to find the ratio of output to input voltages or currents using any of the known methods of circuit analysis (by the computation of nodal voltages using the Gauss elimination method, for example) at a single prescribed frequency, and then repeat the process for all the frequencies of interest. This is a time consuming method, since the analysis of the whole circuit must be done for each frequency. Therefore, a more efficient method is used nowadays.

First, the semisymbolic analysis of the circuit is performed leading to the determination of the required transmittance in semisymbolic form, that is, the determination of the numerical values of all the coefficients of the numerator and denominator of the transmittance. Then, simple tabularization of the magnitude and phase of the transmittance for the required frequencies gives us the required frequency characteristics.

Note that to obtain the transmittance in semisymbolic form it is necessary to use the so-called modified admittance (or hybrid) matrix circuit description in which elements of type $1/s$ must be avoided. Therefore, inductive elements must be described by their impedances $Z_L = sL$ (not by admittances $Y_L = 1/sL$) leading to the mixed description: nodal, for current sources, conductances and capacitances, that is, using Kirchhoff's current law KCL; and mesh, for voltage sources and inductances, that is, using Kirchhoff's voltage law KVL. As will be explained in detail later in this chapter, the modified admittance (hybrid) matrix will be generated based on the so-called 'data matrix' describing the topology of the analyzed circuit together with the element values.

Thus, the following steps should be performed to find the frequency characteristics of the analyzed (or simulated) circuit numerically:

1. reading in circuit data and the creation of the data matrix containing the type of element (R, L, C, independent and controlled sources);
2. the generation of the modified admittance (hybrid) matrix, based on the data matrix;
3. the repetitive solution of a set of hybrid equations (usually using the *LU* method) for the prescribed set of complex frequencies s_j, $j = 1, 0, ..., n$, where n is the order of the network (highest power of s of the denominator of the transmittance) in order to find the semisymbolic form of the transmittance;
4. tabulation of the semisymbolic transmittance for frequencies in the required range to obtain the following characteristics: magnitude $|T(\omega)|$ and phase $\phi(\omega)$.

All these steps are described in detail in the subsequent sections.

11.2 Generation of the 'data matrix'

Information about the topology and element values of the analyzed circuit can be entered into a computer program in several different ways. In this chapter one of the most popular methods is used (sometimes referred to as the SPICE-like method), in which the type of element (that is R, C, etc.), the number of nodes between which the element is located (and the nodes from which is controlled − for controlled sources) and the element value are entered.

The data matrix, denoted here as $[A]$, is composed of *ne* (number of elements) rows and six columns. In each *i*th row particular positions correspond to the following:

1. $A(i, 0)$ and $A(i, 1)$ − number of nodes between which the *i*th element is located,
2. $A(i, 2)$ − type of the *i*th element (see Table 11.1),
3. $A(i, 3)$ − the value of the *i*th element (real part),
4. $A(i, 4)$ − the imaginary part value of the independent voltage E_S or current source J_S,
5. $A(i, 4)$ and $A(i, 5)$ − the number of controlling nodes for the voltage controlled current source − VCCS,
6. $A(i, 5)$ − the value of the source resistance R_S for the independent voltage source E_S.

Note that the ideal independent voltage source E_S and its internal resistance R_S form a single branch (single element) as shown in Fig. 11.3. When zero is declared (entered into the computer) as $R_S = 0$, then, to avoid division by zero, a small value $R_S = 10^{-3}\ \Omega$ is assigned automatically by the program.

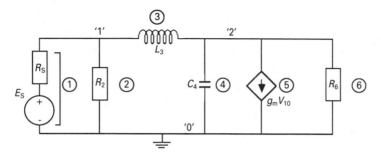

Figure 11.3 The RLC circuit with six elements (E_S and R_S form one branch) and two independent nodes; the node 0 is a reference node

To simplify the program for the generation of the data matrix, the number code (with numbers from 1 to 6) is assigned to particular circuit elements as shown in Table 11.1. The imaginary parts of the independent current source $\text{Im}(J)$ and the independent voltage source $\text{Im}(E)$ are stored in the data-matrix elements $A(i, 4)$. In most practical cases, however, the values of E and J are treated as real (with the imaginary parts equal to zero).

The example program 'DATA_MAT' for the generation of the data matrix is shown in Program 11.1.

See the listing of the program DATA_MAT Directory of Chapter 11 on attached diskette.

Running the program 'DATA_MAT' for the circuit shown in Fig. 11.4 will create the data matrix shown in Table 11.2. The data for particular elements are entered into the program on one line; for example, to enter the data for the VCCS

Table 11.1   Number code for circuit element types

Type of element	Circuit notation	Code assigned to matrix element	Values of matrix elements	
		$A(i, 2)$	$A(i, 4)$	$A(i, 5)$
Independent current source	J_S	1	$\text{Im}(J_S)$	0
Resistor	R	2	0	0
Inductor	L	3	0	0
Capacitor	C	4	0	0
Independent voltage source	E_S	5	$\text{Im}(E_S)$	R_S
Voltage controlled current source, VCCS*	G_m (s)	6	N_s	N_t

*Note: Controlling voltage V_{st} for VCCS is between the nodes N_s and N_t; in the program 'DATA_MAT', the VCCS is denoted by S

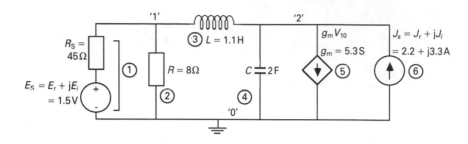

Figure 11.4 Example of active RLC circuit with two independent sources having the same angular frequency ω

of the circuit of Fig. 11.4 the following steps are necessary:

1. for the program query: enter the kind of element (5): S; $\leftarrow$ S – for VCCS,
2. for the program query: for the source (5) enter $i\ j\ g_m$ controlled by $s\ t$ –
    2 0 5.3 1 0.

Note that for the circuits excited by more than one independent source the frequencies of all the sources must be the same. In other cases separate analysis must be performed for sources of different frequencies.

Table 11.2   Data matrix for the circuit of Fig. 11.4

Number of branches	Type of element	Nodes		Element code	Element value	Control node or $Im(J)$, $Im(E)$	Control node or R_S	Comments
		i	j					
k		$A(k,0)$	$A(k,1)$	$A(k,2)$	$A(k,3)$	$A(k,4)$	$A(k,5)$	
1	E	1	0	5	1.5	0	45	$R_S = A(1,5)$
2	R	1	0	2	8	0	0	
3	L	1	2	3	1.1	0	0	
4	C	2	0	4	2	0	0	
5	S	2	0	6	5.3	1	0	$V_{st} = V_{10}$
6	J	2	0	1	2.2	3.3	0	$Im(J_S) = A(6,4)$

11.3 Generation of the modified admittance (hybrid) matrix

To make the computation of the frequency characteristics more efficient we are first going to find the transmittance $T(s)$ of the simulated circuit in semisymbolic form and then tabularize it to find $|T(\omega)|$ and $\phi(\omega)$. An example of the transmittance in symbolic form is the following second-order RLC low pass function:

$$T(s) = \frac{\dfrac{1}{LC}}{s^2 + \dfrac{R}{L} s + \dfrac{1}{LC}} = \frac{\omega_0^2}{s^2 + \dfrac{\omega_0}{Q} s + \omega_0^2} \tag{11.3}$$

When we assume, for example, that the values of the elements are

$$L = 0.5 \text{ H}, \quad C = 0.5 \text{ F}, \quad R = 2 \, \Omega$$

then the semisymbolic form of this transmittance is

$$T(s) = \frac{4}{s^2 + 4s + 4} \tag{11.4}$$

where all coefficients are real numbers.

   As was mentioned in Section 11.1, automatic generation of the semisymbolic form of the transmittance, that is, the numerical computation of all the coefficients of $T(s)$, is possible only when the equations describing the circuit do not contain terms of the type $1/s$. Therefore, we have to apply a mixed method using simultaneously the nodal (Kirchhoff's current law) and the mesh or loop (Kirchhoff's voltage law) descriptions of the analyzed circuit. Kirchhoff's voltage law is applied to branches containing inductances and/or the independent voltage source. The set of coefficients of constitutive equations forms the modified admittance (or hybrid) matrix. As examples, let us create modified admittance matrices for several simple circuits shown in Fig. 11.5. At first we write the voltage and then the current equations.

   For the circuit of Fig. 11.5(a):

$$-sLI_L + V_1 - V_2 = 0$$

$$I_L + G_1 V_1 = J \tag{11.5a}$$

$$-I_L + G_2 V_2 + g_m V_1 = 0$$

or, in matrix form,

$$\begin{bmatrix} -sL & 1 & -1 \\ 1 & G_1 & 0 \\ -1 & g_m & G_2 \end{bmatrix} \begin{bmatrix} I_L \\ V_1 \\ V_2 \end{bmatrix} = \begin{bmatrix} 0 \\ J \\ 0 \end{bmatrix} \tag{11.5b}$$

Note that due to the presence of one inductance the dimensions of the hybrid matrix are larger by one in comparison with the two-node admittance matrix.

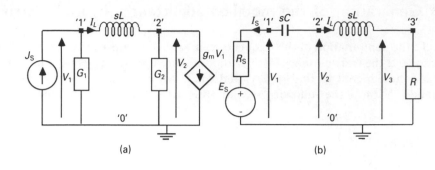

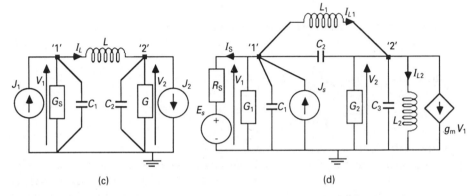

Figure 11.5   (a–d) Examples of simple RLC circuits

For the circuit of Fig. 11.5(b):

$$-R_S I_S + V_1 = E_S$$

$$-sL I_L + V_2 - V_3 = 0$$

$$I_S + sC V_1 - sC V_2 = 0 \tag{11.6a}$$

$$I_L - sC V_1 + sC V_2 = 0$$

$$-I_L + \frac{1}{R} V_3 = 0$$

or, in matrix form,

$$
\begin{bmatrix}
-R_S & 0 & 1 & 0 & 0 \\
0 & -sL & 0 & 1 & -1 \\
1 & 0 & sC & -sC & 0 \\
0 & 1 & -sC & sC & 0 \\
0 & -1 & 0 & 0 & \dfrac{1}{R}
\end{bmatrix}
\begin{bmatrix}
I_S \\ I_L \\ V_1 \\ V_2 \\ V_3
\end{bmatrix}
=
\begin{bmatrix}
E_S \\ 0 \\ 0 \\ 0 \\ 0
\end{bmatrix}
\tag{11.6b}
$$

Due to the presence of the inductance and independent voltage sources the dimensions of the hybrid matrix are larger by two in comparison with the three-node admittance matrix.

For the circuit of Fig. 11.5(c):

$$-sLI_L + V_1 - V_2 = 0$$

$$I_L + (G_S + sC_1)V_1 = J_1 \tag{11.7a}$$

$$-I_L + (G + sC_2)V_2 = -J_2$$

or, in matrix form,

$$\begin{bmatrix} -sL & 1 & -1 \\ 1 & G_S + sC_1 & 0 \\ -1 & 0 & G + sC_2 \end{bmatrix} \begin{bmatrix} I_L \\ V_1 \\ V_2 \end{bmatrix} = \begin{bmatrix} 0 \\ J_1 \\ -J_2 \end{bmatrix} \tag{11.7b}$$

For the circuit of Fig. 11.5(d):

$$-R_S I_S + V_1 = E_S$$

$$-sL_1 I_{L1} + V_1 - V_2 = 0$$

$$-sL_2 I_{L2} + V_2 = 0 \tag{11.8a}$$

$$I_S + I_{L1} + (G_1 + sC_1 + sC_2)V_1 - sC_2 V_2 = J_S$$

$$-I_{L1} + I_{L2} + (g_m - sC_2)V_1 + (G_2 + sC_2 + sC_3)V_2 = 0$$

or, in matrix form,

$$\begin{bmatrix} -R_S & 0 & 0 & 1 & 0 \\ 0 & -sL_1 & 0 & 1 & -1 \\ 0 & 0 & -sL_2 & 0 & 1 \\ 1 & 1 & 0 & G_1 + s(C_1 + C_2) & -sC_2 \\ 0 & -1 & 1 & g_m - sC_2 & G_2 + s(C_2 + C_3) \end{bmatrix} \begin{bmatrix} I_S \\ I_{L1} \\ I_{L2} \\ V_1 \\ V_2 \end{bmatrix} = \begin{bmatrix} E_S \\ 0 \\ 0 \\ J_S \\ 0 \end{bmatrix} \tag{11.8b}$$

In general, the n-node circuit with n_l inductances and n_e independent voltage sources will be described by the $(n + n_l + n_e) \times (n + n_l + n_e)$ modified admittance (hybrid) matrix. Using the modified admittance matrix description we can describe the circuit as

$$\mathbf{YV = J} \tag{11.9}$$

where

1. $\mathbf{Y}$ = the modified admittance (hybrid) matrix containing the real and imaginary admittances, impedances and dimensionless unity terms,
2. $\mathbf{J}$ = the excitation vector containing the real and imaginary parts of the independent voltage sources E_S, and the current sources J_S,
3. $\mathbf{V}$ = the response vector containing the real and imaginary parts of the node voltages V_n, inductance currents I_L, and voltage source currents I_S.

The real and imaginary parts of the **Y** matrix:

$$\mathbf{Y} = \mathbf{Y_r} + j\mathbf{Y_i} \tag{11.10}$$

and the real and imaginary parts of the **J** and **V** vectors:

$$\mathbf{J} = \mathbf{J_r} + j\mathbf{J_i} \tag{11.11}$$

$$\mathbf{V} = \mathbf{V_r} + j\mathbf{V_i} \tag{11.12}$$

will be stored in the computer memory separately as real part and imaginary part arrays. For example, the real and imaginary parts of **Y** and **J** for the circuit of Fig. 11.4, for $\omega = 2$ rad/s, are

$$\mathbf{Y} = \mathbf{Y_r} + j\mathbf{Y_i} = \begin{bmatrix} -45 & 0 & 1 & 0 \\ 0 & 0 & 1 & -1 \\ 1 & 1 & 0.125 & 0 \\ 0 & -1 & 5.3 & 0 \end{bmatrix} + j\begin{bmatrix} 0 & 0 & 0 & 0 \\ 0 & -2.2 & 0 & 0 \\ 0 & 0 & 0 & 0 \\ 0 & 0 & 0 & 4 \end{bmatrix}$$

$$\mathbf{J} = \mathbf{J_r} + j\mathbf{J_i} = \begin{bmatrix} 1.5 \\ 0 \\ 0 \\ 2.2 \end{bmatrix} + j\begin{bmatrix} 0 \\ 0 \\ 0 \\ 3.3 \end{bmatrix}$$

Computer generation of the modified admittance matrix is more complicated than the creation of the normal admittance matrix described in Chapter 3. Here, additional indices (variables) must be introduced to increase the dimension of the **Y** matrix by the number of inductances and independent voltage sources and for proper localization of the voltage source internal resistances and inductive impedances, and the localization of currents flowing through inductances and voltage sources.

The example program 'MOD_YMAT' for the generation of the modified admittance matrix is shown in Program 11.2.

See the listing of the program MOD__YMAT Directory of Chapter 11 on attached diskette.

Note that in the program 'MOD_YMAT', grounded elements must be entered in a specific way:

1. the first node entered must be an ungrounded node,
2. the second node entered must be a grounded (reference) node.

Additionally, the convention for entering the positive values of the source parameters is explained in Fig. 11.6(a, b, c).

To see the print-outs, run the program 'MOD_YMAT' for the two simple circuits shown in Fig. 11.7.

After running the program we have to enter

```
number of nodes: 2
number of elements: 3
number of inductances: 1
number of independent voltage sources: 0
```

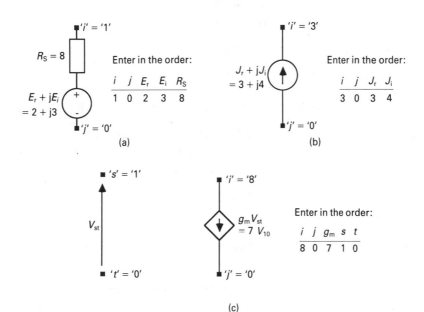

(a)

Enter in the order:

i	j	E_r	E_i	R_S
1	0	2	3	8

(b)

Enter in the order:

i	j	J_r	J_i
3	0	3	4

(c)

Enter in the order:

i	j	g_m	s	t
8	0	7	1	0

Figure 11.6 Conventions for entrance positive values of source parameters. (a) Independent voltage source. (b) Independent current source. (c) Voltage controlled current source

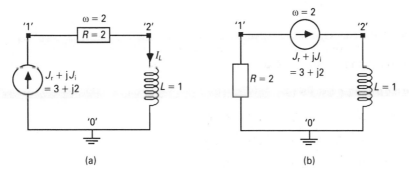

Figure 11.7 Simple circuits excited by (a) grounded current source, (b) ungrounded current source

Then, for the circuit of Fig. 11.7(a):

1. for current source (1): 1 0 3 2
2. for resistance (2):   1 2 2 (or 2 1 2, since it is ungrounded)
3. for inductance (3):   2 0 1

```
The data matrix is:

    1   0   1   3.0   2.0   0.0

    1   2   2   2.0   0.0   0.0

    2   0   3   1.0   0.0   0.0

Enter the value of OMEGA (si): 2
```

```
    The real part of the modified admittance matrix [Yr] is:

        [Yr]            **          [Jr](here, besides [Yr] the real
                                            part of [J] is also printed)

    0.0   0.0   1.0 **          0.0

    0.0   0.5  -0.5 **          3.0

    1.0  -0.5   0.5 **          0.0

    The imaginary part of the modified admittance matrix [Yi]
is:

        [Yi]                  **   [Ji] (here the imaginary part of
                                            [J] is also printed)

   -2.0   0.0   0.0 0.0 **      0.0

    0.0   0.0   0.0 0.0 **      2.0

    0.0   0.0   0.0 0.0 **      0.0
```

For the circuit of Fig. 11.7(b):

1. for resistance (1):   1 0 2
2. for current source (2):   2 1 3 2 (see the convention explained in Fig. 11.6)
3. for inductance (3):   2 0 1

```
The data matrix is:

    1   0   2   2.0   0.0   0.0

    2   1   1   3.0   2.0   0.0

    2   0   3   1.0   0.0   0.0


Enter the value of OMEGA (si): 2.
```

The real part of the modified admittance matrix [Y_r] *is:*

[Y_r]			**	[J_r] (here, besides [Y_r] the real part of [J] is also printed)
0.0	0.0	1.0	**	0.0
0.0	0.5	0.0	**	-3.0
1.0	0.0	0.0	**	3.0

The imaginary part of the modified admittance matrix [Y_i] *is:*

[Y_i]				**	[J_i] (here, the imaginary part of [J] is also printed)
-2.0	0.0	0.0	0.0	**	0.0
0.0	0.0	0.0	0.0	**	-2.0
0.0	0.0	0.0	0.0	**	2.0

Again, let us run the program for the circuit of Fig. 11.5(d), with the following parameter values:

$$E_S = 2 + j3 \text{ V} \quad J_S = 4 + j5 \text{ A} \quad R_s = 15 \, \Omega \quad R_1 = \frac{1}{G_1} = 4 \, \Omega \quad R_2 = \frac{1}{G_2} = 2 \, \Omega$$

$$C_1 = 1 \text{ F} \quad C_2 = 2 \text{ F} \quad C_3 = 3 \text{ F} \quad L_1 = 5 \text{ H} \quad L_2 = 7 \text{ H} \quad g_m = 0.6 \text{ S} \quad \omega = 2 \text{ rad/s}$$

and compare with equation (11.8b).

The modified admittance matrix generated by the program is:

		[Y_r]			**	[J_r]
-15.0	0.0	0.0	1.0	0.0	**	2.0
0.0	0.0	0.0	1.0	-1.0	**	0.0
0.0	0.0	0.0	0.0	1.0	**	0.0
1.0	1.0	0.0	0.25	0.0	**	4.0
0.0	-1.0	1.0	0.6	0.5	**	0.0

		[Y_i]			**	[J_i]
0.0	0.0	0.0	0.0	0.0	**	3.0
0.0	-10	0.0	0.0	0.0	**	0.0
0.0	0.0	-14	0.0	0.0	**	0.0
0.0	0.0	0.0	6.0	-4.0	**	5.0
0.0	0.0	0.0	-4.0	10.0	**	0.0

which agrees with the form of [$\mathbf{Y}$] of equation (11.8b).

Now, after generation of the **Y** matrix, we can solve the matrix equation (11.9) to find the currents flowing through the independent voltage sources and inductance, and the node voltages of the analyzed circuit, for a given frequency ω.

11.4 Solution of the matrix equation with modified admittance matrix **Y**

11.4.1 *LU* decomposition of the **Y** matrix and computation of the unknown vector **V**

Here we present the method for solution of the matrix equation

$$\mathbf{YV} = \mathbf{J}$$

which is based on the *LU* decomposition described in Section 4.6 of Chapter 4. The difference is only in the fact that the matrix equation discussed here is hybrid and the computed unknown vector **V** contains currents flowing through the voltage sources and inductances, and the node voltages of the circuit. The program 'LUCOMPSO' for the solution of the modified admittance matrix equation using the *LU* method is given in Program 11.3.

> **See the listing of the program LUCOMPSO Directory of Chapter 11 on attached diskette.**

Running this program for the simple circuit shown in Fig 11.7(a), we obtain the following results (for $\omega = 2$ rad/s):

$$I(1) = 3 + j2$$

$$V(1) = 2 + j10$$

$$V(2) = -4 + j6$$

This agrees with hand computation, since

$$I(1) = I_L = J = 3 + j2$$

$$V(1) = J(R + j\omega L) = (3 + j2)(2 + j2) = 2 + j10$$

$$V(2) = J(j\omega L) = (3 + j2)(j2) = -4 + j6$$

In a similar way much more complex circuits can be simulated or analyzed.

The output of the program is organized in such a way that all currents flowing through voltage sources and inductors are printed out first, and then the node voltages are printed out.

Note that these computations are performed for a single angular frequency ω. To obtain the results for different frequencies the **Y** matrix generation, *LU* decomposition, and forward–backward substitution procedures could be placed in a loop in which the frequency ω is changed by $\Delta\omega$. Such an approach is not effective for a large number of frequencies ω (computation). However, it will be used in Section 11.5 for the computation of repetitive circuit response for complex frequency $s_i = \sigma_i + j\omega_i$ leading to generation of the semisymbolic transfer function.

11.4.2 Computation of the transfer function (for single ω)

In practical cases of circuit analysis or simulation we use the single excitation source, with real values of E_S or J_S, respectively. Excitation of any N-node circuit by the independent current or voltage source, as shown in Fig. 11.8, allows us, using the results of the program 'LUCOMPSO', to determine all possible transfer functions (transmittances).

The voltage gains between any nodes V_k and V_i are computed directly as

$$T_v = \frac{V_k}{V_i} \tag{11.13}$$

and voltage transducer gains

$$T_{vt} = \frac{V_k}{E_S} \tag{11.14}$$

Similarly, the transducer transimpedance is computed directly as

$$T_{zt} = \frac{V_k}{J_S} \tag{11.15}$$

To find other transmittances it is necessary to compute the branch current of interest and then use the definition of the required current transmittance. To find the transducer current gain T_{it}:

$$T_{it} = \frac{I_L}{J_S} = \frac{\dfrac{V_N}{R_L}}{J_S} \tag{11.16}$$

it is necessary in addition to compute $I_L = V_N/R_L$ (see Fig. 11.8(b)).

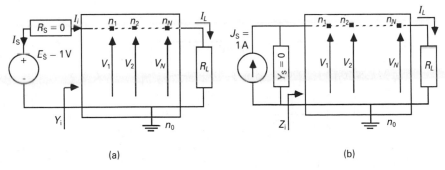

Figure 11.8 The N-node circuit excited by (a) voltage source E_S, (b) current source J_S

As an example let us compute the transmittances for the transistor amplifier of Fig. 11.9(a) having the simplified equivalent circuit shown in Fig. 11.9(c). Here we will use normalized values: J in milliamps, R in kiloohms, g_m in millisiemens, C in nanofarads; therefore voltages are in volts, time constant RC is in microseconds, and angular frequency ω is in megaradians per second.

First, we use the 'MOD_YMAT' program to obtain the real and imaginary matrices for three frequencies: $\omega = 0$, $\omega = 0.01$ Mrad/s and $\omega = 0.5$ Mrad/s (we have to enter ω in megaradians per second: $\omega = 0$, $\omega = 0.01$ and $\omega = 0.5$). The real part Y_r is (the same for all ω)

$$Y_r = \begin{bmatrix} 0.3125 & 0 & 0 \\ 20 & 0.4167 & 0 \\ 0 & 20 & 0.5 \end{bmatrix} \text{ (in millisiemens)}$$

The imaginary parts Y_i are

for $\omega = 0$: $Y_i = 0$

for $\omega = 0.01$ (that is 10 000 rad/s) $Y_i = \begin{bmatrix} 0.02 & 0 & 0 \\ 0 & 0.03 & 0 \\ 0 & 0 & 0 \end{bmatrix}$

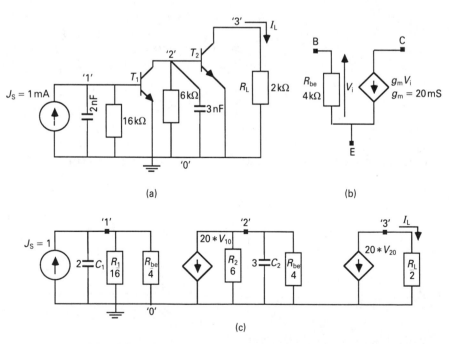

(a)

(b)

(c)

Figure 11.9   Transistor amplifier. (a) Schematic diagram. (b) Simplified model of a transistor. (c) Simplified equivalent circuit (J_S in mA, R in kΩ, g_m in mS, C in nF, V in V)

for $\omega = 0.5$ (that is 500 000 rad/s) $\mathbf{Y}_i = \begin{bmatrix} 1 & 0 & 0 \\ 0 & 1.5 & 0 \\ 0 & 0 & 0 \end{bmatrix}$

Now, using the program 'LUCOMPSO' we obtain the results shown in Table 11.3. Then we can easily find any transmittance of interest; for example,

$$T_{v32} = \frac{V_3}{V_2}, \qquad T_{v31} = \frac{V_3}{V_1}, \qquad T_{zt3} = \frac{V_3}{J_S}, \qquad T_{it3} = \frac{I_L}{J_S} = \frac{\dfrac{V_3}{R_L}}{J_S}$$

as shown in Table 11.4.

11.4.3 Computation of immittance (impedance or admittance)

The program 'LUCOMPSO' can be used directly for computing the input and output immittances (see Fig. 11.2), that is,

$$\text{impedances: } Z_i = \frac{V_i}{I_i} \text{ and } Z_o = \frac{V_o}{I_o} \tag{11.17}$$

or

$$\text{admittances: } Y_i = \frac{I_i}{V_i} = \frac{1}{Z_i} \text{ and } Y_o = \frac{I_o}{V_o} = \frac{1}{Z_o} \tag{11.18}$$

If we assume that the two-port network of Fig. 11.2 is excited by the ideal voltage or current sources (with zero source immittances, that is, $Z_S = 0$, or $Y_S = 0$) with unity values as shown in Fig. 11.8, then we can observe (see Fig. 11.8(b)) that

Table 11.3   Voltages for the circuit of Fig. 11.9

For $\omega = 0$	For $\omega = 0.01$	For $\omega = 0.5$
$V_1 = 3.2$	$V_1 = 3.18695 - j0.20396$	$V_1 = 0.28470 - j0.91103$
$V_2 = -153.6$	$V_2 = -151.47129 + j20.69458$	$V_2 = 10.29786 + j6.65671$
$V_3 = 6144$	$V_3 = 6058.85175 - j827.78306$	$V_3 = -411.91436 - j266.2685$

Table 11.4   Transmittances for the circuit of Fig. 11.9

	$\omega = 0$	$\omega = 10\,000$ rad/s	$\omega = 500\,000$ rad/s
T_{v32}	-40	$-40 + j0$	$-40 + j0$
T_{v31}	1920	$1910.9446 - j137.5079$	$137.5439 - j495.1243$
T_{zt3}	6.144×10^6	$(6058.85175 - j827.783)10^3$	$-(411.91436 + j266.2685)10^3$
T_{it3}	3.072×10^3	$3029.4258 - j413.8915$	$-(205.95718 + j133.1342)$

the input impedance Z_i is numerically equal to the input voltage V_1:

$$V_1 = J_S Z_i = 1 \times Z_i = Z_i \tag{11.19}$$

Similarly, based on the circuit of Fig. 11.8(a), we can state that the input admittance Y_i is numerically equal to the input current $I_i = -I_S$ (in the program 'LUCOMPSO' the direction of I_S is opposite to the direction of the input current I_i):

$$I_i = -I_S = E_S Y_i = 1 \times Y_i = Y_i \tag{11.20}$$

To compute the output immittances, that is, impedance or admittance, it is necessary to modify the circuit by removing the load resistance R_L and to apply the independent source, voltage or current to the output port, as shown in Fig. 11.10.
    Based on Fig. 11.10(a), we can write

$$V_o = J_S Z_o = 1 \times Z_o = Z_o \tag{11.21}$$

that is, the output impedance is numerically equal to the voltage V_o and, based on Fig. 11.10(b), we can write

$$-I_S = I_o = E_S Y_o = 1 \times Y_o = Y_o \tag{11.22}$$

that is, the output admittance is numerically equal to the current $-I_S = I_o$.
    For example, the input impedance Z_i for the circuit of Fig. 11.9(c) corresponds to the value of V_1 at a given frequency; that is,

1. for $\omega = 0$, $Z_i = 3.2$ kiloohms,
2. for $\omega = 10\,000$ rad/s, $Z_i = 3.18695 - j0.20396$ kiloohms,
3. for $\omega = 500\,000$ rad/s, $Z_i = 0.28470 - j0.91103$ kiloohms.

However, to compute the output impedance Z_o for the same circuit, it is necessary to find the value of V_3, for excitation $J_S = 1$ applied to node '3'; that is, the excitation vector $\mathbf{J_r}$ entered into the program 'LUCOMPSO' should be

$$J(1) = 0, \quad J(2) = 0, \quad J(3) = 1$$

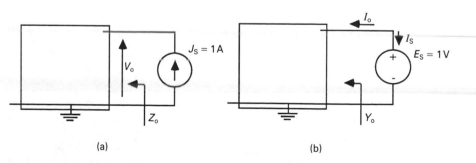

(a)   (b)

Figure 11.10 Circuits explaining the computation of (a) output impedance Z_o, (b) output admittance Y_o

Running the program 'LUCOMPSO' for this case we obtain the following results:

$$V(1) = 0$$

$$V(2) = 0$$

$$V(3) = 2$$

According to relation (11.21), the output impedance Z_o is

$$Z_o = V(3) = 2 \text{ k}\Omega$$

which can be proved by inspection of Fig. 11.9.

As another example, let us compute an input admittance Y_i for the RLC circuit of Fig. 11.11, for which the resonance frequency $\omega_r = 1/\sqrt{6} = 0.408$ rad/s. Using the program 'MOD_YMAT' we find the modified admittance matrix for the circuit of Fig. 11.11(a):

1. for $\omega = 0.2$ rad/s, and $\omega = 2$ rad/s: $\mathbf{Y}_r = \begin{bmatrix} -0.0001 & 0 & 1 & 0 \\ 0 & 0 & 0 & 1 \\ 1 & 0 & 1.5 & -0.5 \\ 0 & 1 & -0.5 & 0.5 \end{bmatrix}$

2. for $\omega = 0.2$ rad/s: $\mathbf{Y}_i = \begin{bmatrix} 0 & 0 & 0 & 0 \\ 0 & -0.6 & 0 & 0 \\ 0 & 0 & 0.4 & -0.4 \\ 0 & 0 & -0.4 & 0.4 \end{bmatrix}$

3. for $\omega = 2$ rad/s: $\mathbf{Y}_i = \begin{bmatrix} 0 & 0 & 0 & 0 \\ 0 & -6 & 0 & 0 \\ 0 & 0 & 4 & -4 \\ 0 & 0 & -4 & 4 \end{bmatrix}$

Running the program 'LUCOMPSO' we obtain the following $I_S = -I_i$ values:

1. for $\omega = 0.2$ rad/s: $I_S = -I(1) = -1.74865 - \text{j}0.2306$, therefore, the input admittance is capacitive:

$$Y_i = 1.74865 + \text{j}0.2306 \ S$$

since $\omega = 0.2$ rad/s < 0.408 rad/s $= \omega_r$,

2. for $\omega = 2$ rad/s: $I_S = -I(1) = -1.00083 + \text{j}0.17376$, therefore, the input admittance is inductive:

$$Y_i = 1.00083 - \text{j}0.17376 \ S$$

since $\omega = 2$ rad/s > 0.408 rad/s $= \omega_r$.

To compute the output admittances Y_o for $\omega = 0.2$ rad/s and $\omega = 2$ rad/s, we have to apply the voltage source E_S between nodes 2 and 0, as shown in Fig. 11.11(b). For this case, we obtain the print-outs of the real part of the modified admittance

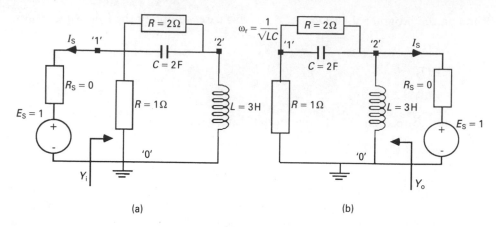

Figure 11.11   RLC circuit with resonance frequency $\omega_r = \dfrac{1}{\sqrt{6}} = 0.408$ rad/s. (a) For Y_i computation. (b) For Y_o computation

matrix (enter E_S as the first element):

[Y_r]				**	[J_r]
0.0001	0.0	0.0	1.0	**	1.0
0.0	0.0	0.0	1.0	**	0.0
0.0	0.0	1.5	-0.5	**	0.0
1.0	1.0	-0.5	0.5	**	0.0

Note the difference of the $\mathbf{Y}_r$ obtained for E_S applied between nodes 1 and 0. The excitation vector in this case is the same as before, that is $J(1) = 1$ and the remaining terms of the vector are zeros. Running the program 'LUCOMPSO' we obtain the following $I_S = -I_o$ values:

1. for $\omega = 0.2$ rad/s: $I_S = -I(1) = -0.3778 + j1.50058$ therefore, the output admittance Y_o is inductive ($\omega < \omega_r$):

    $$Y_o = 0.3778 - j1.50058$$

2. for $\omega = 2$ rad/s: $I_S = -I(1) = -0.91772 - j0.05250$ therefore, the output admittance Y_o is capacitive ($\omega > \omega_r$):

    $$Y_o = 0.91772 + j0.05250$$

11.5 Determination of transmittance $T_{zt} = V_o/J_S$, $T_{vt} = V_o/E_S$, $T_v = V_o/V_i$ and input immittances in semisymbolic form

To find the numerical values of the semisymbolic transfer function $T(s)$ it is necessary to create the modified admittance matrix $\mathbf{Y}$ for the selected set of complex variables s_k:

$$s_k = \sigma_k + j\omega_k, \qquad k = 0, 1, \ldots, N_0 \tag{11.23}$$

where N_0 is the order of the circuit (see Chapter 8), and to solve a matrix equation:

$$\mathbf{Y}(s_k)\mathbf{V} = \mathbf{J} \tag{11.24}$$

for unity current $J_S = 1$ (for $T_{zt} = V_o/J_S$) or unity voltage $E_S = 1$ (for $T_{vt} = V_o/E_S$) excitation source applied to the input port of the circuit. Therefore, the programs described earlier in this chapter form the basis for the determination of transmittances in semisymbolic form.

11.5.1 Determination of semisymbolic $T_{zt} = V_o/J_S$ transmittance

Solution of equation (11.24) for unity excitation and for selected complex frequency $s_k = \sigma_k + j\omega_k$, and for selected output voltage V_o yields

$$T(s_k) = \frac{V_o(s_k)}{J_S} = \frac{V_o(s_k)}{1} = \frac{N(s_k)}{D(s_k)} \tag{11.25}$$

Since the modified admittance matrix has no $1/s$ terms, the denominator $D(s_k)$ of the transmittance (11.25) is equal to the determinant of the matrix $[\mathbf{Y}(s_k)]$, $\det\{[\mathbf{Y}(s_k)]\}$.

Since, the solution of equation (11.24) is performed using the LU decomposition, as described in Section 11.4, we can easily find the value of $\det\{[\mathbf{Y}(s_k)]\}$ as a product of the main diagonal terms of the lower triangular matrix $[\mathbf{L}(s_k)]$:

$$\det\{\mathbf{Y}(s_k)\} = D(s_k) = \det\{\mathbf{L}(s_k)\} = \prod_{k=1}^{nt} l_{kk} \tag{11.26}$$

where nt is the dimension of the modified $\mathbf{Y}$ matrix, that is, the number of nodes plus the number of inductances in the analyzed circuit.

After solution of equation (11.25) using forward–backward substitution, we find the value of the prescribed output voltage $V_o(s_k)$:

$$T(s_k) = \frac{V_o(s_k)}{J_S} = \frac{N(s_k)}{D(s_k)} \tag{11.27}$$

and from previous knowledge of $D(s_k) = \det\{[\mathbf{Y}(s_k)]\}$ we compute the value of the numerator $N(s_k)$ of the transfer function $T(s_k)$:

$$N(s_k) = V_o(s_k)D(s_k) \tag{11.28}$$

Repeating N_0 (N_0 is the order of the circuit, that is the degree of the denominator $D(s)$ polynomial of $T(s)$) computations of $D(s_k) = \det\{[\mathbf{Y}(s_k)]\}$, $V_o(s_k)$ and $N(s_k)$ for different s_k values we can find the set of pairs

$$\{s_k, D(s_k)\} \text{ and } \{s_k, N(s_k)\} \tag{11.29}$$

and, based on these, we can find the values of the coefficients of the denominator $D(s)$ and $N(s)$ using the polynomial interpolation method.

Although the selection of s_k values is almost arbitrary, the best results in the computation of the polynomial $D(s)$ and $N(s)$ coefficients are obtained when the s_k values correspond to equally spaced points located at the unity circle on the complex plane s, as shown in Fig. 11.12.

For equally spaced points on unit circles, the values of s_k are given by the relation

$$s_k = \exp\left\{ j\, \frac{2\pi k}{1+N_0} \right\} = \cos\left\{ \frac{2\pi k}{1+N_0} \right\} + j\, \sin\left\{ \frac{2\pi k}{1+N_0} \right\} \tag{11.30}$$

where $k = 0, 1, 2, \ldots, N_0$; $j = \sqrt{-1}$.

For example, for $N_0 = 1$, we obtain

$$s_0 = \cos(0) + j\, \sin(0) \qquad\quad = 1 + j0 = 1$$

$$s_1 = \cos(\pi) + j\, \sin(\pi) \qquad\quad = -1 + j0 = -1$$

and for $N_0 = 2$, we obtain

$$s_0 = \cos(0) + j\, \sin(0) \qquad\quad = 1 + j0 = 1$$

$$s_1 = \cos\left(\frac{2}{3}\pi\right) + j\, \sin\left(\frac{2}{3}\pi\right) = -0.5 + j0.866$$

$$s_2 = \cos\left(\frac{4}{3}\pi\right) + j\, \sin\left(\frac{4}{3}\pi\right) = -0.5 - j0.866$$

etc., as shown in Fig. 11.12.

The interpolation procedure leading to the determination of the polynomial coefficients of the denominator $D(s)$ can be explained as follows. The relation for

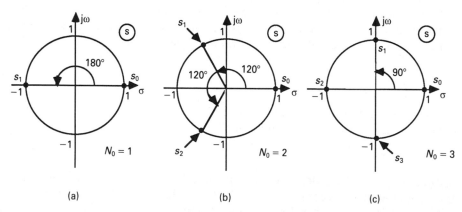

Figure 11.12 Locations of selected s_k points at unity circle on the complex s plane. (a) For $N_0 = 1$; (b) For $N_0 = 2$. (c) For $N_0 = 3$

the polynomial $D(s)$ written as

$$D(s) = d_0 s_0 + d_1 s_1 + \cdots + d_{N_o} s_{N_o} = \sum_{i=0}^{N_o} d_i s_i \tag{11.31a}$$

must be satisfied for the set of doublets $(s_k, D(s_k))$, that is,

$$d_0 s_k^0 + d_1 s_k^1 + \cdots + d_i s_k^i + \cdots + d_{N_o} s_k^{N_o} = D(s_k) \tag{11.31b}$$

for $k = 0, 1, ..., N_0$.

Therefore, we obtain the set of $N_0 + 1$ equations, which can be written in the matrix form

$$
\begin{bmatrix}
s_0^0 & s_0^1 & . & s_0^i & . & s_0^{N_o} \\
. & . & . & . & . & . \\
s_k^0 & s_k^1 & . & s_k^i & . & s_k^{N_o} \\
. & . & . & . & . & . \\
s_{N_o}^0 & s_{N_o}^1 & . & s_{N_o}^i & . & s_{N_o}^{N_o}
\end{bmatrix}
\begin{bmatrix}
d_0 \\
. \\
d_i \\
. \\
d_{N_o}
\end{bmatrix}
=
\begin{bmatrix}
D(s_0) \\
. \\
D(s_k) \\
. \\
D(s_{N_o})
\end{bmatrix}
\tag{11.32}
$$

Solution of the matrix equation (11.32) gives us the set of coefficients $d_i, i = 0, 1, 2, ..., N_0$. For example, for the first-degree polynomial, that is, $N_0 = 1$, we obtain the following set of equations:

$$d_0 + s_0 d_1 = D(s_0) \tag{11.33a}$$

$$d_0 + s_1 d_1 = D(s_1) \tag{11.33b}$$

Insertion of d_1 from (11.33b) into (11.33a) gives

$$d_0 = \frac{D(s_0)}{1 - \dfrac{s_0}{s_1}} - \frac{s_0}{s_1} \frac{D(s_1)}{1 - \dfrac{s_0}{s_1}} \tag{11.34a}$$

Since, for $N_0 = 1$ we have $s_0 = 1$, $s_1 = -1$; then

$$d_0 = \frac{D(s_0)}{2} + \frac{D(s_1)}{2} = \frac{1}{2}\{D(s_0) + D(s_1)\} \tag{11.34b}$$

Insertion of (11.34b) into (11.33) gives (with $s_0 = 1$, $s_1 = -1$)

$$d_1 = \frac{D(s_0)}{2} - \frac{D(s_1)}{2} = \frac{1}{2}\{D(s_0) - D(s_1)\} \tag{11.34c}$$

Generally, the values of the d_i coefficients, with s_k given by equation (11.30), are given by the relation

$$d_i = \frac{1}{N_0 + 1} \sum_{k=0}^{N_o} D(s_k) \exp\left\{ -j \frac{2\pi ki}{N_0 + 1} \right\}, \; i = 0, 1, 2, ..., N_0 \tag{11.35}$$

Knowing the values of the d_i coefficients, we can find the value of the polynomial $D(s_k)$, for a given s_k at the unit circle from the relation:

$$D(s_k) = \sum_{i=0}^{N_o} d_i \exp\left\{ j \frac{2\pi ki}{N_0 + 1} \right\}, \; k = 0, 1, 2, ..., N_0 \tag{11.36}$$

Relations (11.35) and (11.36) are referred to as the discrete Fourier transform (DFT). The coefficients n_i of the numerator $N(s)$ of the transfer function $T(s)$ are computed in exactly the same way.

To show all the steps of computation in detail let us consider a simple first-order RC circuit of Fig. 11.13, described by the matrix equation

$$\begin{bmatrix} 1 + 4s & -4s \\ -4s & 1 + 4s \end{bmatrix} \begin{bmatrix} V_1 \\ V_2 \end{bmatrix} = \begin{bmatrix} 1 \\ 0 \end{bmatrix}$$

The determinant of the $\mathbf{Y}$ matrix is

$$\det(\mathbf{Y}) = 1 + 8s$$

Let us find the semisymbolic form of the transfer function $T_{zt}(s)$:

$$T_{zt}(s) = \frac{V_2(s)}{J_S} = \frac{V_2(s)}{1} = V_2(s)$$

Since the circuit is of the first order, the number $N_0 = 1$. Thus, according to relation (11.30) we select $s_0 = 1$, and $s_1 = -1$.

The relation for $V_2(s)$ is (obtained by using the determinant method)

$$V_2(s) = \frac{1 \times 4s}{\det(\mathbf{Y})} = \frac{4s}{1 + 8s}$$

Then for $s_0 = 1$, and $s_1 = -1$, we obtain $V_2(s)$ and $D(s) = \det(\mathbf{Y})$:

$$V_2(s_0) = \frac{4}{9}, \quad V_2(s_1) = \frac{-4}{-7} = \frac{4}{7}$$

$$D(s_0) = 9, \quad D(s_1) = -7$$

Now we can compute $N(s_0)$ and $N(s_1)$ using formula (11.28):

$$N(s_0) = V_2(s_0)D(s_0) = \frac{4}{9} \times 9 = 4$$

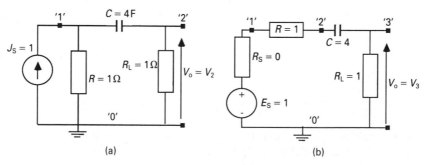

Figure 11.13 Simple first-order RC circuit. (a) Excited by the ideal current source J_S. (b) Excited by the ideal voltage source E_S

$$N(s_1) = V_2(s_1)D(s_1) = \frac{4}{7}(-7) = -4$$

Now we can compute the coefficients d_i and n_i using the general expression (11.35), or the simple expressions (11.34b and c) derived for $N_0 = 1$:

$$d_0 = \frac{1}{2}\{D(s_0) + D(s_1)\} = \frac{1}{2}(9 - 7) = 1$$

$$n_0 = \frac{1}{2}\{N(s_0) + N(s_1)\} = \frac{1}{2}(4 - 4) = 0$$

$$d_1 = \frac{1}{2}\{D(s_0) - D(s_1)\} = \frac{1}{2}(9 + 7) = 8$$

$$n_1 = \frac{1}{2}\{N(s_0) - N(s_1)\} = \frac{1}{2}(4 + 4) = 4$$

Thus, the transfer function $T_{zt}(s) = V_2(s)/J_s$ is $T_{zt}(s) = 4s/(8s + 1)$.

The computer program 'SEMISYMB' for the generation of the semisymbolic transfer function is shown in Program 11.4.

See the listing of the program SEMISYMB Directory of Chapter 11 on attached diskette.

This program is somewhat lengthy, but it contains all the steps to obtain the semi-symbolic transfer function, starting from entering the circuit elements. To see the results of the program let us run it for the data of the circuit of Fig. 11.13(a). We also have to enter the numbers of the output nodes – in this case out1 = 2, out2 = 0. The results are as follows.

The transmittance $T(s) = N(s)/D(s)$ is

```
N(s) = + 4.000*s^1 + 0.000*s^0
D(s) = + 8.000*s^1 + 1.000*s^0
```

This is equivalent to

$$T(s) = \frac{N(s)}{D(s)} = \frac{4s}{8s + 1}$$

11.5.2 Determination of semisymbolic $T_{vt} = V_o/E_s$ transmittance

The same program 'SEMISYMB' can be used for the determination of $T_{vt} = V_o/E_s$ transmittance in semisymbolic form. Let us prove this for the circuit of Fig. 11.13(b), which is equivalent to the circuit of Fig. 11.13(a). In this case, the result should be the same as that obtained in Section 11.5.1 (with V_{out} – voltage on R_L), that is,

$$T_{vt} = \frac{V(3)}{E_S} = T_{it}$$

Since the circuit has three independent nodes, we have to enter out1 = 3 and out2 = 0 as the output nodes. The matrix equation for this circuit is

$$
\begin{bmatrix}
R_S & 1 & 0 & 0 \\
1 & \dfrac{1}{R} & -\dfrac{1}{R} & 0 \\
0 & -\dfrac{1}{R} & \dfrac{1}{R} + sC & -sC \\
0 & 0 & -sC & \dfrac{1}{R_L} + sC
\end{bmatrix}
\begin{bmatrix}
I_S \\ V_1 \\ V_2 \\ V_3
\end{bmatrix}
=
\begin{bmatrix}
E_S \\ 0 \\ 0 \\ 0
\end{bmatrix}
$$

After running the program 'SEMISYMB' we obtain the following results.
    The transmittance $T(s) = N(s)/D(s)$ is

```
N(s) = + -4.000*s^1 + -0.000*s^0
D(s) = + -8.000*s^1 + -1.000*s^0
```

This is equivalent to the result obtained in Section 11.5.1:

$$T(s) = \frac{N(s)}{D(s)} = \frac{-4s}{-8s - 1} = \frac{4s}{8s + 1}$$

Note that for stable circuits the coefficients of the denominator $D(s)$ of the transmittance must be real and positive.

11.5.3 Determination of voltage transmittance $T_v = V_o/V_i$ in semisymbolic form

Determination of the symbolic form of the voltage transmittance $T_v = V_o/V_i$ between any node pairs of the circuit can also be performed using the 'SEMISYMB' program. In this case it is necessary to apply the excitation voltage source E_S with zero source resistance $R_S = 0$ to the node pair selected as input nodes. Consider, for example, the fourth-order RLC circuit shown in Fig. 11.14(a).
    The voltage gain $T_v = V_o/V_i = V_4/V_1$ of this circuit equals the transducer voltage gain $T_{vt} = (V_4/E_S)|_{R_S = 0}$ with $R_S = 0$, and can easily be computed manually, since the circuit is composed of two unilaterally coupled sections:

$$T_v = \frac{V_4}{V_2} \frac{V_2}{V_1} = \frac{1}{s^2 + s + 1} \frac{-2}{0.25s^2 + s + 2}$$

or

$$T_v = \frac{V_4}{V_1} = \frac{-2}{0.25s^4 + 1.25s^3 + 3.25s^2 + 3s + 2}$$

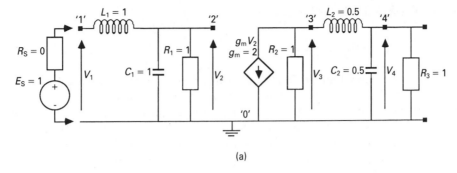

(a)

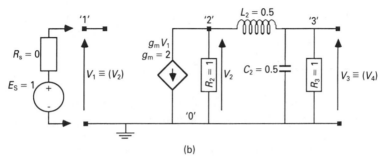

(b)

Figure 11.14   RLC circuit for determination of voltage gain $T_v = \dfrac{V_o}{V_i}$:

(a) $T_v = \dfrac{V_4}{V_1} = \dfrac{V_4}{E_S}\bigg|_{R_S = 0}$.   (b) $T_v = \dfrac{V_3}{V_1} = \dfrac{(V_4)}{(V_2)} = \dfrac{V_3}{E_S}\bigg|_{R_S = 0}$

Running the program 'SEMISYMB' with data corresponding to the circuit of Fig. 11.14(a), that is, number of nodes = 4, number of elements = 9 and output nodes out1 = 4, out2 = 0, we obtain the following results:

```
N(s) = + 2.0*s^0
D(s) = -0.25*s^4 -1.25*s^3 -3.25*s^2 -3*s^1 -2*s^0
```

which corresponds to

$$T_v = \frac{V_4}{V_1} = \frac{N(s)}{D(s)} = \frac{-2}{0.25s^4 + 1.25s^3 + 3.25s^2 + 3s + 2}$$

since all the coefficients of $D(s)$ must be positive for stable circuits.

Now, if we wish to determine the voltage transmittance $T_v = V_3/V_1$ of the circuit of Fig. 11.14(a), we have to select the output nodes out1 = 3, out2 = 0, and enter all the elements of the circuit into the program:

```
N(s) = + 0.5*s^2 + 1.0*s^1 + 2.0*s^0
D(s) = -0.25*s^4 -1.25*s^3 -3.25*s^2 -3.0*s^1 -2.0*s^0
```

which corresponds to

$$T_v = \frac{V_3}{V_1} = \frac{-(0.5s^2 + s + 2)}{0.25s^4 + 1.25s^3 + 3.25s^2 + 3s + 2}$$

Of course, the denominator $D(s)$ is the same as in the previous case. If we wish, however, to determine the voltage transmittance $T_v = V_4/V_2$ of the circuit of Fig. 11.14(a), we apply the input source E_S between the nodes '2' and '0'. However, all the elements of the circuit connected in parallel with E_S have no influence on the transfer function, since $R_S = 0$. Therefore, the circuit (with E_S connected to nodes '2–0') is simplified, as shown in Fig. 11.14(b). In this case the circuit is reduced to three nodes and six elements. To run the program we have to enter three nodes, six elements and the new notation for the nodes (out1 = 3, out2 = 0). The results in this case are

$$T_v = \frac{V_3}{V_1} = \frac{(V_4)}{(V_2)} = \frac{N(s)}{D(s)} = \frac{-2}{0.25s^2 + s + 2}$$

To find the transfer function $T_v = V_4/V_3$, we have to apply the voltage source E_S between nodes '3' and '0'. In this case the circuit is reduced to a two-node circuit with four elements.

   Changing the enumeration of nodes '3' → '1' and '4' → '2', and selecting the output nodes out1 = 2, out2 = 0, we obtain the results of the program:

$$T_v = \frac{V_2}{V_1} = \frac{(V_4)}{(V_3)} = \frac{1}{0.25s^2 + 0.5s + 1}$$

11.5.4 Determination of input (output) impedances in semisymbolic form

Again, the 'SEMISYMB' program can be used for the determination of the input and output, or, generally speaking, the driving-point, impedances. As was discussed in Section 11.4.3 the driving-point impedance Z_i can be obtained directly by application of the excitation current source J_S between selected nodes 'i' and '0' and selection of the same nodes as output nodes. For example, for the circuit of Fig. 11.14(a), the input impedance Z_{10} is obtained by the introduction of the current source $J_S = 1$ between nodes '1' and '0' and selection of the output nodes as out1 = 1, out2 = 0. Running the program 'SEMISYMB' we obtain the following results (in this case it is sufficient to enter only the input section of the circuit with nodes '1' and '2' because the rest of the circuit is isolated by the controlled sources):

```
N(s) = - 1.0*s^2 - 1.0*s^1 - 1.0*s^0
D(s) = 0.0*s^2 - 1.0*s^1 - 1.0*s^0
```

which corresponds to

$$Z_{in} = \frac{V_1}{J_S} = \frac{1 \times s^2 + 1 \times s + 1}{1 \times s + 1} = \frac{s^2 + s + 1}{s + 1}$$

since all the coefficients of the numerator and denominator of the stable impedance must be positive. Running the program 'SEMISYMB' again for the circuit of Fig. 11.13(a), we obtain the input impedance

$$Z_{in} = \frac{V_1}{J_S}\bigg|_{J_S = 1} = \frac{N(s)}{D(s)}$$

with program results

```
N(s) = 4.0*s^1 + 1.0*s^0
D(s) = 8.0*s^1 + 1.0*s^0
```

what corresponds to

$$Z_{in} = \frac{V_1}{J_S} = \frac{4 \times s + 1}{8 \times s + 1}$$

To find the output impedance of the circuit of Fig. 11.14(a), with the input short-circuited, that is, $V_1 = V_2 = 0$, and $g_m V_2 = 0$, we have to enter into the program the part of the circuit with nodes '3' and '4', change the node enumeration '3' → '1', '4' → '2' and apply the current source $J_S = 1$ between nodes '2' and '0', as shown in Fig. 11.15.

Running the program 'SEMISYMB' for the data of the circuit of Fig. 11.15, we obtain the following results:

```
N(s) = 0.0*s^2 - 0.5*s^1 - 1.0*s^0
D(s) = -0.25*s^2 - 1.0*s^1 - 2.0*s^0
```

which corresponds to

$$Z_{out} = \frac{0.5s + 1}{0.25s^2 + s + 1}$$

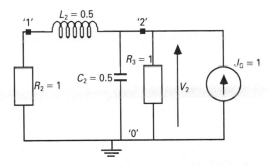

Figure 11.15 Output part of the circuit of Fig. 11.14(a), for Z_{out} computation (with shorted input)

11.6 Computation of frequency characteristics

As was mentioned in Section 11.1, the frequency characteristics of a transfer function $T(s)$, with $s = j\omega$, are presented most often as the Bode plots, that is, in rectangular coordinates as magnitude:

$$\text{Mag}\{T(\omega)\} = |T(\omega)| \tag{11.37a}$$

or in decibels:

$$|T(\omega)|_{\,\text{dB}} = 20 \, \log_{10}[\,|T(\omega)|\,] \tag{11.37b}$$

and phase:

$$\text{phase}\{T(\omega)\} = \arg\{T(\omega)\} \tag{11.37c}$$

with linear or geometric (logarithmic) frequency scale. The frequency characteristics of an immittance, that is, impedance $Z(j\omega)$ or admittance $Y(j\omega)$, and also an 'open-loop gain' of feedback amplifiers, are frequently represented as Nyquist plots, that is, in polar coordinates, with frequency ω as the parameter:

$$Z(j\omega) = \text{Re}\{Z(\omega)\} + j\text{Im}\{Z(\omega)\} \tag{11.38a}$$

$$\text{Re}\{Z(\omega)\} = |Z(\omega)| \cos\{\arg\{Z(\omega)\}\} \tag{11.38b}$$

$$\text{Im}\{Z(\omega)\} = |Z(\omega)| \sin\{\arg\{Z(\omega)\}\} \tag{11.38c}$$

Since the transmittance $T(j\omega)$ and impedance $Z(j\omega)$ (admittance) of the circuits discussed are rational functions, that is, the ratios of two polynomials, $T(j\omega) = N(j\omega)/D(j\omega)$, thus to find the frequency characteristics of $T(j\omega)$, or $Z(j\omega)$ it is necessary to first compute the values of the real and imaginary parts of $N(j\omega_k)$ and $D(j\omega_k)$ for the single frequency ω_k:

$$T(j\omega_k) = \frac{N(j\omega_k)}{D(j\omega_k)} = \frac{\displaystyle\sum_{i=0}^{dn} c_i \{j\omega\}^i}{\displaystyle\sum_{i=0}^{dd} d_i \{j\omega\}^i} = \frac{nr(\omega_k) + jni(\omega_k)}{dr(\omega_k) + j\,di(\omega_k)} \tag{11.39}$$

where dn and dd are the degrees of the numerator and denominator, $nr(\omega_k)$, $dr(\omega_k)$ are the real parts of $N(j\omega_k)$ and $D(j\omega_k)$ and $ni(\omega_k)$, $di(\omega_k)$ are the imaginary parts of $N(j\omega_k)$ and $D(j\omega_k)$ at a given angular frequency ω_k.

Then, it is necessary to compute the magnitude

$$|T(\omega_k)| = \frac{\sqrt{\{nr(\omega_k)\}^2 + \{ni(\omega_k)\}^2}}{\sqrt{\{dr(\omega_k)\}^2 + \{di(\omega_k)\}^2}} \tag{11.40}$$

and phase shift

$$\text{phase}\{T(\omega_k)\} = \text{atan}\left(\frac{ni(\omega_k)}{nr(\omega_k)}\right) - \text{atan}\left(\frac{di(\omega_k)}{dr(\omega_k)}\right) \tag{11.41}$$

or the real and imaginary parts of $T(j\omega)$ or $Z(j\omega)$ according to relations (11.38),

if necessary, and to repeat the computations for other frequencies in the frequency range of interest.

Thus, the main problem is in the computation of the polynomial values at a given frequency. It is similar to the problem described in Section 10.3.5, but here the computation techniques differ, since we are only interested in the polynomial values for $s = j\omega$, that is, for the real frequencies ω. Any polynomial $P(j\omega)$ of degree dp can be written as

$$P(j\omega) = \sum_{i=0}^{dp} p_i \{j\omega\}^i = p_0\{j\omega\}^0 + p_1\{j\omega\}^1 + p_2\{j\omega\}^2 + p_3\{j\omega\}^3 + \cdots + p_{dp}\{j\omega\}^{dp}$$

$$(11.42)$$

or

$$P(j\omega) = pr(\omega) + jpi(\omega)$$

with

$$pr(\omega) = p_0 - p_2\{\omega\}^2 + p_4\{\omega\}^4 - \ldots \qquad (11.43a)$$

or

$$pr(\omega) = p_0 + p_2\{\omega\}^2(-1)^{2/2} + p_4\{\omega\}^4(-1)^{4/2} + \cdots + p_{dp}\{\omega\}^{dp}(-1)^{dp/2}$$

$$(11.43b)$$

and:

$$pi(\omega) = p_1\omega^1 - p_3\omega^3 + p_5\omega^5 - \ldots \qquad (11.43c)$$

or

$$pi(\omega) = p_1\omega(-1)^{(1-1)/2} + p_3\omega^3(-1)^{(3-1)/2} + \cdots + p_{dp-1}\omega^{dp-1}(-1)^{(dp-1-1)/2}$$

$$(11.43d)$$

for dp even, and with:

$$pr(\omega) = p_0 - p_2\omega^2 + p_4\omega^4 - \ldots \qquad (11.44a)$$

or

$$pr(\omega) = p_0 + p_2\omega^2(-1)^{2/2} + p_4\omega^4(-1)^{4/2} + \cdots + p_{dp-1}\omega^{dp-1}(-1)^{(dp-1)/2}$$

$$(11.44b)$$

and

$$pi(\omega) = p_1\{\omega\}^1 - p_3\{\omega\}^3 + p_5\{\omega\}^5 - \ldots \qquad (11.44c)$$

or

$$pi(\omega) = p_1\omega(-1)^{(1-1)/2} + p_3\omega^3(-1)^{(3-1)/2} + p_5\omega^5(-1)^{(5-1)/2}$$
$$+ \cdots + p_{dp}\omega^{dp}(-1)^{(dp-1)/2}$$

$$(11.44d)$$

for dp odd.

Therefore the computations of the real and imaginary parts of the polynomial can be performed in two separate loops with the upper limits $dp/2$ and $dp/2 - 1$, respectively, and with steps equal to two. Inside the loops it is necessary to compute the ω to power i, ω^i, and $(-1)^i$.

When the frequency scale is linear, we have to select the frequency range limits f_{min} and f_{max}, and the frequency increment df; then the frequency is changed according to the following relations:

$$f := f_{min}, \quad f := f + df, \quad until \quad f:>= f_{max}$$

When the frequency scale is logarithmic (geometric increase), then the frequency is changed according to the following relations:

$$f := f_{min}, \quad f := f * df, \quad until \quad f:>= f_{max}$$

with $f_{min} > 0$.

For the logarithmic scale the increment df is usually chosen as $df = 2$ − increase in octaves, or $df = 10$ − increase in decades.

The octaves and decades can be divided into equal parts according to the relation (mth part of octave, or decade) $2^{1/m}$ for octaves and $10^{1/m}$ for decades.

An example program 'FREQCHAR' for the computation of the frequency characteristics is shown in Program 11.5.

See the listing of the program FREQCHAR Directory of Chapter 11 on attached diskette.

To make this program as simple as possible it is assumed that the coefficients of the numerator and denominator of $T(s)$ or $Z(s)$ are known, or are found using the program 'SEMISYMB'. The magnitude $|T(j\omega)|_{dB}$, the phase $\arg\{T(j\omega)\}$, the real part $Re\{T(j\omega)\}$ and the imaginary part $Im\{T(j\omega)\}$ are computed for the linear frequency scale. These results can be used for Bode or Nyquist plot drawing.

Let us run the program 'FREQCHAR' for two transfer functions of the circuit composed of two identical RC sections separated by a unity voltage amplifier shown

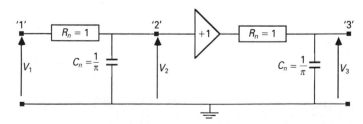

Figure 11.16 Circuit composed of two identical RC sections separated by unity gain voltage amplifier

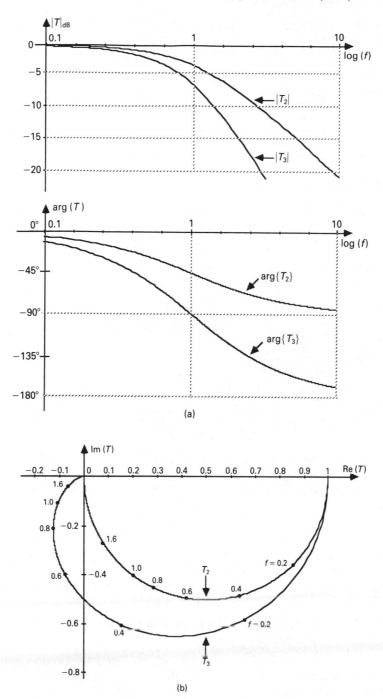

Figure 11.17 Frequency characteristics of transmittances for the circuit of Fig. 11.16: (a) Bode plot of magnitude and phase. (b) Nyquist plot

Table 11.5   Results for frequency range $f_{min} = 0$, $f_{max} = 2$, and frequency increment $df = 0.2$

f	$\lvert T_2 \rvert_{dB}$	$\arg\{T_2\}$	$Re\{T_2\}$	$Im\{T_2\}$	$\lvert T_3 \rvert_{dB}$	$\arg\{T_3\}$	$Re\{T_3\}$	$Im\{T_3\}$
0.0	0.0	0.0	1.0	0.0	0.0	0.0	1.0	0.0
0.2	−0.645	−21.801	0.862	−0.345	−1.289	−43.607	0.624	−0.595
0.4	−2.148	−38.660	0.610	−0.488	−4.297	−77.320	0.131	−0.595
0.6	−3.874	−50.194	0.410	−0.492	−7.748	−100.389	−0.074	−0.403
0.8	−5.515	−57.995	0.281	−0.449	−11.029	−115.985	−0.123	−0.252
1.2	−8.299	−67.380	0.148	−0.355	−16.599	−134.760	−0.104	−0.105
1.6	−10.501	−72.646	0.089	−0.285	−21.015	−145.292	−0.073	−0.051
2.0	−12.304	−75.964	0.059	−0.235	−24.609	−151.928	−0.052	−0.028

in Fig. 11.16:

$$T_2(j\omega) = \frac{V_2}{V_1} = \frac{\dfrac{1}{RC}}{s + \dfrac{1}{RC}} = \frac{3.14159}{1 \times (j\omega) + 3.14159} = \frac{2\pi f_0}{j\omega + 2\pi f_0}$$

and

$$T_3(j\omega) = \frac{V_3}{V_1} = \frac{\left(\dfrac{1}{RC}\right)^2}{s^2 + \dfrac{2}{RC}s + \left(\dfrac{1}{RC}\right)^2} = \frac{9.8696}{1 \times (j\omega)^2 + 6.28318\{j\omega\} + 9.8696}$$

The normalized corner frequency is $f_{0n} = 1/(R_n C_n) 2\pi = 0.5$ in this case.

For the frequency range $f_{min} = 0$, $f_{max} = 2$, and the frequency increment $df = 0.2$, the tabulated results are shown in Table 11.5. The results represented as Bode and Nyquist plots are shown in Fig. 11.17.

Exercises

Exercise 1.   For the circuit shown in Fig. E11.1, generate the data matrix by inspection, and verify the result applying the program 'DATA_MAT'.

Exercise 2.   Construct the admittance matrix defined in Chapter 3 for the examples given in Fig. 11.5(a, b, c, d) and compare with the results of Section 11.3.

Exercise 3.   For the two circuits shown in Fig. 11.7(a, b), change the direction of the current of the current source and run the program 'MOD_YMAT'. Note the differences between the results obtained in Section 11.3.

Exercise 4.   Use the program 'MOD_YMAT' to generate the modified admittance matrix of circuits shown in Fig. E11.1 of exercise 1 for $\omega = 3$ rad/s.

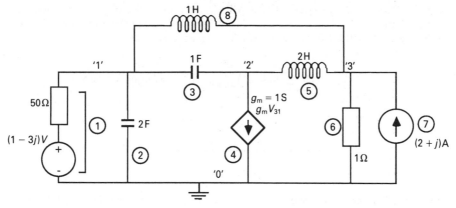

Figure E11.1

Exercise 5. Using the program 'MOD_YMAT', determine the modified admittance matrix corresponding to the circuit shown in Fig. E11.2 for $\omega = 1$ rad/s.

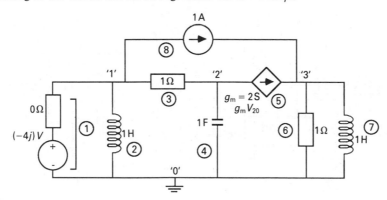

Figure E11.2

Exercise 6. Run the program 'MOD_YMAT' for the circuit shown in Fig. 11.5(c) for the following parameter values: $J_1 = 1 - 4j$ A; $G_S = 1/R_S = 0.15$ S; $C_1 = 1$ F; $C_2 = 2$ F; $G = 1/R = 1$ S; $J_2 = -1 + 3j$ A; $L = 3$ H and $\omega = 1$ rad/s. Verify the results obtained using equations (11.7b).

Exercise 7. In Section 11.1, eight transmittances are defined; indicate the calculation method for these transmittances running the program 'LUCOMPSO'.

Exercise 8. Modify the program 'LUCOMPSO' in order to study the variations of the input impedance of a circuit as a function of frequency. Apply this program to the circuit shown in Fig. E11.3 (piezoelectric quartz model).

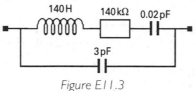

Figure E11.3

Exercise 9. Modify the program 'LUCOMPSO' in order to show the variations of the output impedance of a circuit as a function of frequency. Apply this program to the circuit shown in Fig. E11.4 (third-order Butterworth filter with $E_S = 0$).

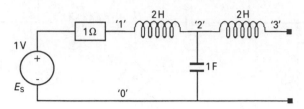

Figure E11.4

Exercise 10. For the circuit shown in Fig. E11.5(a), using the program 'LUCOMPSO', study the transmittance as in Section 11.4.2, for $\omega = 50$ rad/s and $\omega = 5000$ rad/s. The transistor model used is shown in Fig. E11.5(b).

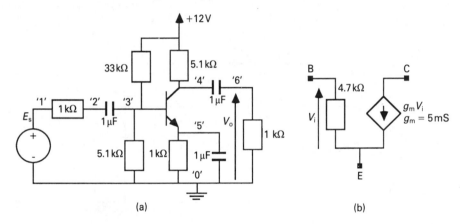

Figure E11.5

Exercise 11. For the circuit shown in Fig. E11.6(a), using the program 'LUCOMPSO', study the transmittance V_o/E_S as in Section 11.4.2, for $\omega = 5000$ rad/s. The transistor model used is shown in Fig. E11.6(b).

Exercise 12. For the circuit shown in Fig. E11.7(a), using the program 'LUCOMPSO', study the transmittance V_6/V_1 as in Section 11.4.2, for $\omega = 50$ rad/s and $\omega = 5000$ rad/s. The transistor model used is shown in Fig. E11.7(b).

Exercise 13. Using the program 'SEMISYMB', determine the impedance of the two-terminal networks shown in Fig. E11.8(a, b).

Exercise 14. Using the program 'SEMISYMB', determine the output impedance of the two-terminal networks shown in Fig. E11.9(a, b), with $E_S = 0$, $J_S = 0$.

Exercise 15. Using the program 'SEMISYMB', determine the transducer voltage gain V_6/E_S of the two-ports shown in Fig. E11.10.

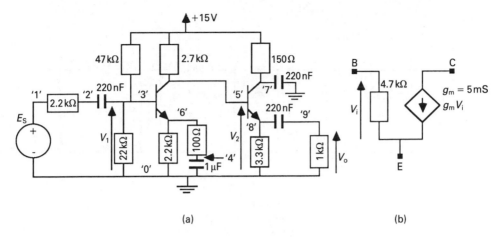

(a)

(b)

Figure E11.6

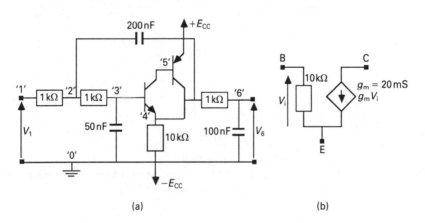

(a)

(b)

Figure E11.7

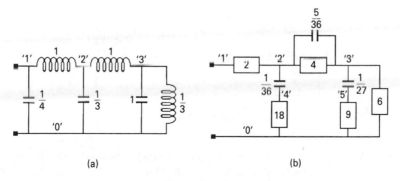

(a)

(b)

Figure E11.8

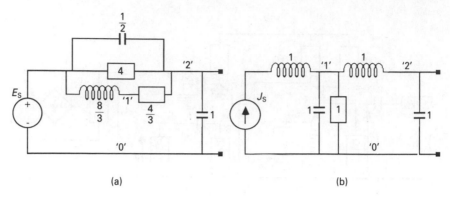

Figure E11.9

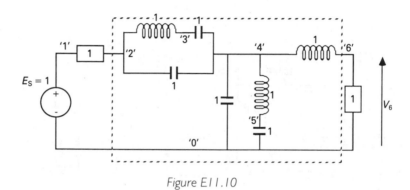

Figure E11.10

Exercise 16. Using the program 'SEMISYMB', obtain the two-port y_{ij} parameters in semi-symbolic form. Apply this program to the circuit shown in Fig. E11.11.

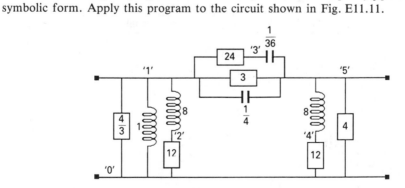

Figure E11.11

Exercise 17. Consider the circuit shown in Fig. E11.12(a) (transistor active filter). Using the program 'SEMISYMB', determine the ratio V_6/V_1. Fig. E11.12(b) shows the small signal equivalent circuit used for transistors.

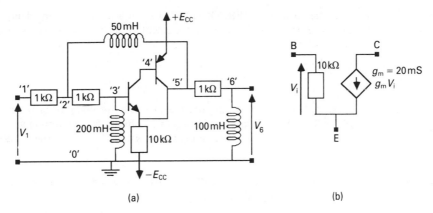

Figure E11.12

Exercise 18.   The circuits shown in Fig. E11.13(a, b) represent an *n*-element normalized low-pass filter (*n* odd in Fig. E11.13(a) and *n* even in Fig. E11.13(b)). The value g_i of elements ($g_i = L_i$ or C_i) satisfies the following relation:

$$g_i = 2\,\sin(\theta_i)\ i = 1, 2, ..., n \quad \text{with} \quad \theta_i = \frac{2i-1}{2n}\,\pi$$

Use the program 'SEMISYMB' in order to determine the transmittance V_o/E_S for $n = 1, 2,$ 3 and 4.

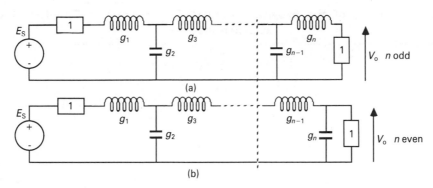

Figure E11.13

Exercise 19.   The circuit shown in Fig. 11.16 has a unit gain voltage amplifier. Use the program 'FREQCHAR' for the cascade connection of the two RC cells without a buffer amplifier.

Exercise 20.   Modify the program 'FREQCHAR' in order to find Bode plots for the following three second-order transfer functions:

$$\frac{\omega_0^2}{s^2 + \dfrac{\omega_0}{Q}\,s + \omega_0^2} \; ; \quad \frac{\dfrac{\omega_0}{Q}\,s}{s^2 + \dfrac{\omega_0}{Q}\,s + \omega_0^2} \; ; \quad \frac{s^2}{s^2 + \dfrac{\omega_0}{Q}\,s + \omega_0^2}$$

It is possible to use a frequency ω normalized with regard to ω_0.

Exercise 21. Modify the program 'FREQCHAR' to obtain the group delay time $\tau = -d\phi/d\omega$ as output. Apply this modification to the calculation of the group delay time of the following transfer function:

$$\frac{4 - 6s + 4s^2 - s^3}{4 + 6s + 4s^2 + s^3}$$

Exercise 22.   The program 'BODE' shown in program 11.6 is a graphic plot program using the program 'FREQCHAR' procedure and permitting us to obtain the magnitude and phase of the Bode diagrams for any transfer function; apply the program 'BODE' to the circuit analyzed in Section 11.6.

See the listing of the program BODE, Directory of Chapter 11 on attached diskette.

Exercise 23.   The program 'NYQUIST' shown in program 11.7 is a graphic plot program using the program 'FREQCHAR' as a procedure and permitting us to obtain the Nyquist diagram for any transfer functions; apply the program 'NYQUIST' to the following transfer functions:

$$\left\{\frac{1}{s+1}\right\}^n \ (n = 1, 2, 3, \ldots); \qquad \frac{1}{s^2 + \dfrac{s}{Q} + 1} \ (Q = 0.1, \ 1, 10); \qquad \frac{(s^2 + 1)(s^2 + 3)}{s(s + 2)(s + 4)}$$

See the listing of the program NYQUIST Directory of Chapter 11 on attached diskette.

---

12 Steady State Analysis for Non-sinusoidal Periodic Signals: Fourier Analysis

12.1 Trigonometric and exponential forms of the Fourier series

Non-sinusoidal periodic voltage or current signals $f(t)$, with period T, described generally by the relation (see Fig. 12.1)

$$f(t + T) = f(t) \tag{12.1}$$

can be decomposed into an infinite (or finite) number of sinusoidal signals, with harmonically related frequencies, using the trigonometric Fourier series

$$f(t) = a_0 + \sum_{n=1}^{\infty} \{a_n \cos(n\omega_0 t) + b_n \sin(n\omega_0 t)\} \tag{12.2}$$

where

$$\omega_0 = 2\pi f = \frac{2\pi}{T} = \text{fundamental (basic) frequency}$$

and

$$a_0 = \frac{1}{T} \int_0^T f(t) \, dt \tag{12.3a}$$

$$a_n = \frac{2}{T} \int_0^T f(t) \cos(n\omega_0 t) \, dt, \quad n = 1, 2, ..., \infty \tag{12.3b}$$

$$b_n = \frac{2}{T} \int_0^T f(t) \sin(n\omega_0 t) \, dt, \quad n = 1, 2, ..., \infty \tag{12.3c}$$

For example, the signal $f(t)$ shown in Fig. 12.1 is a sum of two cosines with frequencies ω_0 and $3\omega_0$:

$$f(t) = 2 \cos(\omega_0 t) - 0.5 \cos(3\omega_0 t)$$

whereas, the 'triangular wave' shown in Fig. 12.2 with period $T = 2$ can be

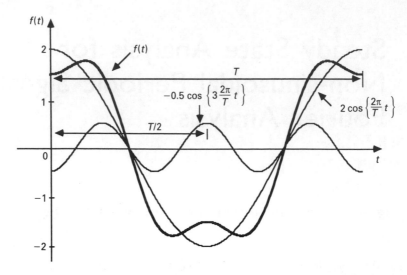

Figure 12.1   Periodic signal $f(t)$ with period T composed of two harmonic signals with angular frequencies $\omega_0 = \dfrac{2\pi}{T}$ and $3\,\omega_0$

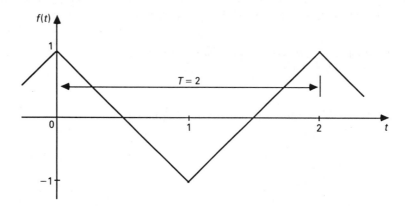

Figure 12.2   'Triangular wave' with period $T = 2$

decomposed using relation (12.2) into the infinite series

$$f(t) = \frac{8}{\pi^2}\left\{\cos(\pi t) + \frac{1}{9}\cos(3\pi t) + \frac{1}{25}\cos(5\pi t) + \frac{1}{49}\cos(7\pi t) + \ldots\right\} \qquad (12.4)$$

where the fundamental frequency ω_0 is $\omega_0 = \pi = 2\pi/T$.

Thus, using the Fourier series we can convert the non-sinusoidal signal into several sinusoidal signals; for example the frequency spectra of signals from Fig. 12.1 and Fig. 12.2 are shown in Fig. 12.3(a) and Fig. 12.3(b), respectively.

Therefore, excitation of a linear circuit with a single non-sinusoidal periodic

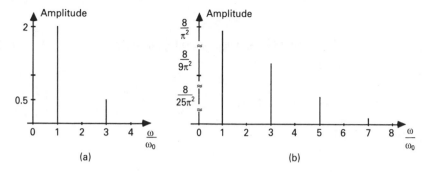

Figure 12.3 Frequency spectrum of signals. (a) Of Fig. 12.1. (b) Of Fig. 12.2

current source can be treated as excitation of the circuit by several current sources with amplitudes and frequencies obtained from relation (12.2), as is shown in Fig. 12.4. Then the response in the time domain can be computed using the super-position method.

The trigonometric form of the Fourier series can be converted into an equivalent exponential form using the Euler identity (valid for all n):

$$\cos(n\omega_0 t) \pm j \sin(n\omega_0 t) = \exp(\pm jn\omega_0 t) \tag{12.5}$$

from which we obtain

$$\cos(n\omega_0 t) = \frac{1}{2} \{\exp(jn\omega_0 t) + \exp(-jn\omega_0 t)\} \tag{12.6a}$$

$$\sin(n\omega_0 t) = \frac{1}{2j} \{\exp(jn\omega_0 t) - \exp(-jn\omega_0 t)\} \tag{12.6b}$$

Insertion of relations (12.6) into (12.2) yields

$$f(t) = a_0 + \sum_{n=1}^{\infty} \left[\frac{a_n - jb_n}{2} \exp(jn\omega_0 t) + \frac{a_n + jb_n}{2} \exp(-jn\omega_0 t) \right] \tag{12.7a}$$

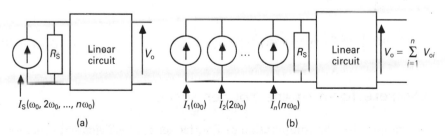

Figure 12.4 Linear circuit excited by (a) single current source with non-sinusoidal periodic current having n components, (b) n sinusoidal sources

or

$$f(t) = c_0 + \sum_{n=1}^{\infty} [c_n \exp(jn\omega_0 t) + c_{-n} \exp(-jn\omega_0 t)] \qquad (12.7b)$$

where

$$c_0 = a_0$$

$$c_n = \frac{a_n - jb_n}{2}$$

$$c_{-n} = \frac{a_n + jb_n}{2} = c_n^* - \text{complex conjugate with } c_n \qquad (12.7c)$$

Changing the limits of summation in (12.7b) we obtain the compact exponential form of the Fourier series

$$f(t) = \sum_{-\infty}^{\infty} [c_n \exp(jn\omega_0 t)] \qquad (12.8)$$

The complex coefficients c_n can be computed from the relation

$$c_n = \frac{1}{T} \int_0^T f(t)\{\cos(n\omega_0 t) - j\sin(n\omega_0 t)\ dt\} \qquad (12.9a)$$

which results from the combination of equations (12.3b), (12.3c) and (12.7c) or the equivalent formula

$$c_n = \frac{1}{T} \int_0^T f(t) \exp(-jn\omega_0 t)\ dt \qquad (12.9b)$$

To eliminate complex values in computer computations we can express the complex c_n coefficients by their magnitude $|c_n|$ and phase ϕ_n:

$$c_n = |c_n| \exp(j\phi_n) \qquad (12.10a)$$

and relate them to the coefficients a_n and b_n:

$$c_0 = a_0$$

$$|c_n| = \frac{1}{2} \sqrt{a_n^2 + b_n^2} \quad n \neq 0$$

$$\phi_n = \text{atan}\left\{-\frac{b_n}{a_n}\right\} = -\text{atan}\left(\frac{b_n}{a_n}\right) \qquad (12.10b)$$

12.2 Discrete form of the Fourier series

To use a computer for the computation of Fourier series coefficients it is necessary to discretize the time t and then to compute numerically definite integrals in equation (12.3) using any method described in Section 2.3.2 (that is rectangular, trapezoidal or Simpson's method). Since the integral limits are 0 and T, it is thus necessary to divide the time interval T for K equal time sections ΔT:

$$\Delta T = \frac{T}{K}$$

or

$$T = K \Delta T \tag{12.11}$$

to obtain the discrete time points

$$t_i = i\Delta T, \quad i = 0, 1, 2, \ldots, K$$

and to compute the discrete values of the function $f(t)$:

$$f_i = f(t_i) = f(i\Delta T), \quad i = 0, 1, 2, \ldots, K-1 \tag{12.12}$$

as shown in Fig. 12.5. Note that $f(K\Delta T) = f(T) = f(0)$, since $f(t)$ is periodic. The considered function $f(t)$ must be a function of time which can have a limited number of discontinuities in one period.

It is also necessary to note that using the computer we can take into account only N lower frequencies of the infinite spectrum. Therefore, the discrete (in time) Fourier series is limited:

$$f(t_i) = f(i\Delta T) = a_0 + \sum_{n=1}^{N} \{a_n \cos(n\omega_0 i\Delta T) + b_n \sin(n\omega_0 i\Delta T)\} \tag{12.13}$$

This has considerable influence on the accuracy of computation, as will be discussed later in this chapter. Since the fundamental frequency ω_0 can be expressed as

$$\omega_0 = \frac{2\pi}{T} = \frac{2\pi}{K\Delta T} \tag{12.14}$$

relation (12.13) thus takes the form

$$f(i\Delta T) = a_0 + \sum_{n=1}^{N} \left\{ a_n \cos\left(2\pi n \frac{i}{K}\right) + b_n \sin\left(2\pi n \frac{i}{K}\right) \right\} \tag{12.15}$$

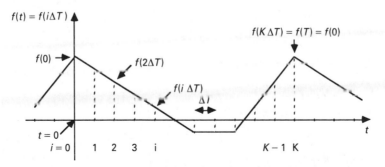

Figure 12.5   Discretization of time in one period T, and $f(i\Delta T)$

When the coefficients a_0, a_n, b_n are computed using the rectangular integration method, we obtain the simple expressions

$$a_0 = \frac{1}{T} \sum_{i=0}^{K-1} f(i\Delta T)\, \Delta T \tag{12.16a}$$

$$a_n = \frac{2}{T} \sum_{i=0}^{K-1} \left(f(i\Delta T)\, \cos\left(2\pi n\, \frac{i}{K}\right) \right) \Delta T \tag{12.16b}$$

$$b_n = \frac{2}{T} \sum_{i=0}^{K-1} \left(f(i\Delta T)\, \sin\left(2\pi n\, \frac{i}{K}\right) \right) \Delta T \tag{12.16c}$$

with $n = 1, 2. \ldots, N$.

Relations (12.16) and (12.15) allow us to find the Fourier coefficients for a given function $f(t_i) = f(i\Delta T)$, and to reconstruct the function $f(t_i)$ based on knowledge of the values of the coefficients a_0, a_n, b_n. Relations (12.15) and (12.16) are referred to as the discrete Fourier transform. When the coefficients a_0, a_n, b_n are computed using the trapezoidal integration method, we obtain the formula

$$a_0 = \frac{\Delta T}{T} \left\{ \frac{f(0) + f(K)}{2} + \sum_{i=1}^{K-1} f(i\Delta T) \right\} \tag{12.17a}$$

or, since $f(0) = f(K)$ and $(\Delta T)/T = 1/K$

$$a_0 = \frac{1}{K} \left\{ f(0) + \sum_{i=1}^{K-1} f(i\Delta T) \right\} \tag{12.17b}$$

$$a_n = \frac{2}{K} \left\{ f(0) + \sum_{i=1}^{K-1} \left(f(i\Delta T)\, \cos\left\{ 2\pi n\, \frac{i}{K} \right\} \right) \right\} \tag{12.17c}$$

$$b_n = \frac{2}{K} \left\{ f(0) + \sum_{i=1}^{K-1} \left(f(i\Delta T)\, \sin\left\{ 2\pi n\, \frac{i}{K} \right\} \right) \right\} \tag{12.17d}$$

In a similar way we can find the formulas for a_0, a_n, b_n for Simpson's integration method. The discrete form of the exponential Fourier series is

$$f(i\Delta T) = \sum_{n=-N}^{N} \left\{ c_n \exp\left(j2\pi n\, \frac{i}{K} \right) \right\} \tag{12.18}$$

with coefficients

$$c_n = \frac{\Delta T}{T} \left\{ \sum_{i=0}^{K-1} \left(f(i\Delta T)\, \exp\left\{ -j2\pi n\, \frac{i}{K} \right\} \right) \right\} \tag{12.19a}$$

or

$$c_n = \frac{1}{K} \left\{ \sum_{i=0}^{K-1} \left(f(i\Delta T)\, \exp\left\{ -j2\pi n\, \frac{i}{K} \right\} \right) \right\} \tag{12.19b}$$

12.3 Computation of the Fourier series coefficients for function $f(t)$ given in analytic form

The simplest method of computation of Fourier series coefficients is direct application of relation (12.16), that is, application of the rectangular integration method. This method is used in the program 'FOUCOREC' shown in Program 12.1.

> **See the listing of the program FOUCOREC Directory of Chapter 12 on attached diskette.**

The function $f(t)$ is predefined as the 'triangular wave' shown in Fig. 12.2, for which the exact Fourier series is given by relation (12.4). In the program the 'triangular' function $f(t)$ with amplitude equal to unity and period $T = tt = 2$ is defined as

```
f := 1 - 2*t;
if t >= tt/2 then
f := - 1 + 2*(t-tt/2)
```

For this function the exact coefficient values are

$$a_0 = 0, b_n = 0$$

$$a_n = \frac{8}{\pi^2 n^2}, \text{ for } n \text{ odd, and } a_n = 0, \text{ for } n \text{ even} \qquad (12.20)$$

Note that for different period T, or different amplitude of this triangular wave, it is necessary to change the function definition; for example for $T = 4$, the function definition should be

```
f := 1 - t;
if t >= tt/2 then
f := - 1 + (t-tt/2)
```

Knowing the values of the coefficients given by relation (12.20) we can compare them with computer computations and see the accuracy of the computations. Running the program 'FOUCOREC' with $N = nn = 7$, $T = tt = 2$, $K = kk = 40$ and $K = kk = 80$, we obtain the results shown in Table 12.1.

As was expected, the a_0, b_n and even a_n coefficients equal zero; the accuracy

Table 12.1 Results of running 'FOUCOREC' for 'triangular wave'

	$K = kk = 40$		$K = kk = 80$	Exact $a_n = 8/\pi^2/n^2$
$n = 1$	$a_n = 0.81224$	$b_n = 0$	$a_n = 0.81099$	0.81057
$n = 3$	$a_n = 0.09175$	$b_n = 0$	$a_n = 0.09048$	0.09006
$n = 5$	$a_n = 0.03414$	$b_n = 0$	$a_n = 0.03284$	0.03242
$n = 7$	$a_n = 0.01831$	$b_n = 0$	$a_n = 0.01697$	0.01654

Table 12.2   Results of running 'FOURCOREC' for 'saw-type' function

	kk = K = 40		kk = K = 80		kk = K = 200		Exact b_n $\dfrac{-1}{\pi n}$
n	a_n $a_0 = 0.48750$	b_n	a_n $a_0 = 0.49375$	b_n	a_n $a_0 = 0.49750$	b_n	$a_n = 0, b_n$ $a_0 = 0.5$
1	−0.025	−0.31766	−0.125	−0.31815	−0.005	−0.31828	−0.31831
2	−0.025	−0.15784	−0.125	−0.15883	−0.005	−0.15910	−0.15915
3	−0.025	−0.10413	−0.125	−0.10561	−0.005	−0.10602	−0.10610
4	−0.025	−0.07694	−0.125	−0.07892	−0.005	−0.07947	−0.07958
5	−0.025	−0.06036	−0.125	−0.06284	−0.005	−0.06353	−0.06366

of computations increases with increasing K, that is, decrease in ΔT. Running the program 'FOUCOREC' for the 'saw-type' function $f(t) = t$, $0 \leqslant t \leqslant T$, with $T = 1$ again, we obtain the results shown in Table 12.2.

An application of the trapezoidal method of integration for the computation of Fourier coefficients is implemented in the program 'FOUCOTRA' shown in Program 12.2.

> **See the listing of the program FOUCOTRA Directory of Chapter 12 on attached diskette.**

Running the program 'FOUCOTRA' for the 'saw-type' function with $T = 1$, we obtain the same values for b_n coefficients as are obtained from the program 'FOU-COREC'; but for trapezoidal integration all the a_n coefficients are equal to zero, and the coefficient a_0 takes the exact value $a_0 = 0.5$, for $kk = 40$, 80 and 200.

It can be seen that the values of the b_n coefficients in the rectangular and trapezoidal methods of integration are equal since the values of $\sin(0)$ and $\sin(2\pi)$ are equal to zero.

12.4 Computation of the Fourier series coefficients for function $f(i \Delta T)$ given as a set of discrete values

In situations when the analyzed function is a result of measurement, that is, it is represented graphically as a function of time on the screen of an oscilloscope, or plotter, it is necessary to create a set of K pairs of equally spaced time points $(i \Delta T)$ and the function $f(i \Delta T)$ values for these time points. Then the function values are entered into the computer program for the computation of Fourier series coefficients. In such cases the rectangular or trapezoidal methods of integration are usually used since the accuracy of the function value readings from graphical display equipment is not very high.

The program 'FOUDISRE' for the computation of Fourier coefficients using

the rectangular integration method, when the analyzed function is given as the set of discrete values $f(i\Delta T)$, is shown in Program 12.3.

> **See the listing of the program FOUDISRE Directory of Chapter 12 on attached diskette.**

This program differs from the program 'FOUCOREC' of Program 12.1 by a different definition of the analyzed function. In this case, instead of the 'Function' in which the analytic definition of the function is given, the 'Procedure readfunction' is used to enter the function $f(i\Delta T)$ values as elements of the vector $f[i]$. The equally spaced time points $(i\Delta T)$ are assured by the statement 'for i: $= 0$ to K $- 1$ do'.

Let us run the program 'FOUDISRE' for the 'saw-type' wave defined as

$$f(t) = t, \ 0 \leqslant t < T$$

with input data $N = nn = 5$, $K = kk = 20$, $T = tt = 1$.

Thus, we have to enter twenty function $f(t) = t$ values:

$$f[0] = 0, \ f[1] = 0.05, \ f[2] = 0.1, ..., f[19] = 0.95$$

Running the program we obtain results, which are compared with the exact values, in Table 12.3.

The program 'FOUDISTR' for computation of Fourier series coefficients using the trapezoidal method, when the analyzed function is given as the set of discrete values $f(i\Delta T)$, is shown in Program 12.4.

> **See the listing of the program FOUDISTR Directory of Chapter 12 on attached diskette.**

Better accuracy of computation of the a_0 and a_n coefficients is obtained by using the trapezoidal method of integration implemented in the program

Table 12.3 Results of running 'FOUDISRE' for 'saw-type' function

	Rectangular integration		Exact values: $b_n = \dfrac{-1}{\pi \times n}$	
n	a_n	b_n	a_n	b_n
1	-0.05	-0.31569	0	-0.31831
2	-0.05	-0.15388	0	-0.15915
3	-0.05	-0.09813	0	-0.10610
4	-0.05	-0.06882	0	-0.07958
5	-0.05	-0.05000	0	-0.06366
	$a_0 = 0.4750$		$a_0 = 0.5$	

'FOUDISTR'. Running this program for the same input data as for the program 'FOUDISRE', we obtain the results shown in Table 12.4.

Again we see that the b_n coefficients have the same values as in the rectangular integration, as was discussed earlier in this section. Better accuracy of computation of Fourier series coefficients is obtained by using Simpson's method of integration. This method is used in the program 'FOUCOSI' shown in Program 12.5.

> **See the listing of the program FOUCOSI Directory of Chapter 12 on attached diskette.**

To compare the accuracy of computation, let us run this program for the 'saw-type' function

$$f(t) = -t, 0 \leqslant t < T, \text{ with } T = 1$$

with initial data $N = nn = 5$, $K = kk = 20$ and $Kkk = 40$, $T = tt = 1$.

The results of the computation and the exact values of the b_n coefficients, $b_n = 1/(\pi n)$, are given in Table 12.5.

Table 12.4 Results of running 'FOUDISTR' for 'saw-type' function

	Trapezoidal integration	
n	a_n	b_n
1	0.0	-0.31569
2	0.0	-0.15388
3	0.0	-0.09813
4	0.0	-0.06882
5	0.0	-0.05000
	$a_0 = 0.50000$	

Table 12.5 Results of running 'FOUCOSI' for 'saw-type' function

		$K = 20$	$K = 40$	exact $-\dfrac{1}{\pi n}$
$n = 0$	$a_n = -0.5$	$b_n = -0.0$	$b_n = -0.0$	$b_n = -0.0$
$n = 1$	$a_n = 0.0$	$b_n = 0.31833$	$b_n = 0.31831$	$b_n = 0.31831$
$n = 2$	$a_n = 0.0$	$b_n = 0.15930$	$b_n = 0.15916$	$b_n = 0.15915$
$n = 3$	$a_n = 0.0$	$b_n = 0.10662$	$b_n = 0.10613$	$b_n = 0.10610$
$n = 4$	$a_n = 0.0$	$b_n = 0.08093$	$b_n = 0.07965$	$b_n = 0.07958$
$n = 5$	$a_n = 0.0$	$b_n = 0.06667$	$b_n = 0.06381$	$b_n = 0.06366$

12.5 Reproduction of the function $f(i\Delta T)$ based on knowledge of Fourier series coefficients

Reproduction of the function $f(i\Delta T)$, which approximates the function $f(t)$ based on the knowledge of the Fourier coefficients for that function, can be used for verification: how many coefficients have we to take into account for a good-quality representation of the function, and how small should the time increments ΔT be, during the determination of N coefficients?

From the sampling theorem we know that the number K of discrete points necessary for proper determination of the Nth harmonic frequency coefficient of the function $f(t)$ is given by the relation

$$K > 2N \qquad\qquad (12.21)$$

The number N of harmonic frequency coefficients a_n and b_n, $n = 1, 2, ..., N$ necessary for reproduction of the function $f(t)$ with proper accuracy depends on the Fourier series convergence for a given function. This convergence is slower for discontinuous functions (for example for square wave functions).

When we know the values of the coefficients a_n, b_n then the values of the $f(i\Delta T)$ function can be computed at discrete time points from relation (12.15) or the equivalent formula

$$f(i\Delta T) = a_0 + \sum_{n=1}^{N} \left\{ d_n \sin\left\{ 2\pi n\, \frac{i}{K} + \theta_n \right\} \right\} \quad i = 0, 1, 2, ..., K-1 \qquad (12.22a)$$

where

$$d_n = \sqrt{a_n^2 + b_n^2} \qquad\qquad (12.22b)$$

$$\theta_n = \operatorname{atan}\left\{ \frac{a_n}{b_n} \right\} \qquad\qquad (12.22c)$$

The program 'FUNREPRO' for the reproduction of the $f(i\Delta T)$ function is shown in Program 12.6. In this program the Fourier series coefficients a_n, b_n are computed using Simpson's method of integration. Insertion of the coefficients a_0, a_n and b_n into formula (12.15) gives us the reproduced $f(t)$ function.

See the listing of the program FUNREPRO Directory of Chapter 12 on attached diskette.

Running the program 'FUNREPRO' for different N and K is very educational, since it gives us some insight into sampling theorem satisfaction or dissatisfaction. Running the program for the function

$$f(t) = -t, \quad 0 \leqslant t < T, \quad \text{with } T = tt = 1$$

we obtain the results for the different $N = nn$ and $K = kk$ shown in Table 12.6.

The original function $f(t) = -t$ and its reproductions, for different N and K, are shown in Fig. 12.6. The deviations of the reproduced function $f(i \cdot \Delta T)$ from the original function $f(t) = -t$ are caused by truncation of the infinite Fourier series, and also by dissatisfaction with the sampling theorem (see relation (12.21)). When the number of harmonic frequency N is closer to $K/2$, half the sampling number in period T, then the deviations are larger, as can be seen from Fig. 12.6, for $N = 6$ and $N = 9$, with $K/2 = 10$. This effect is referred to as the Gibbs effect

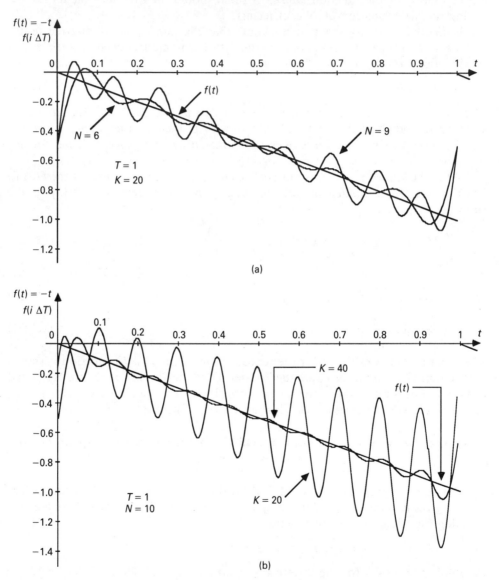

(a)

(b)

Figure 12.6 Reproduction of the function $f(t) = -t$, $0 \leqslant t < T$, $T = 1$, $f(i\Delta T)$: (a) For $K = 20$, $N = 6$ and $N = 9$. (b) For $N = 10$, $K = 20$ and $K = 40$

Table 12.6   Result of the program 'FUNREPRO': reproduction of the function $f(t) = -t$, $0 \leqslant t < T$, $T = 1$

$T = 1$	$K = kk = 20$				$K = kk = 40$	
	$N = nn = 6$	$N = nn = 8$	$N = nn = 9$	$N = nn = 10$	$N = nn = 10$	Exact
$t = i\Delta T$	$f(t) \cong f(i\Delta T)$	$f(t) \cong f(i\Delta T)$	$f(t) \cong f(i\Delta T)$	$f(t) \cong f(i\Delta T)$	$f(t) \cong f(i\Delta T)$	$f(t) = -t$
0.00	-0.50	-0.50	-0.50	-0.166667	-0.50	0.00
0.05	-0.021738	0.065035	0.10	-0.233333	0.04027	-0.05
0.10	-0.047253	-0.166826	-0.233333	0.10	-0.148637	-0.10
0.15	-0.207088	-0.124872	-0.033333	-0.366667	-0.118163	-0.15
0.20	-0.186891	-0.192389	-0.30	-0.033333	-0.222558	-0.20
0.25	-0.221629	-0.279815	-0.166667	-0.50	-0.233529	-0.25
0.30	-0.332957	-0.259056	-0.366667	-0.033333	-0.311999	-0.30
0.35	-0.345285	-0.391539	-0.30	-0.633333	-0.341572	-0.35
0.40	-0.375723	-0.366826	-0.433333	-0.10	-0.405399	-0.40
0.45	-0.475673	-0.468298	-0.433333	-0.766667	-0.447377	-0.45
0.50	-0.50	-0.50	-0.50	-0.166667	-0.50	-0.50
0.55	-0.524327	-0.531702	-0.566667	-0.90	-0.552623	-0.55
0.60	-0.624277	-0.633174	-0.566667	-0.233333	-0.594621	-0.60
0.65	-0.654715	-0.608461	-0.70	-1.033333	-0.658428	-0.65
0.70	-0.667043	-0.740944	-0.633333	-0.30	-0.688001	-0.70
0.75	-0.778371	-0.720185	-0.833333	-1.166667	-0.766471	-0.75
0.80	-0.813109	-0.807611	-0.70	-0.366667	-0.777442	-0.80
0.85	-0.792912	-0.875128	-0.966667	-1.30	-0.881837	-0.85
0.90	-0.952742	-0.833174	-0.766667	-0.433333	-0.851363	-0.90
0.95	-0.978262	-1.065035	-1.10	-1.433333	-1.040207	-0.95
1.00	-0.5	-0.5	-0.5	-1.166667	-0.5	-1.00

and causes the greatest deviation at the vicinity of discontinuity of the original function.

When the inequality (12.21), representing the sampling theorem, is not satisfied then the reproduced function $f(i\Delta T)$ cannot represent the original function $f(t)$, as can be seen from Fig. 12.6(b), for $N = 10$ and $K/2 = 10$. Increasing the number K to 40, again gives the correct reproduction.

Exercises

Exercise 1.   Determine, the time period of the following functions, (if it exists):

$$\sin(3t) - \sqrt{2} \, \cos(4t); \ \cos(t) + \cos\{(\sqrt{3}t)\}; \ \sin(\pi t) - 0.5 \, \cos(t)$$

Exercise 2.   Consider a periodic function $f(t)$ with time period T and the corresponding Fourier series coefficients. Determine the Fourier coefficients of the function $\{f(t) \sin(\omega t)\}$ with $\omega = 2\pi/T$ (if they exist).

Exercise 3.   Determine analytically the coefficients a_n and b_n of the Fourier series for the periodic functions represented in Fig. E12.1 (a, b, c, d).

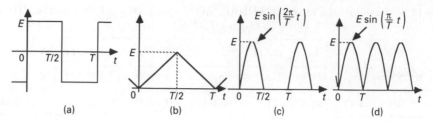

Figure E12.1

Exercise 4. Determine analytically the coefficients a_n and b_n of the Fourier series for the periodic function represented in Fig. E12.2. The function considered is a step approximation of the sine function.

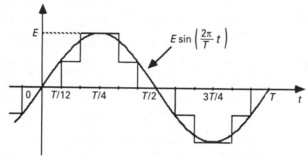

Figure E12.2

Exercise 5. Determine analytically the complex Fourier coefficients c_n of the periodic functions represented in Fig. E12.3(a, b). Use several values of α, for example $\alpha = 0.05, 0.1, 0.2, 0.4, 0.6, 0.8$.

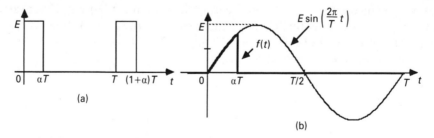

Figure E12.3

Exercise 6. The two periodic functions $f_1(t)$ and $f_2(t)$ shown in Fig. E12.4(a, b) can be obtained one from another by the time shift t_0. The Fourier series expansions of these two periodic functions lead to the set of coefficients (a_{n1} and b_{n1}) and (a_{n2} and b_{n2}). What relationship exists between the two sets of values? Deduce the influence of a time shift t_0 on the Fourier series expansion.

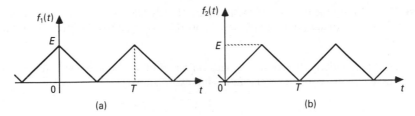

Figure E12.4

Exercise 7. Consider the influence of a time scale modification on Fourier series expansion.

Exercise 8. Consider the influence of a 'vertical translation' of the two signals represented in Fig. E12.5 on Fourier series expansion; find the Fourier coefficients of these two signals.

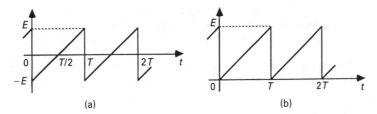

Figure E12.5

Exercise 9. Determine the properties of the Fourier coefficients (a_n, b_n, c_n) for periodic functions having the following properties:

1. $f(t)$ is an even function: $f(t) = f(-t)$
2. $f(t)$ is an odd function: $f(t) = -f(-t)$.

Apply these results to the functions studied in exercise 3.

Exercise 10. Determine the properties of the Fourier coefficients (a_n, b_n, c_n) for periodic functions having the following properties: $f\{t + T/2\} = -f(t); f(T - t) = -f(t)$. Apply these results to the functions represented in Fig. E12.6(a, b).

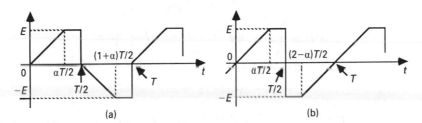

Figure E12.6

Exercise 11. From the programs 'FOUCORE', 'FOUCOTRA' and 'FOUCOSI', develop a program permitting comparison of the values of the Fourier coefficients a_n and b_n. Compare the results with the Fourier coefficients obtained analytically in exercise 4.

Exercise 12. Modify the outputs of the program 'FOUCOSI' in order to obtain the values of the coefficients c_n (magnitude and phase). Apply this program to take into account the time shift properties.

Exercise 13. For periodic function $f(t)$ with time period T, the average value V_{av} and the effective value V_{eff} are defined by the following expressions:

$$V_{av} = \frac{1}{T} \int_{\alpha}^{\alpha + T} (f(t))\, dt \qquad V_{eff} = \sqrt{\frac{1}{T} \int_{\alpha}^{\alpha + T} (f(t))^2\, dt}$$

Develop a program, from the program 'FOUCOSI', giving as an output the average value and the effective value of the analyzed signal.

Exercise 14. Using the program 'FOUDISRE', determine the Fourier coefficients of the functions represented in Figs. E12.6a and b, with $\alpha = 0.5$.

Exercise 15. Use the output data of the program 'HEUN' of Section 7.3 in order to determine the Fourier coefficients using the program 'FOUDISRE'. Vary the time constant τ of the circuit of Fig. 7.6. Consider the first period of the input signal.

Exercise 16. The program 'PLOT' shown in Program 12.7 permits restoration of a periodic signal from Fourier coefficients a_n, b_n. The signal used in this exercise is represented in Fig. E12.7. Its Fourier series expansion is as follows:

$$\left| \cos\left(\frac{2\pi}{T} t\right) \right| = \frac{4}{\pi}\left[\frac{1}{2} + \frac{1}{3}\cos\left(\frac{2\pi}{T} t\right) - \frac{1}{15}\cos\left(\frac{4\pi}{T}t\right) + \cdots + \frac{(-1)^{n+1}}{4n^2 - 1}\cos\left(\frac{n2\pi}{T} t\right)\right]$$

> ### See the listing of program **PLOT** Directory of Chapter 12 on attached diskette.

Using the program 'PLOT', show the influence of the suppression of the average value, fundamental and several harmonics on the restored signal.

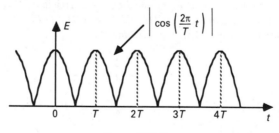

Figure E12.7

Exercise 17. Using program 'PLOT' for the same signal as in exercise 16, show the influence of the phase-shift:

1. Add to every coefficient a constant phase ϕ_0, and show the influence on the restored signal ($\phi_0 = 10°$, $45°$, $90°$ and $180°$).
2. Add to each coefficient a phase $\phi_n = n\phi_0$ ($n = 1, 2, 3, \ldots$) and show the influence on the restored signal ($\phi_0 = 10°$, $45°$, $90°$ and $180°$).

Exercise 18.   A second-order all-pass system is characterized by the following transfer function $T(s)$:

$$T(s) = \frac{s^2 - \dfrac{\omega_0}{Q} s + \omega_0^2}{s^2 + \dfrac{\omega_0}{Q} s + \omega_0^2}$$

If the input signal is periodic (with period T), the output signal is also periodic with the same time period T. Develop a program which, from the Fourier coefficients of the input signal, gives the Fourier coefficients of the output signal. The values of ω_0 and Q are given as parameters. Use the program 'PLOT' to show the results. Consider a square input signal.

Exercise 19.   We want to obtain the steady state response for the circuit shown in Fig. E12.8(a). The input source voltage $e(t)$ is represented in Fig. E12.8(b). Develop a program based on the program 'FOUCOSI' to determine the current $i(t)$ and the voltage $v_C(t)$ on the capacitor in one period T. $R = 1 \text{ k}\Omega$; $C = 1 \text{ } \mu\text{F}$. $T = 0.1 \text{ } \tau, \tau, 10 \text{ } \tau$, where $\tau = RC$.

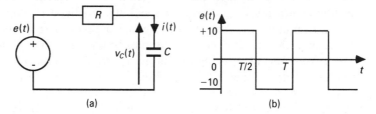

(a)   (b)

Figure E12.8

Exercise 20.   Use the program 'FUNREPRO' with the program 'PLOT' in order to obtain the curves shown in Fig. 12.6(a, b).

Exercise 21.   Modify the program 'FUNREPRO' in order to study the influence of the choice of N and K on the restoration of a square signal. Use the program 'PLOT'.

Exercise 22.   We want to obtain the steady state response for the circuit shown in Fig. E12.9(a). The circuit is characterized by the three parameters R, $\omega_0 = 1/\sqrt{LC}$ and $Q = L\omega_0/R$; the input source voltage is represented in Fig. E12.9(b). Develop a program based on the program 'FOUCOSI' permitting determination of the current $i(t)$ and the voltage $v_C(t)$ on the capacitor in one period. T is constant and ω_0 and Q are variable parameters.

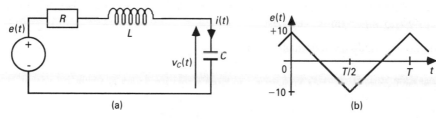

(a)   (b)

Figure E12.9

13 Artificial Intelligence Approach to Electronic Circuit Analysis and Design

13.1 Introduction to artificial intelligence methods

There is no exact definition of 'artificial intelligence' and, also, it is difficult to define human intelligence precisely. Roughly speaking, artificial intelligence concerns the computer simulation of some processes which are considered to be intellectual, such as reasoning, that are connected to human activity and that are attributes of intelligent behaviour. Usually, during reasoning or inference in problem solving, we are not thinking in terms of algorithms, and use symbolic, rather than numeric, concepts and objects. A rough definition of 'artificial intelligence' is expressed as follows: '"Artificial intelligence" is a branch of computer science which concerns symbolic and non-algorithmic methods of problem solving.' What is very important in the area of artificial intelligence is heuristics rather than strict algorithms, that is, procedures based on experience and common sense (so-called 'rules of thumb'), are used, allowing us to reach the solution of a given complex problem in a shorter time.

Good examples of the application of heuristics are chess-playing computer programs based on the knowledge of some chess-masters. To check all the possible movements during play by using strict algorithms, it would be necessary to verify an enormous number of situations (over 10^{100}), which is completely impossible in practice. However, using the experience of chess-masters and their heuristics, some computer chess-playing programs are constructed that beat, in a short time, the majority of good chess-players.

Artificial intelligence methods are used in fields that are not formalized mathematically (no algorithms) such as:

- diagnosis (medical, technical, ...),
- construction (design, configuring, synthesis),
- classification of data,
- planning of actions,
- etc.

Heuristic procedures, or rules, heuristic knowledge, or simply heuristics, are acquired by humans over long-term experience and are developed through personal intuition, judgement and common sense thinking. They depend on the talent and skill of a person and they usually evolve over years of personal activity and experience in a given domain. Heuristics reside in the memory of an individual and are treated as private knowledge – sometimes difficult to express and explain. Some heuristics are described in so-called handbooks, where some guidelines for the repair, design, etc., of some equipment are presented by experienced persons.

Examples of heuristics rules are as follows:

1. switch off your radio-receiver when you see smoke coming out of it,
2. do not use a microwave transistor in an audio amplifier construction,
3. neglect stray inductances and capacitances in the analysis of audio equipment,
4. treat as short-circuits all blocking (electrolytic) capacitors in high-frequency analysis,
5. indicate an error in your computations when one of the coefficients of the denominator of the transfer function of a stable circuit is negative,
6. break or modify the design process of an amplifier after detecting its instability,
7. etc.

One of the important characteristics of heuristics is their ability for filtering or screening. This means that heuristics are usually used to reduce considerably the number of possible ways or alternatives that can be considered in a problem solution. When several heuristics are combined and described as a procedure for problem solution, then we speak of heuristic programming. The main difference between algorithmic and heuristic programs is that in heuristic programs the solution found may not be the best one (optimal, in some sense) , but it may be acceptable. Thus, instead of optimization, we are selecting the solution which meets the requirements, with some acceptable margins.

This situation is frequently met with in the design process. For example, we need an amplifier with a voltage gain of $100 < A_v < 120$, and a bandwidth of $\Delta f = 20$ kHz $\pm$ 500 Hz, etc., but not exactly, say, $A_v = 115$ and $\Delta f = 20$ kHz.

Heuristic knowledge is used extensively in expert systems, mainly as production rules of the form

```
If: <condition> ; Then <action> ;
```

for example:

```
If<analysis of audio equipment> Then <neglect stray
                          capacitances and inductances>
```

Heuristics are also used to solve difficult problems for which there are no simple algorithmic solutions. Usually, artificial intelligence programs are large and are therefore referred to as systems. Additionally, they are organized in such a way that the symbolic information, that is, the knowledge about the problem to be solved, is put together in a separate part of the system, called the knowledge base. Therefore, such computer programs are referred to as 'knowledge base systems' (see Fig. 13.1).

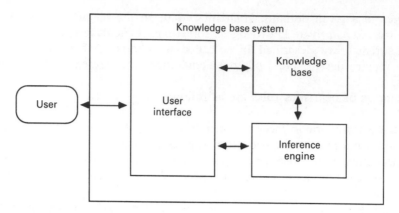

Figure 13.1 Block diagram of the knowledge base system

13.2 Knowledge base systems and expert systems; comparison with classical algorithmic programs

A knowledge base system is a computer program that incorporates knowledge, usually in symbolic form, obtained from various different sources about a specific domain; this knowledge is located in a special program block called the knowledge base, as is shown in Fig. 13.1. Being a separate part of the program, the knowledge base can easily be modified (i.e. expanded or shrunk) without interfering with the rest of the program blocks. The other main blocks of the knowledge base system are as follows:

1. The inference engine, that is, the algorithmic part of the system that controls the operation of the overall system and performs reasoning (based on knowledge located in the knowledge base) leading to conclusions (problem solution).
2. The user interface, which allows the user to enter initial (or starting) data regarding the problem to be solved, to ask some questions in a similar way to natural language, and to obtain results from the system concerning the solution of the problem. This interaction between the user and the knowledge base system is called a consultation.

The expert system is a knowledge base system in which the heuristic knowledge in the knowledge base comes mainly from an expert or group of experts in some narrow, specialized subject area. Thus, the output of the expert system during the consultation (i.e. when running the program) can substitute discussions with human experts.

There is no precise definition of an expert system, but we can give the following rough descriptions to explain its main features:

1. The expert system is a complex computer program that is able to simulate the

actions of human experts during the problem solution in a relatively narrow knowledge domain.

2. The expert system is a computer program that aids the solution of non-algorithmic problems that are normally solved by human experts who have specialized heuristic knowledge gained over long-term experimental practice and studies in the weakly formalized domains.

3. The expert system is an 'intelligent' computer program that uses the knowledge and reasoning (inference) procedures of human experts for solving these problems that require a deep human expertise acquired over long-term scientific and experimental activity in a given narrow domain.

One of the best known and most successful expert systems is called MYCIN, which provides advice on antimicrobial therapy during the medical diagnosis of infectious blood diseases. Running this program, or performing the consultation, make it possible for an inexperienced medical physicist to substitute oral consultations with expert medical doctors.

In speaking about expert systems, a narrow domain knowledge is usually emphasized; nevertheless, the expert system should be equipped, to some extent, with more general knowledge about the domain considered. For example, in an expert system for electronic active RC filter design some general knowledge about linear circuits and systems, including mathematical methods, should be included.

The expertise gained by human experts during their long-term activities leads to fast solutions due to heuristic inference, that is informal short-cuts during reasoning. For example, an experienced electronic engineer (electronic expert) can quickly find the approximate value for the quiescent point of a single-stage transistor amplifier by inspection, rather than by the solution of a set of non-linear equations (which can be quite lengthy without a computer program). In real life, it is quite common for the fast solution of a problem, even not very precise, to be more valuable than the exact solution obtained over a long time. Expert systems are usually provided with such inference mechanisms and knowledge that allow the solution of complex problems in the shortest time.

One of the main differences between the expert system and the conventional algorithmic program is that in the expert system, the data and knowledge are mostly symbolic (non-numeric) and are symbolically processed; also, knowledge about the problem to be solved is given in a declarative (descriptive), not procedural (algorithmic) form. Thus, in an expert system the description of the problem and the goal to be reached are given, without the details of each computation step – as in conventional algorithmic programs.

Expert systems, similar to specialized books, can be used for the transfer of knowledge and reasoning about a given domain from few experts to many users. The majority of existing expert systems are applied in areas that are connected to the concept of analysis, such as diagnosis (medical and technical), interpretation of large numbers of measurement results, classification of objects and division into subclasses, etc.

Expert systems are also used in those areas connected to the concept of construction, such as configuring, design or synthesis and planning. In this area, a good example is the well-known expert system XCON used by the DEC Company for configuring VAX computers in accordance with user demand.

Though the possibilities of applications of expert systems in different areas of human activity seem to be unlimited, at present expert systems are mainly used to solve difficult problems in the domains of chemistry, medicine, geology, law, computer science, electronics, military, agriculture, meteorology, etc.

The application of an expert system in practice is useful in those cases where the cost of the system and the maintenance of the computer are lower than the employment of a human expert in a given domain.

Comparison of expert systems with conventional algorithmic programs to show the similarities and differences can be performed from different points of view. In

Table 13.1 Comparison of expert systems with conventional algorithmic programs

Conventional programs	Expert systems
-numerical data processing	-symbolic knowledge and data processing
-procedural (algorithmic) description of knowledge and processes ('how to do')	-declarative (descriptive) description of knowledge ('what to do'); algorithmic part concerns only the 'inference engine'
-knowledge is mixed with computer action statements in the program code	-knowledge is collected in the separate part of the system and described declaratively
-to change the knowledge it is necessary to modify large part, or all program	-easy change of knowledge by modification of the knowledge base without change of the rest of program
-program is 'closed' (completed) for given problem	-system is 'open' (not completed – as human knowledge) and can be extended easily
-explanations of operation are not used	-explanations: How a goal has been reached, and Why the system is asking a question
-with data not complete, program stops	-with not complete data or knowledge, it is possible to obtain approximate results
-program is used for single purpose	-by changing the knowledge in the knowledge base, the system can be used for several similar purposes (for ex. diagnosis, medical or technical)
-there exists precise 'flow-graph' of the program action	-it is difficult to create a 'flow-graph' since the action depends on facts (data) and answers to questions
-etc....	-etc....

Table 13.1, several attributes and remarks concerning both types of program are collected to show some differences in the two approaches.

13.3 Internal structure of an expert system

13.3.1 Introduction

The most important property of an expert system is the separation of the knowledge base from the rest of the program. Other important properties are as follows:

1. restriction of the knowledge to a narrow specific domain,
2. use of heuristics in problem solving,
3. symbolic description of problems,
4. possibility of explanations as to how the solution has been reached, etc.

A more detailed structure of an Expert System is shown in Fig. 13.2. Due to separation of the knowledge base of the system, the knowledge can be easily modified or even completely changed. Note that an expert system with an empty (cleaned out) knowledge base is referred to as a shell system. This empty base can be filled with the new knowledge from a similar domain type. For example, the expert system for the diagnosis of automobiles, that is, with the knowledge base filled with

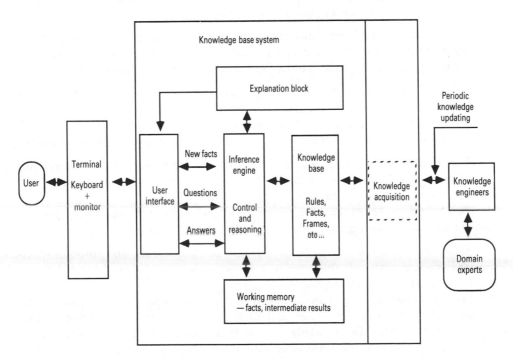

Figure 13.2 Detailed structure of an expert system

knowledge about Ford automobiles, can be used for the diagnosis of agriculture tractors after changing the knowledge about automobiles into tractors.

An expert system is created, maintained and used by different groups of specialists. The overall system, that is, the inference engine, user interface, explanation block, working memory, is created by computer scientists; the knowledge base is filled and updated practically by knowledge engineers, that is, specialists who elicit knowledge from the domain experts and code this knowledge in the knowledge base in different forms; the users are specialists in the given domain who wish to solve problems (usually difficult) using the expert system.

The most popular expert systems are those with a rule-based knowledge base, that is, knowledge about a specific domain is coded in the form of rules and facts.

13.3.2 Knowledge base and knowledge representation

The knowledge that resides in an expert system can be classified in different ways. One of these classifications can be the following:

1. axioms, or facts (that are always true) describing considered objects, concepts and their attributes,
2. inference rules or other forms relating objects and their attributes,
3. control knowledge in the form of rules and/or procedures that is necessary for proper and efficient problem solution; this type of knowledge is called 'meta-knowledge' or 'knowledge about knowledge'.

The first and second types of knowledge are contained in the knowledge base, and the metaknowledge is placed in the inference engine of an expert system.

Another type of classification distinguishes between the following:

1. *a priori* knowledge, composed of facts and rules about a specific domain prior to running an expert system (prior to consultation with the expert system),
2. inferred knowledge, composed of facts concerning a specific problem, derived during the run-time of the expert system, that is, during the problem solution.

The *a priori* knowledge is located in the knowledge base of an expert system, while the inferred knowledge is dynamic, as it changes for each problem considered, and resides in the working memory of the system.

Depending on the source from which the knowledge is obtained, it can be also divided into the following:

1. public domain knowledge, obtained from books, manuals, and through specified studies at school; this type of knowledge is termed deep knowledge and forms part of the knowledge base,
2. private knowledge, or heuristic knowledge developed by a human expert through long-term experience and skill (intuition, ability to make proper choices); this heuristic knowledge, most often described as heuristic rules, forms the essential part of the knowledge base of an expert system; it is often referred to as shallow knowledge.

The knowledge base of an expert system must be stored in the memory of a digital computer. Therefore, its representation should be formalized in order to:

1. provide a format that is compatible with the digital computer and establish knowledge representation that can easily be processed, that is, accessed, modified, updated, etc. Additionally, it would be highly profitable to:
2. use a representation format that can be easily read and understood by humans; this property is termed the transparency of the knowledge representation.

Several different modes of knowledge representation are used in expert systems. The most popular are:

1. facts and rules,
2. semantic networks,
3. frames.

Examples of facts are the statements

> 'amplifier μA741 has gain $= 100\,000$'

> 'amplifier μA741 has input resistance > 1 M ohms'

or, in an abbreviated form,

> 'μA741, gain, $100\,000$'

> 'μA741, $R_{in,} > 1\,000\,000$'

Such representation of facts is known as the 'object–attribute–value' or OAV triplet. In this case, we have

> Object: μA741

> Attribute: gain, or input resistance

> Value: $= 100\,000$, or $> 1\,000\,000$, respectively.

OAV triplets can be represented graphically using the network representation shown in Fig. 13.3. The attribute values can be numeric (gain $= 100\,000$) and symbolic ($R_{out} =$ low).[1]

Another fact concerning the same object may be

> 'amplifier μA741 is a linear active circuit'

and an additional fact with 'linear active circuit' as an object may be

> 'linear active circuit is an analog circuit'

Here the relationship of the type 'is–a' shows the hierarchical dependence between different objects and is used in the representation of semantic network and frame knowledge.

Besides facts, the most popular mode of knowledge representation is in terms of IF–THEN rules. A typical rule consists of at least two statements or clauses,

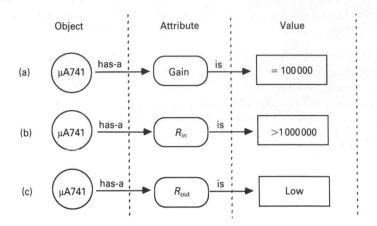

Figure 13.3 OAV networks for μA741 with attributes (a) gain, (b) input resistance, (c) output resistance

called premise and conclusion, or condition and action:

`IF <premise> THEN <conclusion>`

or

`IF <condition> THEN <action>.`

Examples of the IF-THEN rules are

`IF diode forward voltage > 0.7 V`
`THEN diode is conducting.`

`IF battery voltage is lower than nominal`
`THEN charge up battery.`

The first rule is an example of the inference rule since it allows us to conclude, or infer, that the diode is conducting, while the second rule is the production rule since it suggests taking concrete action.

The premise or condition part of the rule, also named LHS (left-hand side) of the rule, and conclusion or action part of the rule, also named RHS (right-hand side) of the rule, may be composed of several clauses connected by AND as well as OR operators. However, it is advisable to avoid using the OR operators, since this often leads to difficulties in interpretation of the inference process. Clauses connected by AND operators are referred to as conjunctive clauses.

Examples of rules with conjunctive clauses are

`IF emitter-base transistor junction is forward biased`
`AND collector-base transistor junction is reverse biased`
`THEN transistor is in active-forward mode.`

IF emitter-base transistor junction is reverse biased
AND collector-base transistor junction is reverse biased
THEN transistor is in cut-off mode.

It is necessary to distinguish between IF–THEN rules and IF...THEN... statements in algorithmic programs. The IF–THEN rules are mostly used to infer new facts (for example knowing the condition of transistor junctions we can conclude the mode of operation of the transistor), while IF...THEN... statements in algorithmic programs are mostly used to control the flow of consecutive computer operations.

Putting together the above facts concerning the μA741 amplifier and associated objects, we can obtain a graphical representation known as the semantic network, shown in Fig. 13.4.

The semantic network represents several objects with their attributes and the relations between the objects. Note that the relation 'has-a' concerns the attributes of a given object, while the relation 'is-a' shows the hierarchical associations between objects.

A very important property of hierarchical structures is the inheritance of properties. For example, in Fig. 13.4, when we define a linear active circuit as one composed of transistors operating in linear mode, resistors and capacitors, having input and output ports, requiring supply power, etc., then it is not necessary to repeat these properties in the definition of the μA741 amplifier, since it is a linear

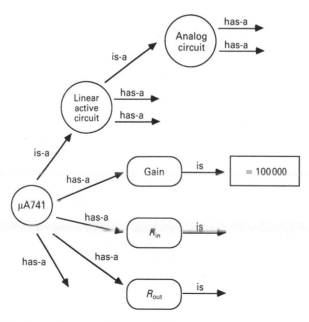

Figure 13.4 Semantic network for μA741 amplifier and associated objects

active circuit and inherits the properties of this circuit. The inheritance property saves a considerable amount of memory in object description.

Like semantic networks, more general knowledge representation is obtained using the concept of frames. A frame contains an object together with all the information related to that object. This information concerns the attributes and their values, default values when the concrete value is not known, procedures for the computation of the values if this is necessary and pointers to other frames – creating in this way a hierarchical frame structure with inheritance properties, and sets of rules that may be implemented. The properties of an object in a frame representation are located in so-called slots. A graphical representation of the frame for the μA741 amplifier is shown in Fig. 13.5.

The versatility of frame representation of knowledge is rather obvious. However, the inference process in this representation is more difficult to control and understand in comparison with rule-based expert systems.

13.3.3 Inference engine and basic inference strategies

The inference engine is an algorithmic block of an expert system and serves as the control and inference mechanism for the system. In this block, inference or problem solution strategies are coded. Inference is the process of drawing final or intermediate conclusions based on knowledge existing in the knowledge base (described by means of a set of rules, facts or other knowledge representation). The inference engine is also called a knowledge interpreter or knowledge processing algorithm; in the case of rule-based knowledge the name 'rule interpreter' is also used.

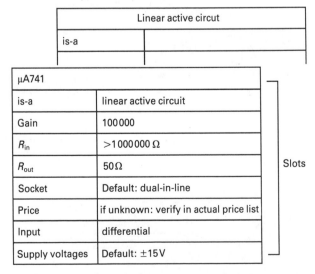

Figure 13.5 Frame based representation of μA741 amplifier

The control processes of the inference engine include starting the inference, selecting the method by which the search for a problem solution is conducted, determination of which rule to execute (or to fire) when more than one is selected for execution, deciding whether to ask questions of the user, etc.

The most popular inference method used in rule-based expert systems is based on the *modus ponens* rule of inference. The *modus ponens* rule states that if the premise P of a rule is true, and if P infers the conclusion C, then C is true, which can be represented as

$$[P \text{ and } (P \rightarrow C)] \rightarrow C \tag{13.1}$$

However, when the conclusion C is true, we cannot always say that the premise P is true. For example, we can state

IF a linear circuit is reciprocal[2]

THEN it is passive

However, we cannot state that if a circuit is passive then it is reciprocal, since the gyrator circuit is known to be passive and not reciprocal.

To accomplish development of a recommended solution to a given problem, the expert system must conduct a search for the solution, that is, select proper intermediate solution steps in order to reach the final goal in an efficient and effective way. There are two basic search strategies, or knowledge processing mechanisms, based on rule and fact evaluations:

1. forward chaining, or forward reasoning,
2. backward chaining, or backward reasoning.

In the forward chaining method, the premises or LHSs of rules are compared with existing facts, or data residing in the working memory of an expert system. If the premise, or premises, of a specified rule or rules agree with existing facts then the rules are activated and can be fired (executed). If several rules are activated at the same time, then the control mechanism of the inference engine decides which rule is fired first. Execution of the activated rules results in new facts that are put into the working memory. Since the activation of rules depends on facts (or data) provided during consultation, forward chaining is also termed data-driven reasoning as the inference proceeds from premises (or data) to conclusions (or goals).

In the backward chaining method the conclusions, or RHSs of rules, are compared with a goal to be proven which is located in the working memory. Those rules are activated whose RHSs agree with the goal provided. After firing a specified active rule, its premises are the new subgoals which are put into the working memory and which again have to be proven.

In this case the inference process proceeds from conclusion (or goals) to premises; it is also called goal-driven inference.

In the backward chaining method, only one fact as a goal to be proven is provided for the expert system during the consultation, while in the forward chaining process there are two different approaches. In the first approach, all known facts (or data) are delivered to the expert system at the beginning of the consultation. In the second approach, only those data are delivered (usually at the demand of expert system) which are necessary to reach intermediate and final conclusions.

As an example let us consider the set of rules concerning applications of transistors, with which an expert system is equipped (B, E, C stand for base, emitter and collector, respectively):

```
Rule 1 :  IF transistor-junction-B-E is forward-biased
          AND transistor-junction-B-C is forward-biased
          THEN transistor is saturated

Rule 2 :  IF transistor is saturated
          THEN transistor-application is not-recommended

Rule 3 :  IF transistor-junction-B-E is forward-biased
          AND transistor-junction-B-C is reverse-biased
          THEN transistor-application is amplification

Rule 4 :  IF transistor-application is amplification
          AND transistor has positive-feedback
          THEN transistor can generate-periodic-waves

Rule 5 :  IF transistor-application is amplification
          AND transistor has negative-feedback
          THEN transistor can
                       amplify-with-increased-stability
```

Now, let us assume forward chaining and provide three known facts to the working memory of the expert system:

```
F1 : transistor-junction-B-E is forward-biased
F2 : transistor-junction-B-C is reverse-biased
F3 : transistor has positive-feedback
```

Then the inference engine will start to check the LHSs of the rules and will activate Rule 3 since its two premises correspond to facts F1 and F2. No other rules can be activated. Thus, Rule 3 will be fired and its conclusion:

```
transistor-application is amplification
```

will be added as a new fact, F4, to the working memory.

The inference engine again checks the knowledge base to activate those rules LHSs of which correspond to facts in the working memory. Since Rule 3 has already been fired, it cannot be activated again. But, Rule 4 will be activated since its two premises correspond to facts F4 and F3. Thus, Rule 4 will be fired and its conclusion:

`transistor can generate-periodic-waves`

will be added as a new fact, F5, to the working memory.

The inference engine again checks the knowledge base to activate the rules. Since Rule 3 and Rule 4 have been fired, they cannot be activated again, and there are no other rules to be activated by the facts from working memory. Therefore, the inference process is stopped and the inference engine 'tells' the user interface to display information for the user (for example):

`It has been concluded that the transistor can`
`                                        generate-periodic-waves`

and closes the consultation session.

In the second approach of the forward chaining no facts are delivered *a priori* to the expert system (empty working memory). In this case the inference engine takes the rules one after the other and asks the user about the facts. Let us assume that the known facts are F1, F2, F3, as before. Now the consecutive steps are as follows:

1. Inference engine gets Rule 1 and checks working memory for its premises; not knowing this, the inference engine 'tells' the user interface to ask the user.
2. The user interface displays (for example):

   `is the transistor-junction-B-E forward-biased? (Yes/No)`
   `user:              yes`
   `is the transistor-junction-B-C forward-biased? (Yes/No)`
   `user:              no`

3. The working memory now has fact:

   `transistor-junction-B-E is forward-biased`

   This is insufficient to fire Rule 1.
4. The inference engine gets Rule 2 and causes display:

   `is the transistor saturated? (Yes/No)`

   user :__ (no answer because it is not known)
5. The inference engine gets Rule 3 and causes display:

   `is the transistor-junction-B-E forward-biased? (Yes/No)`
   `user:yes`
   `is the transistor-junction-B-C reverse-biased? (Yes/No)`
   `user:yes`

6. The working memory now has facts:

   `transistor-junction-B-E is forward-biased`
   `transistor-junction-B-C is reverse-biased`

7. The inference engine fires Rule 3 to conclude that:

    ```
    transistor-application is amplification
    ```

    and puts this to the working memory as a new fact.
8. The working memory has new facts:

    ```
    transistor-junction-B-E is forward-biased
    transistor-junction-B-C is reverse-biased
    transistor-application is amplification
    ```

9. The inference engine gets the Rule 4 and checks the working memory for:

    ```
    transistor-application is amplification
    transistor has positive-feedback
    ```

10. The first premise of Rule 4 agrees with the fact in the working memory and the second does not, therefore the inference engine causes to display question:

    ```
    has the transistor positive-feedback? (Yes/No)
    user:yes
    ```

11. The working memory has now facts as in step 8, and additionally:

    ```
    transistor has positive-feedback
    ```

12. The inference engine fires Rule 4 to conclude that:

    ```
    transistor can generate-periodic-waves
    ```

    and puts this as a new fact to the working memory.
13. The inference engine gets Rule 5 and checks the working memory for:

    ```
    transistor-application is amplification
    transistor has negative-feedback
    ```

14. The first premise of Rule 5 agrees with the fact in the working memory and the second does not; thus, the inference engine causes display:

    ```
    has the transistor negative-feedback? (Yes/No)
    user:no
    ```

15. The working memory has not changed, since Rule 5 has not been fired.
16. The inference engine checks again all rules starting from Rule 1 and, since no new rules can be fired, the inference engine 'tells' the user interface to display:

    ```
    It has been concluded that the transistor can
                                     generate-periodic-waves
    ```

    and closes the consultation session.

Let us now briefly consider the backward chaining method with the same set of rules R1–R5 in the knowledge base as in the previous example. Also, let us assume that three facts, F1, F2, F3, are known and entered into the working memory of the expert system.

The backward chaining begins with the specification of the goal; let us assume that we wish to know whether

`"the transistor can generate-periodic-waves"`

based on information given by facts F1, F2 and F3.

The inference engine first checks whether this goal is reached by examining the facts in the working memory. Since such facts are not known, the inference engine checks the RHSs of rules in the knowledge base and finds Rule 4 which has the goal as its conclusion. Then the inference engine checks the premises of this rule if they are satisfied by the facts in the working memory.

The second premise of Rule 4:

`"transistor has positive-feedback"`

is satisfied by fact F3; however, the first premise of Rule 4:

`"transistor-application is amplification"`

is not known.

Therefore, the inference engine makes this premise a subgoal and checks again the RHSs of the rules. Now it finds Rule 3 which has the subgoal as its conclusion. Then the inference engine checks the premises of Rule 3 to see if they are satisfied by the facts in the working memory. In this case, both premises are satisfied by facts F1 and F2, respectively.

Since there are no other subgoals to be satisfied, the backward chaining process stops and the inference engine may 'tell' the user interface to inform the user that the goal is TRUE, that is, 'the transistor can generate periodic-waves' when

1. transistor-junction-B-E is forward-biased, (F1),
2. transistor-junction-B-C is reverse-biased, (F2),
3. transistor has positive-feedback,   (F3).

Note that in the backward reasoning process only Rule 4 and Rule 3 have been used. Generally, unlike the forward chaining method, with backward chaining reasoning only the relevant rules (to reach an assumed goal) are used.

To examine the relationships among rules we can draw an inference network or inference tree for the rules of interest. The inference tree for Rule 1–Rule 5, for the example considered earlier with forward chaining is shown in Fig. 13.6(a). Here, starting from facts F1–F3, we 'go-up' the tree to obtain some new intermediate facts until we reach the final conclusion. Thus forward chaining is sometimes called the 'bottom-up' method. The inference tree for backward chaining is simpler (see Fig. 13.6(b)), since only two rules, R4 and R3, were considered in the previous examples.

Here, starting from the assumed goal, we 'go-down' the tree to form some subgoals, until we prove that all the subgoals are satisfied by the known facts. Therefore, backward chaining is sometimes called the 'top-down' method. Some expert systems use both forward and backward methods at the same time.

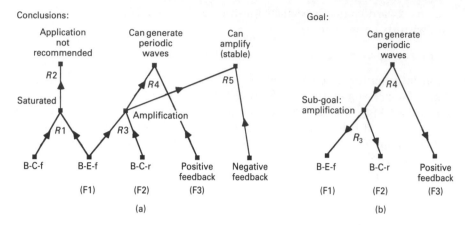

Figure 13.6 Inference tree. (a) For forward chaining. (b) For backward chaining
(B-C-f -junction B-C is forward biased, B-C-r -junction B-C is reverse biased)

To reduce the number of rules to be examined in the forward chaining method, the inference engine is sometimes equipped with metarules, that is, rules about rules. The metarule applied to the example considered earlier, where we are interested in the active operation of a transistor, might be:

```
IF transistor-operation is active
THEN avoid rules where both junctions are equally biased
AND other rules being consequences of these rules.
```

Thus, in our example, Rule 1 and Rule 2 could be omitted and the search space reduced.

13.3.4 User interface and explanation block

The user interface enables user communication with the inference engine and the knowledge base of an expert system during consultation. During realization of the user interface much attention is paid to the so-called user-friendliness of the inter-face, which means that the user of the expert system need not know much about the operation of a computer and expert system construction. The user interface inter-prets, displays on the graphical monitor, prints, plots and transmits the necessary information to the user in a manner that can be easily understood by those humans who are not computer specialists. Usually, the dialog between the expert system and the user is carried on in simplified language that is close to natural language.

At the very beginning of the consultation session, the initiative is taken by the expert system, which explains what the user has to do in order to proceed with the dialog.

If the initial data concerning the problem solution delivered by the user are insufficient, then the expert system prompts the user for additional data. The choice of options presented to the user by the system is often exposed by means of a multi-level menu, sometimes using window facilities. The interface should also have the possibility of expressing the intermediate and final results of the consultation in a form that is acceptable to humans.

The typical properties of the user-friendly interface may be summarized as follows:

1. operation should be simple and not require long training,
2. errors during data (or facts) introduction should be detected immediately and improvements suggested,
3. intermediate and final results should be in a form that is acceptable to the user (sentences close to natural language), graphical interpretations should be understandable, etc.
4. prompts to the user during interactive consultation should be clear,
5. explanations (if any) of user questions should also be clear and acceptable to the user.

The last property requires an additional system block with which some expert systems are equipped.

The explanation block is not necessary for proper operation of an expert system. However, its existence facilitates understanding of the operation of the expert system since, usually, the explanation block can explain each step of the consultation process. It is quite important that users who may not be outstanding experts in the domain can have the possibility of understanding the line of reasoning and the recommended solutions proposed by the expert system.

Typical discussion between the user and the explanation block (via the user interface) concerns questions of the type WHY and HOW, that is, why a particular question has been asked of the user, and how the system came to a specific conclusion.

For example, if during the forward chaining consultation of Section 13.3.3, after the final conclusion 'transistor can generate periodic-waves', we have asked the system HOW ?, the explanation block could respond (for example) 'because of Rule 4' or 'because transistor can amplify and has positive feedback', or in any more sophisticated way.

13.4 Heuristic programming and propagation of constraints

As has already been mentioned in this chapter, in expert systems, heuristic knowledge is used intensively and the knowledge is located in a separate part of the system – the knowledge base. Thus, the use of heuristics and the separation of the knowledge base from the rest of the program, especially from the control block, are the main features or properties of an expert system.

When, however, we combine several heuristic rules and facts in a logical format and use the control statements of classical algorithmic programs, then we obtain a computer program which can be called a heuristic program. For some specific problems, when we have no access to an expert system shell (that is, system with empty knowledge base), we can use the heuristic programming approach to construct an expert system to solve these problems.

Heuristic programs may be employed in cases where exact algorithmic methods cannot be used efficiently. Since in this approach approximate rules are applied, heuristic programs can provide a problem solution which may be approximate (not optimal), but acceptable.

A typical field in the area of electronics for the application of heuristic programming, as well as expert systems, is in the design of electronic circuits and equipment. In this area, to reach an optimal (the best in some sense) solution is very difficult, since there are no exact general algorithms (exceptions are synthesis methods of passive idealized circuits, for example filters) and we are looking for circuit realization that fulfil the requirements demanded with given *a priori* acceptable margins (approximate solution).

Heuristic programs can also be used in configuration, diagnosis and repair of electronic equipment; here, the experience of an expert can be utilized successfully. In the analysis of electronic circuits neither expert systems nor heuristic programming are used, since there exist numerous efficient algorithmic methods for the exact solution of analytic problems. However, even in analysis, heuristic programs can be applied when we wish to emphasize not only the final result of the analysis, but also the details of the operation of the analyzed circuit, that is, its behaviour during the operation. In this case, the approach known as constraints propagation can be used successfully both in the analysis and design of electronic circuits.

The constraints propagation method is a generalization of the informal Guillemin method of ladder circuit analysis. Guillemin's method is illustrated by analysis of the circuit of Fig. 13.7. The analysis starts from the assumption that the last node voltage V_3 has a symbolic value, that is, $V_3 = a$.

Then, single-step deduction is carried out:

$$\text{if } V_3 = a \text{ and } V_3 = R_6 I_6 \text{ then } I_6 = \frac{a}{R_6}$$

The KCL for node 3 states that $I_6 - I_5 = 0$

$$\text{if } I_6 - I_5 = 0 \text{ and } I_6 = \frac{a}{R_6} \text{ then } I_5 = \frac{a}{R_6}$$

The KVL for last mesh states that $-V_2 + R_5 I_5 + V_3 = 0$

$$\text{if } -V_2 + R_5 I_5 + V_3 = 0 \text{ and } V_3 = a \text{ and } I_5 = \frac{a}{R_6} \text{ then } V_2 = \frac{a}{R_6} R_5 + a$$

The KCL for node 2 states that $I_4 + I_5 - I_3 = 0$ and Ohm's law states that $I_4 = V_2/R_4$ giving as a result

$$I_3 = \frac{V_2}{R_4} + \frac{V_3}{R_6} \text{ or } I_3 = a\left\{\frac{R_5}{R_6 R_4} + \frac{1}{R_4} + \frac{1}{R_6}\right\}, \text{ and so on}$$

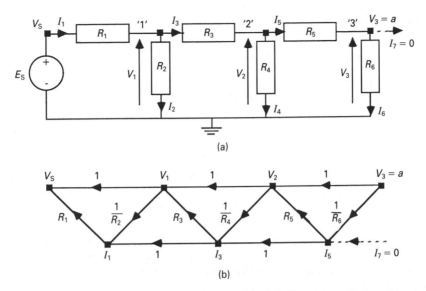

Figure 13.7   (a) Resistive ladder circuit for illustration of Guillemin's analysis method. (b) Its
directed flow graph or constraints propagation diagram

Finally, the voltage V_S is determined:

$$V_S = af(R_i), \qquad i = 1, 2, ..., 6$$

which must be equal to E_S; hence, the assumed unknown symbolic value a is obtained:

$$a = \frac{E_S}{f(R_i)}$$

Knowing a, we can easily compute all the node and branch voltages of the circuit. In numerical computer programs, we can assume that $a = 1$ for simplicity. Then, the actual node voltages V_{ak} are computed as

$$V_{ak} = V_k \frac{E_S}{V_S}, \qquad k = 1, 2, 3$$

In this example, we see directly how the constraint imposed on the last voltage $V_3 = a$ is propagated step by step, according to one-step deduction, backward through the circuit until the input voltage E_S is reached; the intermediate constraints are Ohm's and Kirchhoff's laws.

Note that the circuit description is not in terms of nodal equation and matrix description, but, rather, it is in terms of the directed flow graph as shown in Fig. 13.7(b). The compact description of this circuit is by means of the following six simple equations:

$$I_5 = \frac{V_3}{R_6}, \qquad V_2 = V_3 + I_5 R_5$$

$$I_3 = I_5 + \frac{V_2}{R_4}, \qquad V_1 = V_2 + I_3 R_3$$

$$I_1 = I_3 + \frac{V_1}{R_2}, \qquad V_S = V_1 + I_1 R_1$$

This type of circuit description has properties similar to the sparse matrix method since only small fragments of the circuit are described and proper variables are computed step by step, avoiding multiplication by zeros, as happens in other computation methods based on matrix description of circuits.

The constraints propagation method has an educational advantage since, besides simple analysis of a circuit, it provides behavioural information, that is, it shows how the circuit operates. It has been used for years by many teachers in the classroom and it is sometimes called step-by-step description of circuit operation.

Consider, for example, the transistor circuit of Fig. 13.8 for which all resistances are known and assume that the transistor is in forward active (linear) operation mode. We can find the DC properties of the circuit using the constraints propagation method starting from different nodes of the circuit.

A – constraints formulation starting from the input (V_B – assumed).

assume that V_{CC} is known

assume that $V_B = aV_{CC}$, usually $a < 1/2$ (heuristic assumption)

assume that $I_B \ll I_D$, $I_B \ll I_E$, thus $I_C = I_E$ (approximately satisfied with $\beta > 100$)

$$V_B = aV_{CC} = \frac{R_1}{R_1 + R_2} V_{CC}, \qquad a = \frac{R_1}{R_1 + R_2}$$

$V_E = V_B - V_{BE} = V_B - 0.7$ ($V_{BE} = 0.7$ V – heuristic knowledge for active silicon transistor)

$$I_E = \frac{V_E}{R_E}, \qquad I_C = I_E$$

$V_C = V_{CC} - I_C R_C$

V_C must be $V_C > V_B + 0.8$ (for active linear transistor operation (heuristics))

After constraints formulation (description of the circuit), we can start the constraints initialization by assuming that the resistance and V_{CC} values are

$$R_1 = 10 \text{ k}\Omega, \ R_2 = 40 \text{ k}\Omega, \ R_E = 1.15 \text{ k}\Omega, \ R_C = 5 \text{ k}\Omega, \ V_{CC} = 15 \text{ V}$$

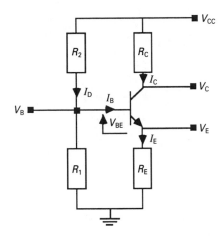

Figure 13.8 Simple transistor circuit with linear transistor operation

hence

$$a = \frac{R_1}{R_1 + R_2} = 0.2$$

This starts constraints propagation

$$V_B = 3 \text{ V} \qquad V_E = 2.3 \text{ V}$$

$$I_E = 2 \text{ mA} \qquad I_C = 2 \text{ mA}$$

$$V_C = 5 \text{ V}$$

checking for proper active transistor operation

$$V_C > V_B + 0.8 \text{ V} = 3.8 \text{ V}$$

B – constraints formulation starting from the middle point of the circuit (I_E assumed).

Assume that I_E is known, for example:

$$I_E = 1 \text{ mA}$$

$$V_E = R_E I_E, \qquad I_C = I_E$$

$$V_B = V_E + V_{BE}$$

Assume a is selected

$$V_{CC} - a V_{BE}, \qquad V_C = V_{CC} - I_C R_C$$

for given a,

$$R_2 = \left\{ \frac{1}{a} - 1 \right\} R_1$$

For the same data as earlier (except the R_2 value), we obtain the constraints propagation

$$I_E = 1 \text{ mA}, \qquad I_C = 1 \text{ mA}$$

$$V_E = 1.15 \times 10^3 \times 10^{-3} = 1.15 \text{ V}$$

$$V_B = 1.15 + 0.7 = 1.85 \text{ V}$$

assume $a = 0.12333$ (to have $V_{CC} = 15$ V)

$$V_{CC} = 0.12333 \times 1.85 = 15 \text{ V}$$

$$V_C = 15 - 5 \times 10^{-3} \times 10^3 = 10 \text{ V}$$

$$R_2 = 71.08 \text{ k}\Omega$$

Checking for proper active operation of the transistor:

$$V_C > V_B + 0.8 \text{ V} = 2.65 \text{ V}$$

Constraints propagation diagrams for the DC properties of the circuit of Fig. 13.8 with two different starting points are shown in Fig. 13.9(a, b), respectively.

As we can see, operations on constraints include the following:

1. constraints formulation,
2. constraints initialization,
3. constraints propagation,
4. constraints satisfaction.

1. Constraints formulation corresponds to circuit description using the directed graph method; therefore, relations between voltages, currents and circuit elements form constraints and correspond to the flow of signals. These constraints include the Kirchhoff current and voltage laws and Ohm's law of local parts of the circuit. This type of circuit description gives much more insight into circuit operation (behaviour) as compared to matrix or any other formal descriptions.

2. Constraints initialization corresponds to the assignment of values to circuit elements and some initial current or voltage values that will be propagated through the circuit. If the problem concerns an analysis, then all circuit elements are given *a priori*.

Much more interesting for the application of constraints propagation is the design of electronic circuits; there, we have to apply several heuristic rules since there are no general algorithms for design. Even for design of such a simple transistor circuit as that shown in Fig. 13.8, we are free to choose many parameters based on designer experience and to combine these with the requirements. For a given supply voltage V_{CC} and total dissipation power P_d, we can select the DC I_E current, for example $I_E = 1$ mA. Then, using heuristics, we can say that the current I_D through the voltage divider composed of R_1 and R_2 should be of order 0.1 I_E, since at the same time this current should be much larger than the base current I_B. Again, using heuristics we can state that V_E should be several times larger than V_{BE} to obtain small dependency of V_E from V_{BE} (V_{BE} varies with temperature). Also, parallel connection of R_1 and R_2 should be significantly larger than R_E, for example $R_1R_2/(R_1 + R_2) > 8R_E$ (this also results from thermal stabilization).

Additionally, if we wish to have a large output voltage without significant distortion, then the operating point Q should have the voltage coordinate $V_{CQ} = (V_{CC} - V_E)/2$; but if we wish to have low noise operation then the current coordinate I_{CQ} should have the proper computed value based on noise factor minimization. For more complicated circuits, many more heuristic rules can be used (when we could have access to an expert circuit designer). These heuristics considerations could also be taken into account during constraints formulation.

3. Constraints propagation starts after constraints initialization; in the most simple case, only one constraint value is propagated in one direction, as happens in a ladder network or as shown in Fig. 13.9(a), or in many directions, as shown in Fig. 13.9(b).

4. Constraints satisfaction corresponds to checking whether initial circuit demands or requirements during design process are satisfied, or whether we come during constraints propagation to some conditions that cannot be accepted, for example collector–base voltage $V_{CB} < 0$ V cannot be accepted for the required forward active operation of an NPN transistor.

In most simple cases, only a single numerical constraint is propagated, as was explained for the circuits of Fig. 13.7 and Fig. 13.8. For more complicated circuits such as parallel ladder or feedback circuits it is also possible to apply the constraints propagation method for the simultaneous propagation of two constraints. Consider, for example, a double-ladder circuit shown in Fig. 13.10(a), and its directed graph or constraints diagram shown in Fig. 13.10(b).

Here, we can choose for example, $V_3 = d$ and then obtain

$$I_6 = \frac{V_3}{R_6} = \frac{d}{R_6}, \qquad I_5 = I_6 - I_9$$

and also assume that

$$I_9 = b$$

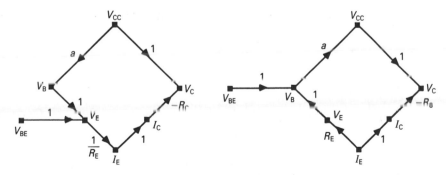

Figure 13.9 Constraints diagrams for the circuits of Fig. 13.8 with starting points: (a) from input, (b) from I_E

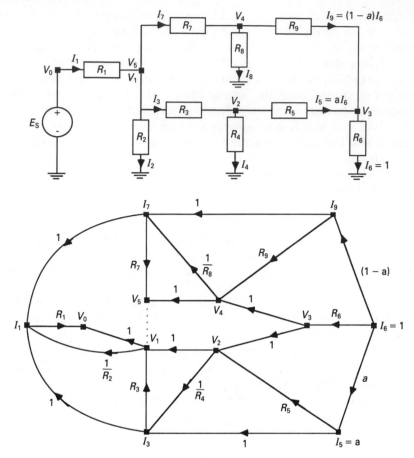

Figure 13.10   (a) Double-ladder resistive circuit. (b) Constraints diagram of the double-ladder circuit

in order to have the two constraints propagation that are d and b

$$I_5 = \frac{d}{R_6} - b$$

However, it is much simpler to choose

$$I_5 = aI_6, \qquad \text{and } I_9 = (1-a)I_6$$

to have

$$I_6 = I_5 + I_9 = 1, \qquad V_3 = I_6 R_6 = R_6$$
$$V_2 = V_3 + I_5 R_5, \qquad V_4 = V_3 + I_9 R_9$$

$$I_3 = I_5 + \frac{V_2}{R_4}, \qquad I_7 = I_9 + \frac{V_4}{R_8}$$

$$V_1 = V_2 + I_3 R_3, \qquad V_5 = V_4 + I_7 R_7$$

The voltages V_1 and V_5 must be equal.

Constraints initialization could be with any value of a, with $0 \leqslant a \leqslant 1$. However, based on heuristics we can assume the value of a in the following way:

1. If resistance values in the upper and lower parts of the double ladder are of the same order, then $a = 0.5$.
2. If resistance values in the upper part are higher, than $a > 0.5$.
3. If resistance values in the upper part are lower, than $a < 0.5$

If after constraints initialization V_1 and V_5 are not equal then we can begin an iterative, or interactive, process by changing the a value to obtain V_1 and V_5 values close to each other and with the required accuracy. The automatic change of the a value could be, for example,

$$a = \frac{a}{\left\{ \left(\frac{V_1 + V_5}{2} \right) \middle/ V_5 \right\}}$$

and the process of the evaluation of V_1 and V_5 could be stopped when

$$1 - acc \leqslant \frac{V_1}{V_5} \leqslant 1 + acc$$

where acc is the required accuracy of the computation. Then, I_1 and V_0 are computed, the value of the source voltage E_S is entered and the node voltages V_n are corrected according to the formula

$$V_{ni} = V_{ni} \times \frac{E_S}{V_0}$$

The example program 'DOUBLLAD' for constraints propagation in the double resistive ladder of Fig. 13.10 is shown in Program 13.1.

See the listing of the program DOUBLLAD Directory of Chapter 13 on attached diskette.

After running the program and entering the resistance values (in ohms or kilohms):

$$R_1 = 1; \; R_2 = 2; \; R_3 = 1; \; R_4 = 2; \; R_5 = 1; \; R_6 = 1; \; R_7 = 2; \; R_8 = 4; \; R_9 = 1$$

and $a = \text{alpha} = 0.5$, we obtain all the intermediate results during constraints propagation and the possibility of entering modified $a = \text{alpha}$ values. Without interactive entering of alpha values we obtain intermediate V_1/V_5 and alpha values during interactions:

$$\frac{V_1}{V_5} = 0.658, \quad alpha = 0.597$$

$$\frac{V_1}{V_5} = 0.843, \quad alpha = 0.643$$

$$\frac{V_1}{V_5} = 0.946, \quad alpha = 0.661$$

$$\frac{V_1}{V_5} = 0.986, \quad alpha = 0.665$$

$$\frac{V_1}{V_5} = 0.997, \quad alpha = 0.666$$

$$V_o = 6.995$$

Then entering E_S, for example $E_S = 10$ V, we obtain the node voltages

$$V_1 = 4.522 \text{ V}, \; V_2 = 2.389 \text{ V}, \; V_3 = 1.430 \text{ V}, \; V_4 = 2.386 \text{ V}, \; V_5 = 4.536 \text{ V}$$

In a similar way, active feedback circuits can be analyzed showing the constraints propagation through the circuit. For example, let us consider the circuit with parallel, or voltage–current, feedback shown in Fig. 13.11(a). Its equivalent circuit and constraints diagrams are shown in Fig. 13.11(b and c), respectively. Here, it is assumed that the load current $I_L = I_5$ equals 1 and the constraint variable α is defined to satisfy

$$I_3 = \alpha I_5, \qquad I_4 = (1 - \alpha)I_5,$$
$$I_L = I_5 = I_3 + I_4 = 1$$

From the constraints diagram of Fig. 13.11(c) we obtain directly

$$V_4 = I_5 R_5, \qquad V_3 = V_4 + I_3 R_3,$$
$$V_2 = -\frac{V_3}{A_V}, \qquad V_5 = V_4 + I_4 R_4$$

If a proper α value has been selected at constraints initialization, then voltages V_2 and V_5 should be almost equal, with the prescribed accuracy; or else it is necessary to start the iteration process to change the α value to obtain

$$1 - acc \leqslant \frac{V_2}{V_5} \leqslant 1 + acc$$

where *acc* is the desired accuracy. After obtaining $V_2 \cong V_5$, we can compute

$$I_2 = \frac{V_2}{R_2}, \qquad I_1 = I_2 + I_4$$

$$V_1 = V_2 + I_1 R_1,$$

and compute the voltage gain V_4/V_1.

The 'first guess' of the value of α can be obtained from heuristic consideration based on the assumption that the output resistance $R_O = R_3$ of an operational amplifier is low; thus, $V_3 \cong V_4$. To obtain $V_2 \cong V_5$ we can write

$$-\frac{V_4}{A_V} = V_4 + (1 - \alpha)R_4$$

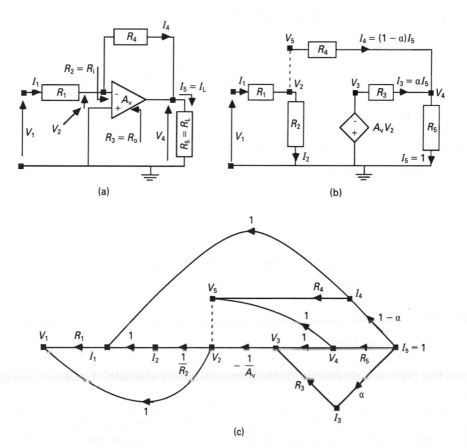

Figure 13.11 (a) Circuit with voltage–current feedback. (b) Its low frequency equivalent circuit. (c) Its constraints diagram

and, with $I_L = I_5 = 1$, we have $V_4 = R_5$ or

$$\alpha = 1 + \frac{R_5}{R_4}\left\{1 + \frac{1}{A_V}\right\}$$

The example program 'CONFEEDB' for constraints propagation in the feedback amplifier of Fig. 13.11(a), with interactive correction of the constraint variable α value is shown in Program 13.2.

See the listing of the program CONFEEDB Directory of Chapter 13 on attached diskette.

After running the program and entering the resistance values in kilohms:

$$R_1 = 1; \qquad R_2 = 1000; \qquad R_3 = 0.1; \qquad R_4 = 10; \qquad R_5 = 1$$

and the voltage gain $A_V = 1000$, we obtain

1. the first guess of α: alpha = 1.10010

2. $\dfrac{V_2}{V_5} = 1.11001$

Assuming accuracy $acc = 0.01$, we have the condition for $V_2 \cong V_5$:

$$0.99 \leqslant \frac{V_2}{V_5} \leqslant 1.01$$

Since this condition is not satisfied we change α for

alpha = 1.10011

and obtain

$$\frac{V_2}{V_5} = 1.00910$$

Since now, $V_2/V_5 < 1.01$, the voltage V_1 is computed:

$$V_1 = -0.10122$$

and also the voltage gain V_4/V_1 is printed out

$$\frac{V_4}{V_1} = -9.87936$$

The print-out of the results of the program 'CONFEEDB' for this numerical example is given in Table 13.2.

In a similar way the constraints propagation method can be used for the analysis of more complex circuits. However, as was described earlier in this chapter, heuristic programs and expert systems are not used for the analysis of electronic circuits, since powerful numerical algorithms exist for this purpose. Thus, the constraints propagation method for analysis can only be used in the educational process.

Table 13.2   Print-out of the results of the program 'confeedb'   427

```
Constraints Propagation in the Feedback Amplifier
^^^^^^^^^^^^^^^^^^^^^^^^^^^^^^^^^^^^^^^^^^^^^^^^^^
Enter the number of resistors nr : 5
Enter R[1] : 1
Enter R[2] : 1000
Enter R[3] : 0.1
Enter R[4] : 10
Enter R[5] : 1

Enter the voltage gain Av:1000

The first guess of alpha is :
Alpha = 1.10010
i5 = 1.000
v4 = 1.0000
v3 = 1.1100
v2 = - 0.0011
^^^^^^^^^^^
v5 = - 0.0010
^^^^^^^^^^^
V2/V5 = 1.11001
^^^^^^^^^^^^^^
Change alpha if necessary :
Increase alpha when V2/V5 > 1.01 or else - Decrease :
Previous Alpha = 1.10010 - New : 1.10012
v4 = 1.0000
v3 = 1.1100
v2 = - 0.0011
^^^^^^^^^^^
v5 = - 0.0012
^^^^^^^^^^^
V2/V5 = 0.92501
^^^^^^^^^^^^^^
Change alpha if necessary :
Increase alpha when V2/V5 > 1.01 or else - Decrease :
Previous Alpha = 1.10012 - New : 1.10011
v4 = 1.0000
v3 = 1.1100
v2 = - 0.0011
^^^^^^^^^^^
v5 = - 0.0011
^^^^^^^^^^^
V2/V5 = 1.00910
^^^^^^^^^^^^^^
Change alpha if necessary :
Increase alpha when V2/V5 > 1.01 or else - Decrease :
Previous Alpha = 1.10011 - New :
i2 = - 0.00000
i1 = - 0.10011
v1 = - 0.10122

V[1] = - 0.1012   V[2] = - 0.0011   V[3] = 1.1100   V[4] =
1.0000   V[5] = - 0.0011
Volt gain V4/V1 = - 9.87936


*The end*
```

Different situations exist in the field of electronic circuits and equipment design. Generally, there are no exact numerical algorithms for electronic circuit design – especially for analog circuits (the exceptions are the synthesis methods for passive RLC filters). Thus, success of design in this field depends mainly on knowledge, experience and intuition, that is, on the heuristics of the design expert.

To illustrate the philosophy of a heuristic design program with the use of step-by-step constraints propagation, let us consider the design of the simple single-stage transistor amplifier shown in Fig. 13.12. The circuit has been selected as one of many possible amplifying circuits that can meet some initial requirements. In larger design programs several circuits can be considered (selection depending on input requirements) as potential candidates to fulfil the initial design requirements.

To begin the design we have to assume some requirements to be met. In our example, let the input requirements be

$$E_{CC} = 12 \text{ V}, \qquad V_o = \frac{E_{CC}}{3} \pm 0.5 \text{ V}, \qquad I_E = 2 \times 10^{-3} \pm 1.5 \times 10^{-3} \text{ A}$$

AC (alternating current) voltage gain $= 10 \pm 0.5$

Total dissipation power $\leqslant 0.05$ W

Then, we can create some heuristic rules, such as the AC voltage gain:

$$\text{Gain} \cong -\frac{R_O}{R_E} \text{ (obtained from previous AC analysis)}$$

if $I_E \leqslant 1$ mA, then $R_E = 1 \text{ k}\Omega$,

if $I_E \geqslant 1$ mA, then $R_E = 1000 - I_E \times 0.1$

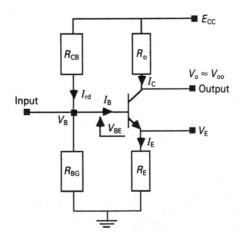

Figure 13.12 Single-stage transistor amplifier

(for larger I_E it is necessary to reduce R_E, to obtain the proper V_o and Gain),

if I_E and R_E are selected, then $V_E = R_E I_E$

if voltage Gain and R_E are selected, then $R_O = \text{abs}(\text{Gain } R_E)$

The computed output DC voltage V_{oo} is

$$V_{oo} = E_{CC} - R_O I_E$$

which must be compared with the required output DC voltage V_o in the range

$$V_{omi} < V_{oo} < V_{oma}$$

$$V_{omi} = V_o - 0.5, \qquad V_{oma} = V_o + 0.5$$

if V_{oo} is outside these limits, then modify the I_E value:

if $V_{oo} < V_{omi}$, then $I_E = I_E \times 0.8$ and repeat computations

if $V_{oo} > V_{oma}$, then $I_E = I_E \times 1.2$ and repeat computations

if proper I_E value has been selected and V_E computed, then
$V_B = V_E + V_{BE}$, $V_{BE} = 0.7$ V (assumed for silicon NPN transistor)

assume $I_B = 0$, that is, $I_B \ll I_{rd}$ (I_{rd} − resistor divider current)

and

assume R_{BG} value, for example, $R_{BG} = 10$ kΩ, or any other value

if R_{BG} is known and $I_B = 0$, then $R_{CB} = R_{BG}\left\{\dfrac{E_{CC}}{V_B} - 1\right\}$

if $I_B \neq 0$, then compute $I_B = \dfrac{I_E}{\beta}$, (β − transistor gain coefficient)

if I_B is known and R_{BG} is known (assumed) then $R_{CB} = R_{BG}\dfrac{(E_{CC} - V_B)}{V_B + I_B R_{BG}}$

check if total actual power dissipated

$$P_{at} = P_{aC} + P_{aB}$$

where

$$P_{aC} = I_E^2(R_E + R_O)$$

and

$$P_{aB} = \frac{E_{CC}^2}{(R_{CB} + R_{BG})} \text{ is lower than the requirement } P_{over}$$

if $P_{at} \leqslant P_{over}$, then design is OK, and print-out circuit resistance, voltage and power values if $P_{at} > P_{over}$, then change the input requirements or change the circuit, that is, select another circuit that will better suit the input requirements

Putting together the above rules we can create a heuristic program. A simple example of the heuristic program 'TRANHEUR' for the single-stage transistor amplifier design of Fig. 13.12 to meet input requirements is shown in Program 13.3.

> **See the listing of the program TRANHEUR Directory of Chapter 13 on attached diskette.**

Let us run this program with the input initial requirements:

$$E_{CC} = 12 \text{ V}, \ I_E = 1.5 \text{ mA (may be changed)}, \ V_o = 4 \text{ V}, \ \text{Gain (AC)} = -8,$$

$$P_{over} \leqslant 0.05 \text{ W; assume also that } R_{BG} = 12 \text{ k}\Omega$$

After several iterations we obtain the results of the design

```
Design = OK!
The circuit elements are
     R_CB = 71720.93,   R_BG = 12000.0,   R_o = 6800.0,
                                                      R_E = 850.0

The voltages are
     E_CC = 12.0,   V_oo = 3.84,   V_B = 1.72,   V_E = 1.02
The actual total power P_at = 0.01274 and required P_t = 0.05.
```

If in above requirements we change the power requirements to $P_{over} = 0.01$ W, then, we obtain the comment:

```
"Power requirement not satisfied-change requirements
                                          or circuit"
```

This example design program can be made more sophisticated in order to design the circuit with better accuracy. In a similar way much larger circuits can be designed using the heuristic knowledge of circuit design experts.

13.5 Expert system shells

Heuristic programs can be written in any of the computer languages. The same is true for expert systems. However, since in expert systems a lot of non-numerical, or symbolic data is processed, the preliminary versions, or prototypes, of expert systems are often written in symbolic or logic oriented languages, sometimes named AI languages, such as LISP (LISt Processing) and/or PROLOG (PROgrammation LOGique). LISP and PROLOG are not very efficient at data processing. Thus, when the prototype written in one of those languages is successful, then to make the system more efficient it may be rewritten in a language such as C, or Pascal.

Easier construction of expert systems is based on the utilization of an expert system shell, which is a software tool that contains all expert system blocks, but with an empty knowledge base. Thus, to build an expert system using the shell, the knowledge engineer or user has to fill the knowledge base with the knowledge of a given domain prepared in an appropriate format, that is, rules and facts, for the rule-based shell. The inference engine in popular shells infers using the forward or backward chaining method. A shell usually contains additional program blocks to facilitate the expert system development process, such as editors and debuggers. Of the more popular expert system shells one can distinguish CLIPS and EXSYS. CLIPS (C Language Intelligent Production System) is written in C language and has the inference engine operating in forward chaining mode. EXSYS (EXpert SYstem Shell) is also written in C language and has the inference engine reasoning in backward and forward chaining modes, but backward is the default mode.

13.5.1 Description and application of CLIPS shell (version 3.2)

The knowledge to be introduced into CLIPS in order to obtain an expert system is represented by means of rules and facts. The syntax of CLIPS is similar to that of the LISP language; all constructs in CLIPS must be surrounded by parentheses (the number of openings must be equal to the number of closings). The CLIPS rule syntax does not use the keywords IF ... THEN, but uses the implication symbol $\Rightarrow$ to distinguish the LHS (left-hand side) and RHS (right-hand side) of the rule.

To enter, or to declare, a rule in the knowledge base of CLIPS the keyword defrule is used, with syntax

```
(defrule name "comment"
    (condition -1) ... (condition -m)
⇒
    (action -1) ... (action -n))
```

The comment is optional and must be enclosed within double quotes. Each rule must have at least one condition and at least one action. The rule can be executed (or fired) when all conditions in LHS are satisfied by the facts existing in the working memory of the system (that is the pattern of each condition corresponds to the pattern of appropriate facts). The rules can be listed on the monitor screen using the keyword (rules).

Each fact represents a piece of knowledge and must be entered, or asserted into the working memory of the system, to form a fact-list. The first fact named (initial-fact) is asserted by the system after the (reset) command. Other facts can be added to the (initial-fact) using the keyword deffacts, with syntax

```
(deffacts name "comment"
    (fact -1) ... (fact n))
```

The comment is optional, but the name is obligatory. The facts from the fact-list are located in the working memory of the system after the (reset) command. Then using the keyword (facts) we can check the facts asserted into the working memory (facts are printed on the monitor screen).

Additional facts can be asserted using the keyword assert, with syntax

```
(assert (fact))
```

The command assert can be used prior to running the system and can also be used in the RHSs of the rules to assert new facts during the run-time.

Table 13.3   Example program 'transist.clp' about transistor properties

```
;;; Program "transist.clp" about transistor properties
(deffacts initiallization
   (transistor junction b-e is forward biased)
   (transistor junction b-c is reverse biased))

;; Rules
(defrule rule01
   (transistor junction b-e is forward biased)
   (transistor junction b-c is forward biased)
   ⇒
   (assert (transistor is saturated)))

(defrule rule02
   (transistor is saturated)
   ⇒
   (printout "Transistor application is not recommended." crlf))

(defrule rule03
   (transistor junction b-e is forward biased)
   (transistor junction b-c is reverse biased)
   ⇒
   (assert (transistor is in active region)))

(defrule rule04
   (transistor is in active region)
   ⇒
   (assert (transistor can amplify)))

(defrule rule05
   (transistor can amplify)
   ⇒
   (printout "It is concluded that transistor can amplify." crlf))

(defrule rule06
   (transistor can amplify)
   (transistor has positive feedback)
   ⇒
   (assert (transistor can generate periodic waves)))
```

(continued)

Table 13.3   (*continued*)

```
(defrule rule07
  (transistor can amplify)
  (transistor has negative feedback)
  ⇒
  (assert (transistor can amplify with increased stability)))

(defrule rule08
  (transistor can generate periodic waves)
  ⇒
  (printout " It is concluded that transistor can " crlf)
  (printout "generate periodic waves." crlf))

(defrule rule09
  (transistor can amplify with increased stability)
  ⇒
  (printout " It is concluded that transistor can " crlf)
  (printout "amplify with increased stability." crlf))
```

Table 13.4   Print-out of consultation session with program 'transist.clp'

```
    CLIPS (V3.2a 01/09/87)
CLIPS> (load "a:transist.clp')
$*********
CLIPS> (reset)
CLIPS> (facts)
f-0 (initial-fact)
f-1 (transistor junction b-e is forward biased)
f-2 (transistor junction b-e is reverse biased)
CLIPS> (agenda)
0    rule03    f-1    f-2
CLIPS> (run)
   It is concluded that transistor can amplify.
3 rules fired
CLIPS> (reset)
CLIPS> (assert (transistor has negative feedback))
CLIPS> (facts)
f-0 (initial-fact)
f-1 (transistor junction b-e is forward biased)
f-2 (transistor junction b-c is reverse biased)
f-3 (transistor has negative feedback)
CLIPS> (agenda)
0    rule03    f-1    f-2
CLIPS> (run)
   It is concluded that transistor can amplify.
   It is concluded that transistor can
   amplify with increased stability
5 rules fired
CLIPS>
```

Table 13.5 Example program 'trampde.clp' for simple transistor amplifier design

```
;;; Program "trampde.clp" for transistor amplifier design
(deffacts requirements
   (Ecc 12)  (Ie 1.5e-3)  (Voo 4)  (Av 8)  (Pd 0.05))
;;; Rules

(defrule R0-title
   (declare (salience 10))
   (initial-fact)
   ⇒
   (printout crlf "Transistor single-stage amplifier design" crlf)
   (printout " ^^^^^^^^^^^^^^^^^^^^^^^^^^^^^^^^^^^ " crlf crlf))
(defrule R1-Ie-Re
   (Ie ?Ie)
   (test (<= ?Ie 0.001))
   ⇒
   (printout " Ie = "? Ie crlf)
   (assert (Re =1000))
   (bind ?Re 1000)
   (printout " For Ie <= 1 mA, Re is : " ?Re crlf))
(defrule R2-Ie-Re
   (Ie ?Ie)
   (test (> ?Ie 0.001))
   ⇒
   (printout " Ie = "? Ie crlf)
   (assert (Re = (-1000 (* ?Ie 1.0000e+0.5))))
   (bind ?Re (-1000 (* ?Ie 1.000e+05)))
   (printout " For Ie > 1 mA, Re is Re(1-Ie*100) = " ?Re crlf))
(defrule R3-Av-Rc
   (Av ?Av)
   (Re ?Re)
   ⇒
   (assert (Rc = (* ?Av ?Re )))
   (bind ?Rc = (* ?Av ?Re ))
   (printout " Rc = " ?Rc crlf))
(defrule R4-Vo
   (Ecc ?Ecc) (Ie ?Ie) (Rc ?Rc)
   ⇒
   (assert (Vo = (- ?Ecc (* ?Ie ?Rc ))))
   (bind ?Vo (- ?Ecc (* ?Ie ?Rc)))
   (assert (Vo ?Vo))
   (printout " Output voltage Vo = " ?Vo crlf))
(defrule R5-Vo-Voo
   (Ie ?Ie) (Voo ?Voo) (Vo ?Vo)
   (test ( | | (<= ?Vo ( - ?Voo 0.2)) ( > ?Vo (+ ?Voo 0.5))))
   (Re ?Re) (Rc ?Rc)
   ?Ie <- (Ie ?Ie) ?Re <- (Re ?Re)
   ?Ro <- (Rc ?Rc) ?Vo <- (Vo ?Vo)
   ⇒
   (printout " Vo= " ?Vo "; Voo= " ?Voo crlf)
```

(continued)

Table 13.5   (*continued*)

```
    (printout "if Vo < Voo then lower Ie; - else increase Ie:Ie= ")
    (retract ?Ie) (bind ?Ie (read))
    (printout " Modified Ie = " ?Ie crlf)
    (assert (Ie ?Ie)) (retract ?Re ?Rc ?Vo))
(defrule R6-Ve
    (Ie ?Ie) (Re ?Re) (Vo ?Vo)
    ⇒
    (assert (Ve = (* ?Ie ?Re)))
    (bind ?Ve (* ?Ie ?Re))
    (printout " Voltage Ve = "?Ve crlf "^^^^^^^^^^^^^^^^^^^^^" crlf))
(defrule R7-Vb
    (Ve ?Ve)
    ⇒
    (assert (Vbe = 0.7))
    (bind ?Vbe 0.7)
    (assert (Vb = (+ ?Ve ?Vbe)))
    (bind ?Vb (+ ?Ve ?Vbe))
    (printout " Voltage Vb = "?Vb crlf "^^^^^^^^^^^^^^^^^^^^^" crlf))
(defrule R8-Rbg-Rcb
    (Vb ?Vb) (Ecc ?Ecc)
    ⇒
    (printout " Enter Rbg value, for example Rbg = 12000 : ")
    (bind ?Rbg (read)) (assert (Rbg ?Rbg))
    (assert (Rcb = (- (/ (* ?Rbg ?Ecc) ?Vb) ?Rbg)))
Rcb=Rbg*Ecc/Vb-Rbg
    (bind ?Rcb (- (/ (* ?Rbg ?Ecc) ?Vb) ?Rbg))
    (printout " Rbg = " ?Rbg " ; Rcb = " ?Rcb crlf))
(defrule R9-Pac-Pab-Pat
    (Ie ?Ie) (Re ?Re) (Rc ?Rc) (Ecc ?Ecc) (Rcb ?Rcb) (Rbg ?Rbg)
    ⇒
    (assert (Pac = (* ?Ie (* ?Ie (+ ?Re ?Rc)))))
    (bind ?Pac (* ?Ie (* ?Ie (+ ?Re ?Rc))))
    (assert (Pab = (* ?Ecc (/ ?Ecc (+ ?Rcb ?Rbg)))))
    (bind ?Pab        (* ?Ecc (/ ?Ecc (+ ?Rcb ?Rbg))))
    (assert (Pat =    (+ ?Pac ?Pab)))
    (bind ?Pat        (+ ?Pac ?Pab))
    (printout "Total actual power Pat= Pac+ Pab = " ?Pac "+" ?Pab )
    (printout " = " ?Pat crlf crlf))
(defrule R10-Check-Power
    (Pd ?Pd) (Pat ?Pat)
    (test (< ?Pat ?Pd))
    ⇒
    (printout " OK ! Pat is lower than required Pd " crlf))
(defrule R11-Check-Power
    (Pd ?Pd) (Pat ?Pat)
    (test (> ?Pat ?Pd))
    ⇒
    (printout "Power requirement is not satisfied-Modify design!"
crlf))
```

Facts can also be cancelled or retracted using the keyword retract. This retraction can be made prior to running or during the run-time of the system.

After loading the CLIPS system into the computer and executing it, we obtain on the monitor screen the prompt:

```
CLIPS>
```

Programs, or expert systems in CLIPS can be prepared using any editor and saved in the file 'filename.clp' on diskette or hard disk. Then this program can be loaded into CLIPS using the command (assume the diskette is in drive A)

```
CLIPS>(load "a:filename.clp")
```

As an example let us consider the simple expert system, described in Section 13.3.3, concerning the properties of a transistor. An example program 'transist.clp' is shown in Table 13.3.

After loading this program the fact-list named initialization containing two facts is also loaded. The print-out of the consultation session is shown in Table 13.4.

To run the program it is necessary to enter the command (reset) to put the facts into working memory. Then it is possible to check the facts by entering (facts); as a result three facts are listed (fact f-0 (initial fact) is automatically added). Then we can check which rules are activated, that is, put into the agenda, by entering command (agenda); as a result rule03 is listed together with two facts, f-1 and f-2, causing the activation of rule03. Then executing the (run) command we obtain a print-out of the conclusion of rule05 and the information that '3 rules fired', that is, rule03, rule04, and rule05. After resetting again and asserting the additional fact 'transistor has negative feedback' and running we obtain the print-out of the rule05 and rule09 conclusions with the additional information that '5 rules fired', that is, rule03, rule04, rule05, rule07 and rule09.

As a second example of expert system development using the CLIPS shell let us consider the simple transistor amplifier design described in Section 13.4 as a heuristic program. A listing of a simple expert system or program 'trampde.clp' for this purpose is shown in Table 13.5. In this program application of variables that can store numerical and string values is presented. Performance of some mathematical calculations and data transfer from one to another rule, are also shown. Here, design is performed interactively.

The rule R0-title is triggered by the (initial-fact); in this rule the priority is declared using the fact with the keyword (salience value), where the numeric value can range from $-10\,000$ to $10\,000$. When the rule has no salience declared then CLIPS assumes a salience of 0.

The rule R1-Ie-Re is triggered by the fact (Ie ?Ie) from the fact-list in which ?Ie corresponds to the variable (names of CLIPS variables begin with a question mark); in this rule a test is performed: ?Ie $\Leftarrow 0.001$. Note that CLIPS uses prefix, or Polish reverse notation in comparisons and mathematical computations. To print a value of the variable this value must be bound to the variable name using the keyword (bind) and the structure:

```
(bind variable value)
```

On the RHS of rule R2-Ie-Re the computation of Re is performed:

```
Re = 1000 - Ie*1E5
```

On the RHS of rule R4-Vo the computation of V_o is performed:

```
Vo = Ecc - Ie*Ro
```

The rule R5-Vo-Voo is triggered by the values of I_E, V_o, R_E and R_O passed from previous rules and V_{oo} from the fact-list ; also, the test is performed:

```
(Vo <= (Voo - 0.2)) or (Vo > (Voo + 0.5))
```

where the logical *or* function is denoted by '||'.

If this inequality is satisfied, it is necessary to enter a new value for Ie (this is performed on the RHS using the 'read' command in the structure '(bind ?Ie (read))'), compute new values of R_E, R_O and V_o and to cancel or retract the old values of R_E, R_O and V_o (this is performed on the RHS using the command '(retract ?Re ?Ro ?V_o)'). However, to retract a variable value or fact it is necessary to know the address of the fact, that is, the fact address must first be bound to a variable on the LHS of the rule. This is performed using the 'left arrow' operator '<-' in the rule R5-Vo-Voo:

```
?Ie <- (Ie ?Ie) ?Re <- (Re ?Re)
?Ro <- (Rc ?Rc) ?Vo <- (Vo ?Vo)
```

Note that after retracting R_E, R_O and V_o and asserting new value of Ie a computational loop is formed, since the new fact (Ie ?Ie) can trigger only rules R1-Ie-Re or R2-Ie-Re. When the test in the rule R5-Vo-Voo is not satisfied, rule R5 is not fired. Instead, rule R6-Ve is triggered and fired, and subsequently rules R7-Vb, R8 and R9 are triggered and fired (in the rule R8 Rbg value must be entered). In rules R10 and R11 tests are performed whether the actual computed power P_{at} is lower or larger than the required dissipation power Pd (from fact-list) and the proper comment is printed.

The print-out of the results of the consultation session with the program 'trampdc.clp' is shown in Table 13.6. Due to repetitive computations in the loop, some rules are fired several times; therefore, the number of rules fired is bigger than the number of rules defined in the program.

In a similar way more powerful expert systems using the CLIPS shell can be constructed for the design of electronic circuits or equipment with an application of many heuristic rules obtained from human experts.

Table 13.6 Print-out of results of consultation session with program 'trampde.clp'

```
      CLIPS (V3.2a 01/09/87)
CLIPS> (load "a:trampde.clp")
$ *************
CLIPS> (reset)
CLIPS> (facts)
f-0 (initial-fact)
f-1 (Ecc 12)
f-2 (Ie 0.0015)
f-3 (Voo 4)
f-4 (Av 8)
f-5 (Pd 0.05)
CLIPS> (agenda)
10 R0-title   f-0
0   R2-Ie-Re f-2
CLIPS> (run)


Transistor single-stage amplifier design
^^^^^^^^^^^^^^^^^^^^^^^^^^^^^^^^^^^^^^^^^^^

   Ie = 0.0015
   For Ie > 1 mA, Re is Re(1-Ie*100) = 850
   Rc = 6800
   Output voltage Vo = 1.80000019
   Vo = 1.80000019 ; Voo= 4
   if Vo < Voo then lower Ie ; - else increase Ie : Ie= 1.2e-3
   Modified Ie = 0.0012
   Ie = 0.0012
   For Ie > 1 mA, Re is Re(1-Ie*100) = 880
   Rc = 7040
   Output voltage Vo = 3.55200005
   Vo = 3.552000005 ; Voo= 4
   if Vo < Voo then lower Ie ; - else increase Ie : Ie= 0.2e-3
   Modified Ie = 0.00020000
   Ie = 0.00020000
   For Ie <= 1 mA, Re is : 1000
   Rc = 8000
   Output voltage Vo = 10.39999962
   Vo = 10.39999962 ; Voo= 4
   if Vo < Voo then lower Ie ; - else increase Ie : Ie= 1.1e-3
   Modified Ie = 0.0011
   Ie = 0.0011
   For Ie > 1 mA, Re is Re(1-Ie*100) = 890
   Rc = 7120
   Output voltage Vo = 4.16800022
   Voltage Ve = 0.97899997
   ^^^^^^^^^^^^^^^^^^^^^^^^
   Voltage Vb = 1.6789999
   ^^^^^^^^^^^^^^^^^^^^^^
   Enter Rbg value, for example Rbg = 12000 : 12000
   Rbg = 12000 ; Rcb = 73765.344
   Total actual power Pat= Pac+Pab=0.00969210+0.00167900=
                                          0.01137110


   OK ! Pat is lower than required Pd
21rules fired
CLIPS>
```

13.5.2 Description and application of EXSYS shell

The EXSYS shell enables creation of an expert system with rule-based knowledge base. To fill the knowledge base it is necessary to answer several additional questions posed by the shell and to follow the step-by-step procedure controlled by the EXSYS environment. Explanations of how to use the EXSYS shell are usually included in the EXSYS software package and in the user manual.

The EXSYS shell is menu-driven, which means that during knowledge base creation and consultation the knowledge engineer or user has to answer some questions or select some of several possible choices given by the shell (this will be explained in the course of an example expert system development). During creation of a new knowledge base the editing windows are used and all available actions are listed at the bottom of the rule editing window. After selection of one of the possible actions, another window appears on the terminal screen with new available actions, and so on, until the whole knowledge base is created.

The format of rules in EXSYS is as follows:

```
RULE NUMBER : (number)
    IF :           (conditions)
    THEN :         (choices or conditions)
    ELSE :         (choices or conditions)
    NOTE :         (notes to the user)
    REFERENCE :  (references to the user concerning the rule)
```

The ELSE, NOTE and REFERENCE parts of the rule are optional. To obtain a clearer operation of the expert system created it is advisable not to use the ELSE portion of the rule. The NOTE and REFERENCE parts are useful for better documentation of each rule.

Note that the EXSYS shell uses its own terminology which does not correspond with that used in most texts on rule-based expert systems. According to EXSYS terminology a condition is a complete clause that can be used as a premise of the rule or as an intermediate conclusion. A condition is composed of two parts:

1. a qualifier, which is made up of an attribute and a verb,
2. one or more values.

For example, if the condition of the rule is:

IF the transistor junction b-e is forward biased
the qualifier in this case is:
the transistor junction b-e is
and the value is:
forward biased

Also, according to EXSYS terminology the choices are possible solutions, conclusions, or actions concerning the considered problem.

Before using the EXSYS shell it is recommended to prepare carefully the whole set of rules for the problem under investigation and to prepare the table of attributes (qualifiers) and their values for the knowledge base including the final conclusions, that is, choices. As an example, which will also be used to explain knowledge base creation using the EXSYS shell, let us consider again the knowledge base of the transistor properties described by the following set of rules:

```
Rule 1: IF transistor junction b-e is forward biased
        AND transistor junction b-c is forward biased
        THEN transistor is saturated

Rule 2: IF transistor junction b-e is forward biased
        AND transistor junction b-c is reverse biased
        THEN transistor is in active region

Rule 3: IF transistor is in active region
        THEN transistor can amplify

Rule 4: IF transistor can amplify
        AND transistor has positive feedback
        THEN transistor can generate periodic waves

Rule 5: IF transistor can amplify
        AND transistor has negative feedback
        THEN transistor can amplify with increased
        stability
```

Based on these rules we can construct the table of qualifiers (attributes) and values, including choices, as is shown in Table 13.7. Here we see that we have five qualifiers with different values and five choices, although choices 1, 2 and 3 can be treated as intermediate choices. Enumeration of qualifiers is very important since, during rule edition (creation) we can call defined already qualifiers by their numbers (not by repeating their description – this saves typing and possible incorporation of error); the same concerns the choices – the choices have to be entered before the LHSs of rules are created. Now we can go through the procedures for entry into the knowledge base and run the newly created expert system.

After entering the EXSYS rule editor, the EXSYS logo window appears on the screen and the question

```
Expert System Filename :_
```

is asked. In our case we can name this file 'tran-pro' and type in this name (after typing, press the ENTER key). Then EXSYS asks for confirmation of whether it is a new file:

```
if you wish to start a new file? (Y/N)_
If yes, then press: y key
```

Table 13.7 Table of qualifiers and values of transistor properties

Rule no.	Clause no.	No.	Description	Value
			Qualifier (attribute plus verb)	
1	1	1	transistor junction b-e is	forward biased
2	1	1	transistor junction b-e is	forward biased
1	2	2	transistor junction b-c is	forward biased
2	2	2	transistor junction b-c is	reverse biased
3	1	3	transistor is	in active region
4	1	4	transistor can	amplify
5	1	4	transistor can	amplify
4	2	5	transistor has	positive feedback
5	2	5	transistor has	negative feedback
Choices (conclusions)				
1	3		transistor is	saturated
2	3		transistor is	in active region
3	2		transistor can	amplify
4	3		transistor can	generate periodic waves
5	3		transistor can	amplify with increased stability

The next prompt concerns the description of the expert system:

`Subject of the knowledge base:_`

Here we can enter any text that informs about the subject; in our case we can enter, for example, knowledge base describes properties of transistor. This information will be displayed when the expert system is run.

The next prompt concerns

`Author name :_`

The user can enter his or her own name which will be displayed when the expert system is run. Typing the author's name and pressing the ENTER key leads to a new screen with a prompt about the choice of confidence factors (named probability in EXSYS) for the choices or conclusions of rules.

Three possibilities are given for selection:

1. Rule (1) confidence 1 or confidence 0; confidence 1 corresponds to 'yes' or 100% of confidence, or probability = 1, while confidence 0 corresponds to 'no', or probability = 0,
2. Rule (2) confidence factors on the 0 to 10 scale; any number between 0 and 10 can be used in this case, with 10 corresponding to probability = 1 and 0 corresponding to probability = 0,
3. Rule (3) confidence factors on the -100 to 100 scale, with 100 corresponding to probability = 1 and -100 corresponding to probability = 0.

These three possibilities are displayed on the screen together with the question:

```
Input number of selection or <H> for help :_
```

To start with the simplest situation we can input 1 and press the ENTER key. The next prompt concerns·the number of rules to be used in data derivation — all (1) or just the first (2):

```
Select 1 or 2 (default = 1) :_
```

This means that several rules may lead to the same intermediate (or final) conclusion; in such a situation after firing the first rule the other need not be fired. However, it is safer to use all the rules that are activated. Therefore, let us input 1 (for all rules).

The next prompt concerns the entering of explanatory text to be displayed at the start of a run of the expert system. Here we can enter any information we wish.

The next prompt concerns the entering of explanatory text to be displayed at the end of the consultation session, just before the results are displayed. This text, like the previous text, is optional but it may help the user in understanding the operation and conclusions reached by the system. After the appearance of the next screen the next question will ask whether the rules should be displayed during the consultation session (Y/N); since this is not necessary, we may respond N (for NO).

The next question asks whether or not an external program is to be called at the start of the consultation session. This question refers to the EXSYS capability of calling external algorithmic programs, written in BASIC, Pascal or C languages and using the results of the programs. In this case, our response should be N (for NO) since we are not going to use external programs.

The question at the next screen will ask us to input the Choices or rule conclusions to be considered. In our example we see from Table 13.7 that we have five choices. We enter them one after the other in exactly the same form as they are in the rules; enumeration of choices is done automatically by the shell:

```
1. transistor is saturated                    (press ENTER)
2. transistor is in active region             (press ENTER)
3. transistor can amplify                      (press ENTER)
4. transistor can generate periodic waves      (press ENTER)
5. transistor can amplify with increased
   stability                                   (press ENTER)
6. (press ENTER key)
```

Pressing the ENTER key after the appearance of number 6 leads to the rule editing environment of the EXSYS rule editor. A window composed of three areas will appear: upper left, upper right and lower. The upper areas are empty and in the lower part all the available actions are listed:

```
Add rule <A> or <ENTER>, Edit rule <E>, Delete rule <D>,
                                        Move rule <M>,
```

```
Print rules <P>, Store /exit <S>, Run <R>, Options <O>,
                              DOS <Ctrl-X>, Help <H>:
```

This listing of EXSYS commands represents the MAIN MENU of the rule editor. Now, and later on, it is necessary to read the available actions on each new screen carefully in order to know how to proceed further.

Since we now want to start a new rule, we have to press the A or ENTER key, for: add rule. This causes the screen to change to that shown in Fig. 13.13. The upper left part of the screen shows that EXSYS is ready to build the IF part of Rule 1; the bottom part shows the current menu.

Since we are going to enter a new qualifier we have to press the N key. This results in a change of the contents of the bottom part of the screen where a new prompt appears:

```
Input text of qualifier (ending in verb) :
```

Looking at the table of qualifiers and values (see Table 13.7) we see that our first qualifier is

```
transistor junction b-e is
```

After typing this qualifier we have to press the ENTER key. This moves the text of the qualifier to the upper right part of the screen and the number of the qualifier is given: Qualifier #1; at the same time in the bottom part of the screen a new prompt appears:

```
Input acceptable values. Input just <ENTER> when done.
```

Looking again at Table 13.7, we see that qualifier #1 has one value:

```
1. forward biased
```

So when number '1' appears (automatically) in the upper right part of the screen then we have to type 'forward biased', and press just ENTER when number '2' appears. Pressing ENTER causes the screen to change to that shown in Fig. 13.14.

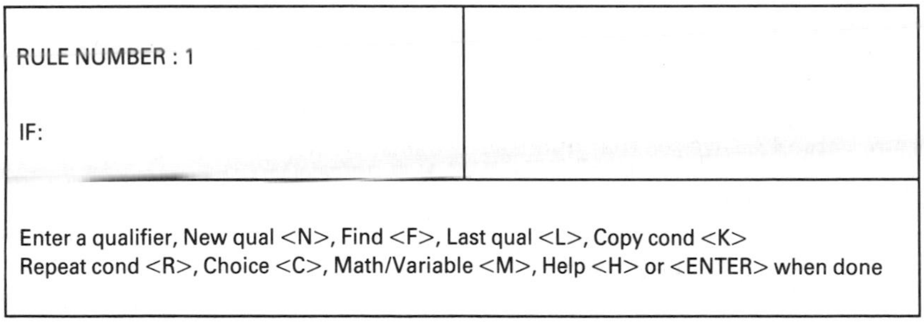

```
RULE NUMBER : 1                |
                               |
IF:                            |
_____|_____

Enter a qualifier, New qual <N>, Find <F>, Last qual <L>, Copy cond <K>
Repeat cond <R>, Choice <C>, Math/Variable <M>, Help <H> or <ENTER> when done
```

Figure 13.13 Screen showing the rule component entry menu

RULE NUMBER : 1	Qualifier #1
IF:	transistor junction b-e is
	1. forward biased
	1
Enter number(s), NOT+number(s), New value <N>, Type correction <T>, Delete <D>, ↑ or ↓ to scroll, Help <H> or just <ENTER> to cancel	

Figure 13.14   Qualifier # I with one value; value I has been selected to the first premise for Rule I

According to the instructions in the bottom part of the screen we have to enter the number corresponding to the proper qualifier value. Since the first premise of Rule 1 is

```
if transistor junction b-e is forward biased
```

we have to enter number '1', which is also shown in the upper right part of the screen of Fig. 13.14. After pressing ENTER, the text of the first premise appears in the upper left part of the screen:

```
RULE NUMBER : 1
IF :
transistor junction b-e is forward
biased
```

The upper right part of the screen becomes clear (empty) and the bottom part becomes as shown in Fig. 13.13, making it possible to input the next qualifier. Since our Rule 1 has two premise clauses, we have to type N to enter the second qualifier:

```
transistor junction b-c is
```

Now the procedure repeats. After typing this qualifier and pressing the ENTER key, the upper right part of the screen will be filled with the following text:

```
Qualifier #2
transistor junction b-c is
1.
```

From Table 13.7 we see that qualifier #2 has two values. Thus, we have to type:

```
1.  forward biased
2.  reverse biased
3.  (press ENTER key)
```

After pressing the ENTER key when number '3' appears, we obtain the upper part of the screen as shown in Fig. 13.15; the lower part is the same as in Fig. 13.14.

RULE NUMBER : 1	Qualifier #2
IF:	transistor junction b-c is
transistor junction b-e is forward biased	1. forward biased 2. reverse biased 1

Enter number(s), NOT+number(s), New value <N>, Type correction <T>, Delete <D>, ↑ or ↓ to scroll, Help <H> or just <ENTER> to cancel

Figure 13.15 Inclusion of qualifier #2 with two values; value 1 has been selected to the second premise for Rule 1

After entering '1' for the first value selection we obtain:

`transistor junction b-c is forward biased`

as is required for Rule 1.

This causes the second premise clause to appear additionally in the left upper part of the screen and the lower part of the screen is as in Fig. 13.13. Since there are no more qualifiers in Rule 1, we press the ENTER key (for done) and the word THEN appears in the left upper part of the screen indicating that EXSYS is ready to construct the RHS of Rule 1, that is, to develop the conclusion (or choice) clause for Rule 1. To do this we have to type 'C' (for choices, as is indicated at the bottom in Fig. 13.13); then all five choices (conclusions) that we entered at the start will appear in the upper right part of the screen and the lower part of the screen displays the possible actions available:

`Select choice number, New value <N> or Typo correction <T>,`
`Delete/reorder <D>, Find <F>, Help <H>, <ESC> or <ENTER>`
`                                        to cancel :_`

To obtain the conclusion of the first rule 'transistor is saturated' we have to select the first choice; thus, we have to type '1'. This causes a new prompt to appear at the bottom part of the screen:

`Enter value 0 (false) or 1 (true)`
`Help <H>, <ENTER> or <ESC> to cancel :_`

Since we have selected 0, 1 confidence (probability) values at the start, we have to enter '1' to assign a probability = 1 to this choice. Now, Rule 1 with both IF and THEN parts appears in the upper left part of the screen, with the available commands in the lower part, as shown in Fig. 13.13.

Pressing the ENTER key again causes the ELSE condition to be displayed in the upper left part of the screen. Next we may enter the NOTE and/or REFERENCE parts of the rule. If not, press the ENTER key for these prompts. The next prompt asks whether the rule is correct. It is always better to respond with a YES, since the response NO erases the entire rule. If there are some mistakes in a rule it is possible to correct them using the editing facility of the rule editor. At this step after the response YES the EXSYS editor returns to the main editing screen, that is, MAIN MENU, and we are ready to enter Rule 2.

Pressing ⟨A⟩ or ⟨ENTER⟩ (for: add rule) we obtain the screen as shown in Fig. 13.13, except that the rule number is now 2. Since the two qualifiers in Rule 2 are the same as in Rule 1, that is, Qualifier #1 and Qualifier #2, we first have to enter 1 (for qualifier number = 1) and we obtain a screen as shown in Fig. 13.14 (for Rule 2). The value for this qualifier is 1, since the junction b-e is forward biased. Then again, to enter Qualifier #2, enter 2; now the screen looks as shown in Fig. 13.15 (for Rule 2). Since in Rule 2 the transistor junction b-c is reverse biased, we have to enter 2 (for the second value of Qualifier #2).

The rest of Rule 2 is entered similarly to Rule 1 (as a conclusion for Rule 2 we select choice #2 for: transistor is in active region). In a similar way Rules 3, 4 and 5 are entered.

After all the rules have been entered correctly, we may start a consultation session by typing R (for RUN) from the MAIN MENU. Then the first question appears on the screen:

```
Do you wish to have the rules displayed as
they are used ? (Y/N) (Default = N) :_
```

Since rule display is not necessary we may respond N (for NO). Then the title of our expert system appears, along with the author's name and the explanatory text that we entered during the creation of our knowledge base. Moving from one to another screen is accomplished by pressing any key.

Subsequent screens will display the qualifiers with their values for our selection. Suppose we wish to know the properties of transistor when

```
junction b-e is forward biased
junction b-c is reverse biased
and it has negative feedback.
```

The first qualifier displayed is

```
transistor junction b-e is
1 forward biased
_1                                      (type 1 and press ENTER)
```

The rest of the qualifiers and responses are

```
transistor junction b-c is
1 forward biased
2 reverse biased
_2                                      (type 2 and press ENTER)
transistor is
1 in active region
_1                                      (type 1 and press ENTER)
```

```
transistor can
1 amplify
_1                                          (type 1 and press ENTER)
transistor has
1 positive feedback
2 negative feedback
_2                                          (type 2 and press ENTER)
```

Since all the qualifiers have been used the explanatory text entered at the start (if any) is displayed, and the prompt appears:

```
Press any key to display results :-
```

Pressing a key we obtain the final results screen shown in Fig. 13.16.

If we wish to know how these conclusions have been reached we may type in the line number corresponding to the particular conclusion; when we type in 3, for example, to determine how the third choice was reached, the rule, or rules associated with this choice will be displayed.

After pressing D (for Done) we obtain a prompt:

```
Run again (Y/N) (Default = N) :-
```

Entering Y makes it possible to run the program again with other qualifier values, while entering N or pressing ENTER brings the system to the MAIN MENU.

To exit to DOS we have to enter S (for Store/exit); again a prompt appears:

```
Done (Y/N) (Default Y) :-
```

Pressing ENTER we exit from EXSYS to DOS.

EXSYS also allows us to use mathematical expressions with variables. The variables of any names must be enclosed in square brackets; they may be both numerical and string variables. As an example let us construct a simple knowledge base concerning feedback amplifiers containing numeric variables both in the IF and THEN

	Values based on 0/1 system	VALUE
1	transistor is in active region	1
2	transistor can amplify	1
3	transistor can amplify with increased stability	1
	All choices <A>, only if value >1 <G>, Print <P>, Change and rerun <C>, rules used <line number>, Quit/save <Q>, Help <H>, Done <D> :	

Figure 13.16 The final results screen

parts of the rules:

```
Rule Number 1
IF amplifier has feedback
THEN
    [loop gain] IS GIVEN THE VALUE [gain]*[feedback factor]
```

(the part of sentence : IS GIVEN THE VALUE is added automatically by EXSYS)

```
Rule Number 2
IF amplifier has positive feedback
THEN
    [pos feedb gain] IS GIVEN THE VALUE [gain]/(1 - [loop gain])
Rule Number 3
IF amplifier has negative feedback
THEN
    [neg feedb gain] IS GIVEN THE VALUE [gain]/(1 + [loop gain])
Rule Number 4
IF [loop gain] >= 1
THEN
    Positive feedback amplifier is unstable
```

At the beginning, the construction of this knowledge base is similar to the previous example until we come to inputting the choices. Here we have to input only one text choice, that is,

```
1. Positive feedback amplifier is unstable
2. (press ENTER key)
```

The RHSs of rules 1 to 3 will be entered during the construction of the THEN parts. During construction of the IF part of Rule 1 we proceed as in the previous example, that is, we press N (for new qualifier) and enter the qualifier:

```
amplifier has
```

and then its three values:

```
1 feedback
2 positive feedback
3 negative feedback
```

After pressing ENTER we select the first value (by pressing 1), and pressing the ENTER key again we obtain in the upper left part of the screen:

```
RULE NUMBER : 1
IF :
amplifier has feedback
THEN :
```

The lower part of the screen is as shown in Fig. 13.13.

Since now we wish to enter a variable [loop gain] we have to press M (for

Math/Variable) to go to the mathematical mode. Then the prompt appears at the bottom:

```
Enter the name of the variable that you wish to use (in [ ])
   [loop gain] (press ENTER key)
```

The next information at the bottom says that it is a new variable and that a description of its meaning is required; we can add, for example:

```
loop gain is a product of gain and feedback factor
```

The next prompt asks

```
Do you wish this variable to be displayed Y/N (Default =N) : Y
```

(answer Yes to check the value at the end of run).
The next prompt asks:

```
Do you wish this variable to be numeric (N) or string (S) ? : N
Do you wish this variable to be initialized to a value at
                                      the start ? : N
Do you wish this variable to be within specified limits ? : N
Enter the expression whose value is to be assigned to
                                      the variable:
```

(we have to enter now :)

```
   [gain]*[feedback factor] (press ENTER key)
```

Now again the information at the bottom of the screen says that [gain] is a new variable, ..., etc., as in the case of the [loop gain] variable; also [feedback factor] is a new variable and we have to answer all questions as in the case of [loop gain].

After this Rule 1 is displayed in the form given at the beginning of this example. In exactly the same way Rule 2 and Rule 3 are entered. Rule 4 is constructed in a different way since it has a variable relation in the IF part. When we proceed to the screen with

```
Rule Number : 4
IF :
```

similar to that shown in Fig. 13.13, we have to press M (for Math/Variable) to enter the mathematical mode. After that, the prompt will be displayed telling us to enter the formula in the IF part; we now have to type in:

```
[loop gain] >= 1 (press ENTER)
```

since the variable [loop gain] has been described earlier EXSYS recognizes it and does not ask for description, ..., etc.

When THEN appears, we have to press C (for Choices); then our choice entered at the start appears:

```
1. Positive feedback amplifier is unstable
```

We select it by entering 1 (the number of this choice). Then we are asked to enter the probability of this choice. Since we are in 0/1 mode the question is

```
Enter value 0 (false) or 1 (true) :_
```

Then the whole of Rule 4 is displayed (we omit the ELSE, NOTE and REFERENCE parts of the rule for simplicity). Thus, the whole knowledge base has been created and we run the system by pressing R (from the MAIN MENU).

At the beginning of the run initial information is displayed (title, authors, ..., etc.), and then the first qualifier appears with its values to choose:

```
amplifier has
1 feedback
2 positive feedback
3 negative feedback
```

We can select one or more values. If we select all three values by typing:

> 1, 2, 3

(numbers must be separated by comma) we will force the system to consider all three situations. Then we shall be asked by our expert system:

```
Please input Gain which is open loop gain : 120
```

(the phrase 'Gain which is open loop gain' was entered during description of the [gain] variable)

```
Please input Feedback factor which is return signal
                                        factor : 0.02
```

the phrase 'Feedback factor ...' was entered during description of [feedback factor]). After entering 120 and 0.02, respectively, as is shown above, we obtain a display of final results:

	Values based on 0/1 system		Value
1	Positive feedback amplifier is unstable		1
2	Gain which is open loop gain	= 120.000 000	
3	Loop gain is gain of feedback loop	= 2.400 000	
4	Feedback factor which is return signal factor	= 0.020 000	
5	Pos feedb gain is positive feedback gain	= -85.714 286	
6	Neg feedb gain is negative feedback gain	= 35.299 118	

Line number 1 of the results is a conclusion of Rule 4, while line numbers 2 to 6 are displays of values of numerical variables [gain], [loop gain], [feedback factor], [pos feedb gain] and [neg feedb gain], respectively, with texts given to corresponding variables during their creation.

In a similar way a more complex expert system can be developed using the EXSYS shell. Further information on more advanced applications may be found in the EXSYS software manual.

Exercises

Exercise 1. Indicate the important differences between an expert system and a classical algorithmic program.

Exercise 2. Describe the similarities and differences between heuristic and algorithmic programs.

Exercise 3. Describe the similarities and differences between expert system and heuristic program.

Exercise 4. An audio stereo amplifier system consists of microphones, low-power amplifier, high-power amplifier, loudspeakers, and two power-supply blocks with fuses (one for each channel). Develop a heuristic program to aid the repair of this audio system in which malfunctions (in the case where an acoustic signal is applied to both microphones) may be as follows: no acoustic signal out of both loudspeakers; noise from both loudspeakers; no signal from left loudspeaker, noise from right loudspeaker; noise from left loudspeaker, normal acoustic signal from right loudspeaker.

Exercise 5. Write IF...THEN rules for the problem described in exercise 4.

Exercise 6. Show the difference between 'if...then' statements in an algorithmic program and IF...THEN rules used in a knowledge base.

Exercise 7. The rule 'IF a THEN b' means: if a is true then b is true. The OR logical operator in the premise parts (LHSs) of rules should be avoided; rewrite the following rules to obtain six rules without the OR operator:

> R1 : IF a OR b THEN c
> R2 : IF (a AND c) OR b THEN d
> R3 : IF d OR (a AND b) THEN f

Exercise 8. Draw the inference tree for the rules of exercise 7.

Exercise 9. Write the rules for the inference tree shown in Fig. E13.1 (the premises connected with an arch correspond to the AND operator, and those not connected correspond to the OR operator or to separate rules).

Exercise 10. Based on the *modus ponens* rule of inference we can say: if A infers B and A is true, then B is also true. In most cases the reverse statement is valid: if B is true then A is true; however, sometimes this statement is not valid:

(a) give an example in the area of electronics;
(b) give an example from animal world.

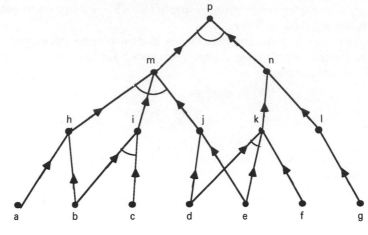

Figure E13.1

Exercise 11. Semiconductor devices are briefly described by means of the object–attribute–value table shown in Table E13.1:

(a) draw the inference tree to identify the devices (start from the number of electrodes),
(b) generate the rules for device identification based on the inference tree.

Table E13.1 Semiconductor device object – attribute – value table

	Semiconductor devices and attribute values							
	Diode		Transistors					
			Bipolar		Field effect			
					Junction		MOS	
Attribute	NP	PN	NPN	PNP	N	P	N	P
Number of electrodes	2	2	3	3	3	3	3	3
Type of charge carriers	bipolar	bipolar	bipolar	bipolar	unipol	unipol	unipol	unipol
Junctions	yes	yes	yes	yes	yes	yes	no	no
Collector/drain polarity for ON, or active operation	+	−	+	−	+	−	+	−

Exercise 12. Give some examples of heuristic rules that can be applied to electronic equipment (or circuit) construction (design), diagnosis and repair.

Exercise 13. Give examples of the object–attribute–value (OAV) triplets describing semiconductor devices: diode, bipolar and MOS transistors.

Exercise 14. Write two or more IF...THEN rules describing the properties of a linear operation of the NPN transistor.

Exercise 15. Draw the constraints diagram for the linear cascode amplifier of Fig. E13.2, and find its approximate DC behaviour; choose the DC voltage V_{o2} to obtain maximum undistorted voltage swing (in such a case an instantaneous voltage $v_{o2}(t)$ can vary between E_{CC} and V_{B2}, i.e. when $V_{CB2} = 0$).

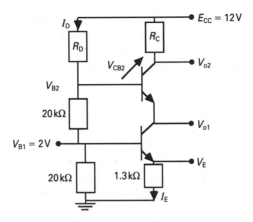

Figure E13.2

Exercise 16. Construct a semantic network relating the differential amplifier and an active RC filter using 'is-a-part-of' relations; for example: transistor 'is-a-part-of' differential amplifier, differential amplifier 'is-a-part-of' ... .

Exercise 17. Construct the frame describing an audio amplifier.

Exercise 18. Suppose that the power transistor in an operating amplifier is too hot; write several IF...THEN rules suggesting actions to avoid its failure.

Exercise 19. Develop the 'knowledge base' using IF...THEN rules to design the cascode amplifier of exercise 15.

Exercise 20. Using CLIPS syntax develop the knowledge base to design the cascode circuit of Fig. E13.2. Use known parameter values as facts in the 'deffacts' statement.

Exercise 21. Modify the 'DOUBLLAD' program to find the behaviour of the circuit shown in Fig. E13.3, for element values $E_S = 10$ V, $R_1 = R_2 = R_3 = R_6 = 1\ \Omega$, $R_4 = 10\ \Omega$, $R_5 = 5\ \Omega$, $R_7 = 7\ \Omega$. Assume that $I_6 = 1$ and start with $a = 0.5$.

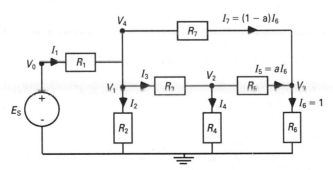

Figure E13.3

Exercise 22.   Draw a constraints diagram for the circuit of Fig. E13.3.

Exercise 23.   Using CLIPS syntax develop a knowledge base to find the voltages in the circuit of Fig. E13.3 for the element values as in exercise 21.

Exercise 24.   Write a heuristic program for the rough design of the two-stage transistor amplifier shown in Fig. E13.4, for the requirements: the DC voltages $E_{CC} = 12$ V, $V_{oDC} = 2/3$ $E_{CC} \pm 2$ V, $V_{iDC} = 1.5$ V, the AC voltage gain $A_V = \Delta V_o / \Delta V_i = 40 \pm 4$, the DC currents $I_{E1} = 0.2$ mA, $I_{E2} \leqslant 1$ mA; use the approximate formula $A_V = A_{V1} A_{V2} = R_{C1}/R_{E1} \ R_{C2}/R_{E2}$, and assume that $\beta > 1000$.

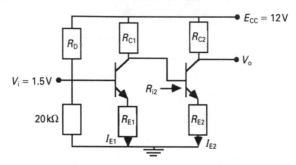

Figure E13.4

Exercise 25.   Find a more precise value of the voltage gain A_V of exercise 24, assuming that $\beta = 200$ and using the formula $R_{i2} = \beta R_{E2}$; write IF...THEN rules for the problem.

Exercise 26.   Using CLIPS syntax develop a knowledge base for the problem described in exercise 24.

Exercise 27.   Write a heuristic program for the rough design of the differential amplifier shown in Fig. E13.5 for the required A_V, R_{in}, CMRR. Use the approximate design formulas

$$I_{ref} = I_0 \exp\left\{\frac{I_0 R_2}{V_T}\right\}; \quad R_1 = \frac{E_{CC} - (E_{EE} + V_{BE})}{I_{ref}}; \quad g_{m2} = \frac{I_C}{V_T} = \frac{I_0}{2V_T}; \quad R_{02} \cong \frac{V_{AN}}{I_C}$$

$$R_{04} \cong \frac{V_{AP}}{I_C}; \quad A_V = g_{m2} R_0; \quad R_0 = \frac{R_{02} R_{04}}{R_{02} R_{04}}; \quad R_{in} = 2 \frac{\beta}{g_{m2}}; \quad g_{m5} = \frac{I_0}{V_T};$$

$$V_{CE5} = -E_{EE} - (I_0 R_2 + V_{BE}); \quad R_{05} \cong \frac{V_{AN}}{I_0}; \quad R_{0CM} = (1 + g_{m5} R_2) R_{05};$$

$$CMRR = 1 + 2 g_{m2} R_{0Cm}; \quad CMRRdB = 20 \log(CMRR)$$

Assume that the following values are given: E_{CC}, E_{EE}, I_0, β. V_{AN} and V_{AP} represent Early Voltage of NPN and DNP transistors.

Notes

1. The meaning of the value 'low' can be additionally expressed by the following rule: IF $R_{out} < 100$ Ohms, THEN $R_{out} <$ is low.
2. The reciprocal circuit has symmetrical immittance matrix, **Y** or **Z**, in respect of the main diagonal.

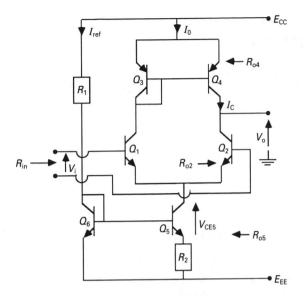

Figure E13.5

References

Antognetti, P., and G. Massobrio, *Semiconductor Device Modeling with SPICE*, New York: McGraw-Hill, 1988.

Bialko, M., 'Analiza ukladow elektronicznych wspomagana mikrokomputerem', *WNT*, 1989 (in Polish).

Calahan, D.A., *Computer-aided Network Design*, New York: McGraw-Hill, 1972.

Chen, U.K., *Linear Networks and Systems*, Monterey: Brooks/Cole, 1983.

Choma, J., *Electrical Networks, Theory and Analysis*, New York: John Wiley, 1985.

Chua, L., and P.M. Lin, *Computer-aided Analysis of Electronic Circuits*, Englewood Cliffs, NJ: Prentice Hall, 1975.

Director, S.W., *Circuit Theory: A Computational Approach*, New York: John Wiley, 1975.

Giarratano, J.C., *CLIPS Users Guide*, Houston, Tex.: NASA, L.B. Johnson Space Center, 1986.

Hayes-Roth, F., *Building Expert Systems*, New York: Addison Wesley, 1983.

Huelsman, L.P., *Digital Computations in Basic Circuit Theory*, New York: McGraw-Hill, 1968.

Huelsman, L.P., *Basic Circuit Theory with Digital Computations*, Englewood Cliffs, NJ: Prentice Hall, 1972.

Huelsman, L.P., *Circuit Theory*, New York: McGraw-Hill, 1975.

Ignizio, J.P., *Introduction to Expert Systems: The development and implementation of rule-based experts systems*, New York: McGraw-Hill, 1991.

Kuo, F.F., and U.G. Magnuson, *Computer Oriented Circuit Design*, Englewood Cliffs, NJ: Prentice Hall, 1969.

McCalla, U.J., *Fundamentals of Computer-aided Circuit Simulation*, Boston, Mass.: Kluster, 1982.

Nagel, L.W., *Spice 2: A computer program to simulate semiconductor circuits*, Berkeley: Electronics Research Laboratory, 1975.

Ruehli, A.E., *Circuit Analysis Simulation and Design*, Amsterdam: North Holland, 1986.

Spence, R., and J.P. Burgess, *Circuit Analysis by Computer from Algorithms to Package*, Hemel Hempstead, Herts: Prentice Hall International, 1986.

Valkenburg, M.E. van, *Network Analysis*, Englewood Cliffs, NJ: Prentice Hall, 1974.

Vlach, J., and K. Singhal, *Computer Methods for Circuit Analysis and Design*, New York: Van Nostrand Reinhold, 1983.

Wong, O., *Circuit Theory with Computer Methods*, Orlando, Fla.: Holt Rinehart & Winston, 1972.

Index